Grundlagen der Math

Foundations of Mathematics I

Part A: Prefaces and §§ 1–2

Part A: Prefaces and §§ 1–2

Part B: §§ 3–5

Parts C and D: §§ 6–8

Grundlagen der Mathematik I

Foundations of Mathematics I

Part A: Prefaces and §§ 1–2

DAVID HILBERT and PAUL BERNAYS

Editors:
CLAUS-PETER WIRTH, JÖRG SIEKMANN,
MICHAEL GABBAY, DOV GABBAY

Advisory Board:
WILFRIED SIEG (chair),
IRVING H. ANELLIS, STEVE AWODEY, MATTHIAS BAAZ,
WILFRIED BUCHHOLZ, BERND BULDT, REINHARD KAHLE,
PAOLO MANCOSU, CHARLES PARSONS, VOLKER PECKHAUS,
WILLIAM W. TAIT, CHRISTIAN TAPP, RICHARD ZACH

First English edition

Commented translation by CLAUS-PETER WIRTH of the
Second German edition of 1968,
including the annotation and translation of all deleted parts of the
First German edition of 1934

© Individual author and College Publications 2011. All rights reserved.

ISBN 978-1-84890-033-2

College Publications
Scientific Director: DOV GABBAY
Managing Director: JANE SPURR
Department of Informatics
King's College London, Strand, London WC2R 2LS, UK

http://www.collegepublications.co.uk

Original cover design by LARAINE WELCH
Printed by Lightning Source, Milton Keynes, UK

Translation from the German language edition:
Grundlagen der Mathematik I by DAVID HILBERT and PAUL BERNAYS.
Copyright © Springer-Verlag Berlin Heidelberg 1968.
Springer-Verlag is a part of Springer Science+Business Media.

All rights reserved. No part of this publication may be reproduced, stored in a retrieval system or transmitted in any form, or by any means, electronic, mechanical, photocopying, recording or otherwise without prior permission, in writing, from the publisher.

Editors' Preface

With great pleasure we introduce this bilingual German–English and annotated edition of HILBERT's and BERNAYS' "Grundlagen der Mathematik". It offers the first English publication of this text and shows the facsimile of the German original on the left of each double page, and its English translation on the right-hand side. Beside commenting on the history and the interpretation of this highly mathematically, logically, and philosophically significant material, we also carefully annotate the differences between the two German editions of this two volume monograph (1st edition 1934/1939; 2nd edition 1968/1970).

For this long overdue translation of one of the most important works in the history of logical foundations of mathematics, we have continued support of some of the most internationally renowned scholars in this field, whose names are listed in the advisory board at the front page of this volume. We are extremely grateful to them for their hard work.

By the 1930s, ground-breaking work had been achieved by German scientists, especially in philosophy, psychology, physics, chemistry, and mathematics. With the Nazis' seizure of power in 1933, the historical tradition of German research was discontinued in most areas, and, as a further consequence, many achievements of German science in the first half of the 20th century have still not been sufficiently recognized. This is the case especially for those developments that had not been completed before the Nazis covered Germany under twelve years of intellectual darkness.

DAVID HILBERT (1862–1943) is one of the most outstanding representatives of mathematics, mathematical physics, and logic-oriented foundational sciences in general [REID, 1970]. From the end of the 19th century to the erosion of the University of Göttingen by the Nazis, HILBERT formed and reshaped many areas of applied and pure mathematics. Most well-known and highly acknowledged are his "Foundations of Geometry" [HILBERT, 1899].

After initial work at the very beginning of the 20th century, HILBERT re-intensified his research into the logical foundations of mathematics in 1917, together with his new assistant PAUL BERNAYS (1888–1977). Supported by HILBERT's PhD student WILHELM ACKERMANN (1896–1962), HILBERT and BERNAYS developed the field of *proof theory* (or *metamathematics*), where formalized mathematical proofs become themselves the objects of mathematical operations and investigations — just as numbers are the objects of number theory. The goal of HILBERT's endeavors in this field was to prove the consistency of the customary methods in mathematics once and for all, without the loss of essential theorems as in the competing intuitionistic movements of LEOPOLD KRONECKER (1823–1891), L. E. J. BROUWER (1881–1966), HERMANN WEYL (1885–1955), and AREND HEYTING (1898–1980). The proof of the consistency of mathematics was to be achieved by sub-division into the following three tasks:

- *Arithmetization* of mathematics.
- *Logical formalization* of arithmetic.
- *Consistency proof* in the form of a *proof of impossibility*: It cannot occur in arithmetic that there are formal derivations of a formula A and also of its negation $\overline{A}$.

The problematic step in this program (nowadays called HILBERT *'s program*) is the consistency proof.

Hilbert's program was nourished by the hope that mathematics — as the foundation of natural sciences, and especially of modern physics — could thus provide the proof of its own groundedness. This was a paramount task of the time, not least because of the foundational crisis in mathematics (which had been evoked among others by Russell's Paradox at the beginning of the 20th century) and the vivid philosophic discussions of the formal sciences stimulated *inter alia* by the Vienna Circle.

It should be recognized that Hilbert's primary goal was neither a reduction of mathematical reasoning and writing to formal logic as in the seminal work of Whitehead & Russell [1910–1913], nor a formalization of larger parts of mathematics as in the publications of the famous French Bourbaki [1939ff.] group of mathematicians. His ambition was to secure — once and for all — the foundation of mathematics with consistency proofs, in which an intuitively consistent, "finitistic" part of mathematics was used for showing that no contradiction could be formally derived in larger and larger parts of non-constructive and axiomatic mathematics.

Hilbert's program fascinated an elite of young outstanding mathematicians, among them John von Neumann (1903–1957), Kurt Gödel (1906–1978), Jacques Herbrand (1908–1931), and Gerhard Gentzen (1909–1945), whose contributions essentially shaped the fields of modern mathematical logic and proof theory.

We know today that Hilbert's quest to establish a foundation for the whole scientific edifice could not be successful to the proposed extent: Gödel's incompleteness theorems dashed the broader hopes of Hilbert's program. Without the emphasis that Hilbert has put on the foundational issues, however, our negative and positive knowledge on the possibility of a logical grounding of mathematics (and thus of all exact sciences) would hardly have been achieved at his time.

The central and most involved presentation of Hilbert's program and Hilbert's proof theory is found in the two-volume monograph "Grundlagen der Mathematik" of Hilbert & Bernays [1934; 1939] to be presented here for the first time in English.

The first volume presents the motivation and philosophical foundation of Hilbert's and Bernays' original view on finitistic mathematics and their methodological standpoint for proof theory (§§ 1–2), a refined introduction to propositional and predicate logic (with equality) (§§ 3–5), and the consistency issues of a variety of (sub-) systems of number theory (including recursion theory, System (Z), and the elimination of the ι-operator) (§§ 6–8).

The second volume of "Grundlagen der Mathematik" is focused on more advanced topics, such as a proof-theoretic sharpening of Gödel's completeness theorem, a full argument for Gödel's second incompleteness theorem, the demarcation of the finitistic standpoint, and especially on Hilbert's ε, some kind of syntactic choice operator: Roughly speaking, "$\varepsilon x.\ A$" denotes some particular object x that renders the formula A as true, in the context of the free variables of A, provided that there are such objects at all.

> "Indeed, the two volumes constitute an encyclopedic synthesis of metamathematical work from the preceding two decades. What is most remarkable, however, is the sheer intellectual force that structures the books"
> [Sieg & Ravaglia, 2004]

After the first edition of the "Grundlagen der Mathematik" had been out of print for a long period of time, Bernays prepared a revised second edition [Hilbert & Bernays,

1968; 1970]. Beside small corrections, the second edition incorporated about a dozen multi-page new parts and some additional supplements; but about 90% of the pages were photo-identical to the first edition (except for the page numbers).

It has been a great impairment to science and foundational research that, although translations into Russian[1] and French[2] are available, there has never been an English edition of these two milestones in the development of modern mathematical logic. One of the reasons for the lack of a translation may have been the copyright issue, but a more important reason is certainly the enormous breadth of the required scholarship of the translators because of the deep insight displayed in the presented mathematical content, but also the very rich and sophisticated German language of the original, mixing philosophical and mathematical concepts and arguments, routed in more than one way in KANTian writings.

Although proof theory is nowadays well represented in excellent English textbooks[3] and the further development of the field after World War II was published in English,[4] and although HILBERT's program and its related form of intuitionism (*finitism*) have been well documented and studied,[5] the international research communities still lack access to the roots of proof theory, especially to the basic philosophical and epistemological conceptions of BERNAYS and HILBERT, as well as HILBERT's ε. The lack of access to the original sources has resulted in a simplistic and sometimes even mistaken evaluation of HILBERT's proof theory, and especially of HILBERT's ε.

We began working on this translation in 2004, when CORINNA ELSENBROICH (then: King's College London) produced a first rough English translation of [HILBERT & BERNAYS, 1934, §§ 1–8] and [HILBERT & BERNAYS, 1939, §§ 1–3]. With the help of KURT ENGESSER (Universität Konstanz), we produced a mathematically and historically more accurate second draft of [HILBERT & BERNAYS, 1934, §§ 1–3] by the end of 2007. When CLAUS-PETER WIRTH became the main translator, commentator, and coordinator, and an outstanding advisory board was formed in 2008, the work gained momentum and professionalism. Since then, the text has been reworked by us many times with the help of the advisory board. In 2010, with the support the International Federation of Computational Logic (IFCoLog), we obtained all formal rights on the English translation and the copyright for the German originals of [HILBERT & BERNAYS, 1934; 1939; 1968; 1970] from Springer-Verlag.

The present translation represents a compromise in several respects:

1. Mathematical logic and proof theory are firmly established today with a pretty uniform set of notions and notation. This was not the case at the time of the origin of the two volumes. We had to find a comprise in translation between modern parlor and the historical flavor of BERNAYS' writing and the historical translations of the technical terms.

---

[1] Cf. [HILBERT & BERNAYS, 1979; 1982].
[2] Cf. [HILBERT & BERNAYS, 2001; 2003].
[3] Cf. e.g. [KLEENE, 1952], [SCHÜTTE, 1977], [TAKEUTI, 1987], and [TROELSTRA & SCHWICHTENBERG, 1996].
[4] Cf. e.g. [KREISEL, 1951; 1952; 1958; 1965; 1982] [PRAWITZ &AL., 1994], [FEFERMAN, 1996] and [DELZELL, 1996].
[5] Cf. e.g. [REID, 1970] [TAIT, 1981], [MANCOSU, 1998], [PARSONS, 1998], [ZACH, 2001] and [SIEG, 2009].

2. Another difficult issue was to assess exactly what was known to HILBERT and BERNAYS at their time and what is our interpretation in the light of what we know today. It is here that the knowledge and the proofreading of the advisory board turned out to be indispensable and crucially valuable.

3. Finally there is the issue of HILBERT's and BERNAYS' writing style, rooted in the tradition of German philosophy with its long and highly grammatically structured sentences. Such German is notoriously difficult to translate into acceptable English while maintaining the full subtlety of the language and content. So when in doubt, while staying as close as possible to the original, we opted for clarity rather than a literal translation (contrary to the French edition)[6] sometimes breaking the long sentences into parts, and translating synonyms with a single word, rather than choosing two not perfectly synonymous English words. It is up to the bilingual reader to judge if and how close we came to an acceptable solution.

Each of the two original German volumes will be published in several commented and bilingual volumes by College Publications with the support of IFCoLog. These books will be printed on demand and offered at a reasonable price, making this textbook also available, once again, to students.

Finally, we would like to express our thanks to the International Federation for Computational Logic (IFCoLog), the German Research Center for Artificial Intelligence (DFKI), and the German Max Planck Institute for Informatics (MPI-INF) for their support. Special thanks are due to CHARLES PARSONS and CHRISTIAN TAPP for their careful proofreading and their magnificent comments on the prefaces of HILBERT and BERNAYS and the first two sections presented in this volume. Further thanks go to IRVING H. ANELLIS, BERND BULDT, JOHN N. CROSSLEY, JUSTUS DILLER, PHILIP EBERT, MURDOCH J. GABBAY, THOMAS HILLENBRAND, ERICA MELIS, MARKUS MOSCHNER, VOLKER PECKHAUS, MARIANEH REZAEI, MARVIN SCHILLER, WILFRIED SIEG, WILLIAM W. TAIT, and RICHARD ZACH.

DOV GABBAY
MICHAEL GABBAY
JÖRG SIEKMANN
CLAUS-PETER WIRTH

---

[6]The emphasis of the French translation on preserving the style of the original finds its expression in the following passage:

"Nous avons donc tenté de respecter la lettre du texte originel, et parfois nous adoptons même le mot à mot (au risque de surprendre le lecteur, en particulier par l'abondance des adverbes). Ce texte originel a ses particularités: c'est, comme le dit HILBERT dans son Introduction, un cours oral; à le lire, on entend parfois littéralement le professeur s'adresser à ses étudiants, depuis sa chaire ou son tableau noir." [HILBERT & BERNAYS, 2001, p.12]

In opposition to the opinion expressed in this passage, we do not take the "Grundlagen der Mathematik" for a "cours oral": The learned language of the "Grundlagen der Mathematik" suits lectures and lecture notes just as poorly as the language of KANT's critiques. In their lectures, great teachers, such as HILBERT and KANT, use a language more appropriate for that special purpose.

Hilbert's Proof Theory

by Wilfried Sieg[7]

> Überhaupt ist es ein wesentlicher Vorzug der Mathematik gegenüber der Philosophie, dass sie sich mit ihrer Spekulation niemals in phantasievolles mystisches Denken hüllt.[8]

1 Changing perspectives

Hilbert gave lectures on the foundations of mathematics throughout his career. Notes for many of them have been preserved and are remarkable treasures of information; they allow us to reconstruct the circuitous path from Hilbert's logicist position, deeply influenced by Dedekind and presented in lectures starting around 1890, to the novel program of finitist proof theory in the 1920s. The general direction of this path in quite unchartered territory was determined by Hilbert's focus on the *axiomatic method* and the associated *consistency problem*. Thus, our journey along that path begins with an analysis of the modern axiomatic method, proceeds to a discussion of the rôle of consistency proofs, and ends with a sketch of the content of Hilbert and Bernays' magisterial work *Grundlagen der Mathematik*. What a remarkable path it is: emerging as it does from the radical transformation of mathematics in the second half of the 19th century, it leads to the new field of metamathematics or mathematical logic, with Hilbert insisting from the start that foundational problems be treated mathematically as far as possible. Examining that path with Hilbert and later on also with Bernays allows us to appreciate Hilbert's changing perspectives on the wide-open mathematical landscape, but it also enriches our perspective on Hilbert's metamathematical work. Though this essay sheds light on contemporary issues, it is intended to be mostly historical. Thus, I will quote extensively in particular from the unpublished source material.

---

[7] We are grateful to Elsevier and Wilfried Sieg for allowing the re-publication of this essay (Elsevier License Number 2750880169006; Sept. 16, 2011), which appeared originally as [Sieg, 2009] and had been dedicated to George Boolos. We have edited the references of this essay so that the new version can share the bibliography and is consistent with the following chapters. Moreover, we have set the relevant names into small capital letters so that they are consistent with our reference style.

[8] Hilbert [1917b] according to [Goeb, 1917, p.102]. The English translation is:

"In fact, it is an essential advantage of mathematics when compared to philosophy that it [mathematics] never wraps itself with its speculation in fanciful mystical thinking."

1.1 A pivotal essay

In some sense, the development towards *Beweistheorie* really begins in late 1917 when HILBERT gave a talk in *Zürich*, entitled *Axiomatisches Denken*. This talk is rooted in the past and points to the future. HILBERT suggested in particular:

> "... we must — that is my conviction — take the concept of the specifically mathematical proof as an object of investigation, just as the astronomer has to consider the movement of his position, the physicist must study the theory of his apparatus, and the philosopher criticizes reason itself."[9]

HILBERT recognized full well, in the very next sentence, that "the execution of this program is at present, to be sure, still an unsolved problem". And yet, initial steps had been taken at the International Congress of Mathematicians in *Heidelberg* (1904); HILBERT presented there a syntactic consistency proof for a purely equational fragment of number theory. [HILBERT, 1905a], on which the talk was based, insisted programmatically on a simultaneous development of logic and number theory. HILBERT expected that his proof could be extended to establish the consistency for such a joint formalism.

Already in 1900, at the previous International Congress of Mathematicians in *Paris*, HILBERT had articulated the need for a consistency proof as his Second Problem [HILBERT, 1901]. However, at that point the consistency of analysis was at stake, and it was to be established by semantic techniques adopted from the theory of real numbers. Neither in his Paris nor in his Heidelberg talk did HILBERT specify a suitable logical calculus; that also holds for all the lectures he gave before the winter term 1917/18. It is only then that the study of *Principia Mathematica* [WHITEHEAD & RUSSELL, 1910–1913] bore fruit; the process of absorbing the core of WHITEHEAD and RUSSELL's work had begun in 1913. It provided a way of formulating joint formalisms with appropriate logicist and inferential principles, thus transcending the restricted systems of the Heidelberg talk. HILBERT had found the means to develop number theory and analysis, strictly formally.

Proof theory, as we understand it, begins only in 1922 when the finitist standpoint provides the methodological grounding for consistency proofs, but the developments between 1917 and 1922 are extremely important. First of all, what was only alluded to at the end of the Zürich talk was actually carried out in HILBERT's remarkable 1917/18 lectures *Prinzipien der Mathematik* [HILBERT, 1917/18]. Secondly, in his lectures from the summer

---

[9][EWALD, 1996, p.1115]. Here is the German text from [HILBERT, 1917a, p.155]:

> "... [wir] müssen — das ist meine Überzeugung — den Begriff des spezifisch mathematischen Beweises selbst zum Gegenstand einer Untersuchung machen, gerade wie ja auch der Astronom die Bewegung seines Standortes berücksichtigt, der Physiker sich um die Theorie seines Apparates kümmern muß und der Philosoph die Vernunft selbst kritisiert."

This programmatic perspective is reiterated in [HILBERT, 1922b, pp. 169–170]:

> "Um unser Ziel zu erreichen, müssen wir die Beweise als solche zum Gegenstande unserer Untersuchungen machen; wir werden so zu einer Art *Beweistheorie* gedrängt, die von dem Operieren mit den Beweisen selbst handelt. Für die konkret-anschauliche Zahlentheorie, die wir zuerst betrieben, waren die Zahlen das Gegenständliche und Aufweisbare, und die Beweise der Sätze über Zahlen fielen schon in das gedankliche Gebiet. Bei unserer jetzigen Untersuchung ist der Beweis selbst etwas Konkretes und Aufweisbares; die inhaltlichen Überlegungen erfolgen erst an dem Beweise. Wie der Physiker seinen Apparat, der Astronom seinen Standort untersucht, wie der Philosoph Vernunftkritik übt, so hat meiner Meinung nach der Mathematiker seine Sätze erst durch eine Beweiskritik sicherzustellen, und dazu bedarf er dieser Beweistheorie."

term of 1920 [HILBERT, 1920], he goes back to the syntactic perspective of the Heidelberg talk and reproves the consistency result I mentioned — in a slightly different way. Last but certainly not least, PAUL BERNAYS joined HILBERT in the fall of 1917 as assistant for work on the foundations of mathematics. He not only helped prepare HILBERT's lectures on foundational issues and wrote the notes for many of them, starting with those of the winter term 1917/18, but he became a crucial collaborator in a joint enterprise and was seen by HILBERT in that light. It is difficult to imagine the evolution of proof theory without his mathematical acumen and philosophical sophistication. Indeed, my systematic presentation of HILBERT's proof theory culminates in a discussion of the second volume of *Grundlagen der Mathematik*, a volume that was written exclusively by BERNAYS.

The volumes of [HILBERT & BERNAYS, 1934; 1939] are milestones in the development of mathematical logic, not just of proof theory. They were at the forefront of contemporaneous research and presented crucial metamathematical results that are still central: from consistency proofs HILBERT and BERNAYS had obtained in weaker forms during the 1920s, through theorems of HERBRAND and GÖDEL, to a sketch of GENTZEN's consistency proof for number theory. The core material is expanded in the second volume through a series of important supplements concerning particular topics. One finds there, for example, an elegant development of analysis and an incisive presentation of the undecidability of the decision problem for first-order logic. In short, the two volumes constitute an encyclopedic synthesis of metamathematical work from the preceding two decades. What is most remarkable, however, is the sheer intellectual force that structures the books: these are penetrating studies concerned with and deeply connected to the foundations of modern mathematics as it emerged in the second half of the 19th century.

1.2 Transformations

Do we have to go back that far in order to obtain a proper perspective on HILBERT's proof theory? Don't we have an adequate grasp of its development in the 1920s? Is there more to be said about the connection to the earlier foundational work than what was indicated above, supplementing the standard view of the evolution of proof theory? According to that common view, endorsed even by BERNAYS [1935a][10] and WEYL [1944], the evolution is divided into two periods. During the first period, roughly from 1899 to 1904, HILBERT is seen as unsuccessfully addressing difficulties for the foundations of arithmetic that had been raised by the set-theoretic paradoxes. The second period begins around 1920, motivated again externally by the emergence of BROUWER's and WEYL's intuitionism, and it is seen as lasting either until 1931 (the publication of GÖDEL's incompleteness paper) or until 1939 (the publication of the second volume of *Grundlagen der Mathematik*). The initial work of the second period is characterized as a refinement of the earlier approach and as leading to HILBERT's *finitist consistency program*. Of course, it is recognized that this

---

[10]In [BERNAYS, 1935a] there is a general smoothing of the developments between 1904/05 and 1917/22: BERNAYS does not mention that HILBERT gave lectures on the foundations of mathematics during that period; the lectures [HILBERT, 1917/18] are not even hinted at. BERNAYS effectively creates the impression that the period is one of inactivity; for example, one finds:

"In diesem vorläufigen Stadium hat HILBERT seine Untersuchungen über die Grundlagen der Arithmetik für lange Zeit unterbrochen[1]. Ihre Wiederaufnahme finden wir angekündigt in dem 1917 gehaltenen Vortrage[2] ‚Axiomatisches Denken'." [BERNAYS, 1935a, p. 200]

The impression is reinforced by Note 1 attached to the first sentence in the above quotation, where BERNAYS points to the work of others who pursued the research direction stimulated by [HILBERT, 1905a].

refinement involves a significant methodological shift and requires novel metamathematical investigations.

For both periods HILBERT's proposals are, however, *more* than defensive ad-hoc reactions against threats to "classical" mathematics and emerge out of long periods of intense work on difficult issues. The issues are connected in both cases to the creative freedom in mathematics celebrated by CANTOR, DEDEKIND, and HILBERT himself, but seen as severely restricted by constructivity requirements that were articulated first by KRONECKER and then by BROUWER. Thus, they are ultimately grounded in the 19th century transformation of classical into modern abstract mathematics. HILBERT's problems and his metamathematical approaches are deeply intertwined with this transformation, and we can better understand it by better understanding HILBERT's perspective on it.[11]

The need for historical study and conceptual analysis is thereby not exhausted, as the precise character of HILBERT's problems and his techniques for solving them developed substantially and often dramatically. The complex evolution is chronicled in the notes for lectures on mathematics and logic that were given by HILBERT between 1890 and 1930. TOEPELL [1986] and PECKHAUS [2003] have made clear how important these notes are for the early part of this long period. HALLETT and MAJER have strengthened that perspective through [HILBERT, 2004]. [SIEG, 1999] investigates the progression from a RUSSELLian logicist program articulated in 1917 to the finitist program of 1922: such a study would have been impossible without access to the lecture notes from those years. The historical evolution shows how deeply connected HILBERT's investigations are. Many fascinating aspects have been detailed or re-examined, for example, by AVIGAD [2006], CORRY [2004], EWALD [1996], FEFERMAN [1988; 1998], FERREIRÓS [2007], GRAY [1992], HALLETT [1994], MAJER [2006], MANCOSU [1998], MOORE [1997], PARSONS [2008], PECKHAUS & KAHLE [2002], RAVAGLIA [2003], ROWE [2000], TAIT [2005], TAPP [2006], and ZACH [1999; 2003].

1.3 Immediate context and overview

HILBERT's synthesizing perspective and the reasons for his insistence on formulating sharp problems when confronting the foundations of mathematics come to life in the long preamble to his Paris talk [HILBERT, 1901]. His account of mathematics emphasizes the subtle and ever recurring interplay of rigorous abstract thought and concrete experience. Turning his attention to the principles of analysis and geometry, he asserts that the "most suggestive and notable achievements of the last century are ... the arithmetical formulation of the concept of the continuum ... and the discovery of non-EUCLIDean geometry ...". Then he states CANTOR's continuum problem as his first and the consistency issue for the arithmetical axioms as his second problem:

> "... I wish to designate the following as the most important among the numerous questions which can be asked with regard to the axioms: *To prove that they are not contradictory, that is, that a finite number of logical steps based upon them can never lead to contradictory results.*"[12]

---

[11] BERNAYS emphasizes in [BERNAYS, 1930/31] (on pp. 17–18 and pp. 19–20) that the changes in the methodological perspective of mathematics during the second half of the 19th century had three characteristic features, namely, the use of the concept of set for the rigorous foundation of analysis, the emergence of existential axiomatics ("existentiale Axiomatik"), and building a closer connection between mathematics and logic. Existential axiomatics, the central concept, will be discussed extensively in §2.

As to the arithmetical axioms, HILBERT points to his paper *Über den Zahlbegriff*, delivered at the Munich meeting of the German Association of Mathematicians in December of 1899 [HILBERT, 1900a]. Its title indicates a part of its intellectual context, as KRONECKER had published only twelve years earlier a well-known paper with the very same title. In his paper, KRONECKER sketched a way of introducing irrational numbers, without accepting the general notion. It is precisely to this general concept HILBERT wants to give a proper foundation — using the axiomatic method and following DEDEKIND.

HILBERT connected the consistency problem also to CANTORian issues. From his correspondence with CANTOR we know that he had been aware of CANTOR's "inconsistent multiplicities" since 1897. He recognized their effect on DEDEKIND's logical grounding of analysis and tried to blunt it by a consistency proof for the arithmetical axioms. Such a proof, he stated, provides "the proof of the existence of the totality of real numbers, or — in the terminology of G. CANTOR — the proof that the system of real numbers is a consistent (finished) set."[13] In the Paris talk, HILBERT re-emphasized and expanded this point by saying:

> "In the case before us, where we are concerned with the axioms of real numbers in arithmetic, the proof of the consistency of the axioms is at the same time the proof of the mathematical existence of the complete system ['complete system' is WILLIAM EWALD's translation for 'Inbegriff', WS] of real numbers or of the continuum. Indeed, when the proof for the consistency of the axioms shall be fully accomplished, the doubts, which have been expressed occasionally as to the existence of the complete system of real numbers, will become totally groundless."[14]

He proposed then to extend this approach to the CANTORian Alephs. Consequently and more generally, the consistency of appropriate axiom systems was to guarantee the existence of sets or, as he put it in [HILBERT, 1905a], to provide an *objective criterion* for the consistency of multiplicities.

I will explore the background for these foundational issues in §2 under the heading *Existential axiomatics*. §3 is concerned with *Direct consistency proofs* for equational calculi. §4 looks at *Finitist proof theory* and its problems, namely, how first to treat quantifier-free systems of arithmetic and then to deal with quantifiers via the ε-substitution method. §5 discusses *Incompleteness* and related mathematical as well as philosophical problems; the essay ends with a very brief outlook beyond 1939.

---

[12] [EWALD, 1996, pp. 1102–1103]. This is part of the German text from [HILBERT, 1901, p. 55]: "Vor allem möchte ich unter den zahlreichen Fragen, welche hinsichtlich der Axiome gestellt werden können, dies als das wichtigste Problem bezeichnen, *zu beweisen, daß dieselben unter einander widerspruchslos sind, d. h., daß man auf Grund derselben mittels einer endlichen Anzahl von logischen Schlüssen niemals zu Resultaten gelangen kann, die mit einander in Widerspruch stehen.*"

[13] [EWALD, 1996, p.1095]. This is part of the German paragraph in [HILBERT, 1900a, p. 261]: "Um die Widerspruchslosigkeit der aufgestellten Axiome zu beweisen, bedarf es nur einer geeigneten Modifikation bekannter Schlußmethoden. In diesem Nachweis erblicke ich zugleich den Beweis für die Existenz des Inbegriffs der reellen Zahlen oder — in der Ausdrucksweise G. CANTORs — den Beweis dafür, daß das System der reellen Zahlen eine konsistente (fertige) Menge ist."

[14] [EWALD, 1996, p.1095]. Here is the German text from [HILBERT, 1901, p. 56]: "In dem vorliegenden Falle, wo es sich um die Axiome der reellen Zahlen in der Arithmetik handelt, ist der Nachweis für die Widerspruchslosigkeit der Axiome zugleich der Beweis für die mathematische Existenz des Inbegriffs der reellen Zahlen oder des Kontinuums. In der That, wenn der Nachweis für die Widerspruchslosigkeit der Axiome völlig gelungen sein wird, so verlieren die Bedenken, welche bisweilen gegen die Existenz des Inbegriffs der reellen Zahlen gemacht worden sind, jede Berechtigung."

2   Existential axiomatics

The axiom system HILBERT formulated for the real numbers in 1900 is not presented in the contemporary logical style. Rather, it is given in an algebraic way and assumes that a structure exists whose elements satisfy the axiomatic conditions. HILBERT points out that this *axiomatic* way of proceeding is quite different from the *genetic method* standardly used in arithmetic; it rather parallels the ways of geometry.[15]

> "Here [in geometry] one begins customarily by assuming the existence of all the elements, i.e. one postulates at the outset three systems of things (namely, the points, lines, and planes), and then — essentially after the model of EUCLID — brings these elements into relationship with one another by means of certain axioms of linking, order, congruence, and continuity."[16]

These geometric ways are taken over for arithmetic or rather, one might argue, are re-introduced into arithmetic by HILBERT; after all, they do have their origin in DEDEKIND's work on arithmetic and algebra. In the 1920s, HILBERT and BERNAYS called this methodological approach *existential axiomatics*, because it assumes the existence of suitable systems of objects. The underlying ideas, however, go back in HILBERT's thinking to at least 1893.

DEDEKIND played a significant rôle in their evolution not only directly through his foundational essays and his mathematical work, but also indirectly, as his collaborator and friend HEINRICH WEBER was one of HILBERT's teachers in Königsberg and a marvelous interpreter of DEDEKIND's perspective. For example, WEBER's *Lehrbuch der Algebra* [WEBER, 1895f.] and some of his expository papers are frequently referred to in HILBERT's lectures around the turn from the 19th to the 20th century. It is a slow evolution, which is impeded by the influence of KRONECKER's views as to how irrational numbers are to be given. The axiomatic method is to overcome the problematic issues KRONECKER saw for the general concept of irrational number. HILBERT's retrospective remarks in the notes to HILBERT's lecture [HILBERT, 1904] by BORN [1904] make this quite clear: this general concept created the "greatest difficulties", and KRONECKER represented this point of view most sharply. HILBERT seemingly realized only in [HILBERT, 1904] that the difficulties are overcome, in a certain way, when the concept of natural number is secured. "The further steps up to the irrational numbers," he claims there, "are then taken in a logically rigorous way and without further difficulties."[17] Let us look at the details of this remarkable evolution, as far as it can be made out from the available documents.

---

[15]In [HELLINGER, 1905] there is a detailed discussion of the genetic foundation (genetische Begründung) with references to textbooks by LIPSCHITZ, PASCH, and THOMAE. HILBERT formulates there also the deeply problematic aspects of this method; see the beginning of § 2.2.

[16][EWALD, 1996, p.1092]. Here is the German text from [HILBERT, 1900a, p. 257]:

> "Hier [in der Geometrie] pflegt man mit der Annahme der Existenz von sämtlichen Elementen zu beginnen, d. h. man setzt von vornherein drei Systeme von Dingen, nämlich die Punkte, die Geraden und die Ebenen, und bringt sodann diese Elemente — wesentlich nach dem Vorbilde von EUKLID — durch gewisse Axiome, nämlich die Axiome der Verknüpfung, der Anordnung, der Kongruenz und der Stetigkeit, miteinander in Beziehung."

HILBERT obviously forgot to mention, between the axioms of order and of congruence, the axiom of parallels.

2.1 Axioms for continuous systems

HILBERT introduces the axioms for the real numbers in [HILBERT, 1900a] as follows:

> "We think of a system of things, and we call them numbers and denote them by $a, b, c, \ldots$. We think of these numbers in certain mutual relations, whose precise and complete description is obtained through the following axioms."[18]

Then the axioms for an ordered field are formulated and rounded out by the requirement of continuity via the Archimedean axiom and the axiom of completeness (Axiom der Vollständigkeit). This formulation is not only in the spirit of the geometric ways, but actually mimics the contemporaneous and paradigmatically modern axiomatic presentation of *Grundlagen der Geometrie* [HILBERT, 1899]:

> "We think of three different systems of things: we call the things of the first system *points* and denote them by $A, B, C, \ldots$; we call the things of the second system *lines* and denote them by $a, b, c, \ldots$; we call the things of the third system *planes* and denote them by $\alpha, \beta, \gamma, \ldots$; .... We think of the points, lines, planes in certain mutual relations ...; the precise and complete description of these relations is obtained by the *axioms of geometry*."[19]

Five groups of geometric axioms follow and, in the first German edition, the fifth group consists of just the Archimedean axiom. In the French edition [HILBERT, 1900b] and the second German edition [HILBERT, 1903], the completeness axiom is included. The latter axiom requires in both the geometric and the arithmetic case that the assumed structure is maximal, i.e., any extension satisfying all the remaining axioms must already be contained

---

[17]These matters are discussed in [BORN, 1904, pp. 164–167]. One most interesting part consists of these remarks on pp. 165–166:

> "Die Untersuchungen in dieser Richtung [foundations for the real numbers] nahmen lange Zeit den breitesten Raum ein. Man kann den Standpunkt, von dem dieselben ausgingen, folgendermaßen charakterisieren: Die Gesetze der ganzen Zahlen, der Anzahlen, nimmt man vorweg, begründet sie nicht mehr; die Hauptschwierigkeit wird in jenen Erweiterungen des Zahlbegriffs (irrationale und weiterhin komplexe Zahlen) gesehen. Am schärfsten wurde dieser Standpunkt von KRONECKER vertreten. Dieser stellte geradezu die Forderung auf: Wir müssen in der Mathematik jede Tatsache, so verwickelt sie auch sein möge, auf Beziehungen zwischen ganzen rationalen Zahlen zurückführen; die Gesetze dieser Zahlen andrerseits müssen wir ohne weiteres hinnehmen. KRONECKER sah in den Definitionen der irrationalen Zahlen Schwierigkeiten und ging soweit, dieselben gar nicht anzuerkennen."

[18][HILBERT, 1900a, pp. 257–258]. The German text is:

> "Wir denken ein System von Dingen; wir nennen diese Dinge Zahlen und bezeichnen sie mit $a, b, c, \ldots$. Wir denken diese Zahlen in gewissen gegenseitigen Beziehungen, deren genaue und vollständige Beschreibung durch die folgenden Axiome geschieht."

[19][HILBERT, 2004, p. 437]. The German text is:

> "Wir denken drei verschiedene Systeme von Dingen: die Dinge des ersten Systems nennen wir *Punkte* und bezeichnen sie mit $A, B, C, \ldots$; die Dinge des zweiten Systems nennen wir *Geraden* und bezeichnen sie mit $a, b, c, \ldots$; die Dinge des dritten Systems nennen wir *Ebenen* und bezeichnen sie mit $\alpha, \beta, \gamma, \ldots$; .... Wir denken die Punkte, Geraden, Ebenen in gewissen gegenseitigen Beziehungen ...; die genaue und vollständige Beschreibung dieser Beziehungen erfolgt durch die Axiome der Geometrie." [HILBERT, 1899, p. 4]

in it. This formulation is frequently criticized as being metamathematical and, to boot, of a peculiar sort.[20] However, it is just an ordinary mathematical formulation, if the algebraic character of the axiom system is kept in mind. In the case of arithmetic we can proceed as follows: call a system A *continuous* when its elements satisfy the axioms of an ordered field and the Archimedean axiom, and call it *fully continuous* if and only if A is continuous and for any system B, if $A \subseteq B$ and B is continuous, then $B \subseteq A$.[21] HILBERT's arithmetic axioms characterize the fully continuous systems uniquely up to isomorphism.

HILBERT thought about axiom systems in this existential way already in *Die Grundlagen der Geometrie* [HILBERT, 1894], his first lectures on the *foundations* of geometry. He had planned to give them in the summer term of 1893, but their presentation was actually shifted to the following summer term. Using the notions "System" and "Ding" — so prominent in [DEDEKIND, 1888] — HILBERT [1894] formulated the central question as follows:

> "What are the necessary and sufficient and mutually independent conditions a system of things has to satisfy, so that to each property of these things a geometric fact corresponds and conversely, thereby making it possible to completely describe and order all geometric facts by means of the above system of things."[22]

At some later point, HILBERT inserted the remark that this axiomatically characterized system of things is a "complete and simple image of geometric reality."[23] The image of "image" is attributed on the next page to HEINRICH HERTZ and extended to cover also logical consequences of the images; they are to coincide with images of the consequences: "The axioms, as HERTZ would say, are images or symbols in our mind, such that consequences of the images are again images of the consequences, i.e., what we derive logically from the images is true again in nature." In the Introduction to the notes [SCHAPER, 1898/99] for the lectures *Elemente der Euklidischen Geometrie* [HILBERT, 1898/99], written by HILBERT's student HANS VON SCHAPER, this central question is explicitly connected with [HERTZ, 1894]:

---

[20]In [TORRETTI, 1984], for example, one finds on p. 234 this observation: "... the completeness axiom is what nowadays one would call a metamathematical statement, though one of a rather peculiar sort, since the theory with which it is concerned includes Axiom V2 [i.e., the completeness axiom, WS] itself."

[21]This is dual to DEDEKIND's definition of simply infinite systems: here we are focusing on maximal extensions, and there on minimal sub-structures. Indeed, the formulation in the above text is to be taken in terms of sub-structures in the sense of contemporary model theory.

[22][TOEPELL, 1986, pp. 58–59] and [HILBERT, 2004, pp. 72–73]. The German text is: "Welches sind die nothwendigen und hinreichenden und unter sich unabhängigen Bedingungen, die man an ein System von Dingen stellen muss, damit jeder Eigenschaft dieser Dinge eine geometrische Thatsache entspricht und umgekehrt, so dass also mittelst obigen Systems von Dingen ein vollständiges Beschreiben und Ordnen aller geometrischen Thatsachen möglich ist."

[23]In [TOEPELL, 1986, p. 59] and [HILBERT, 2004, pp. 73, 74]. The German text is:
  "... so dass also unser System ein vollständiges und einfaches Bild der geometrischen Wirklichkeit werde."
The next German text is:
  "Die Axiome sind, wie HERTZ sagen würde, Bilder oder Symbole in unserem Geiste, so dass Folgen der Bilder wieder Bilder der Folgen sind, d. h. was wir aus den Bildern logisch ableiten, stimmt wieder in der Natur."
On the very page of the last quotation, in note 7 of [HILBERT, 2004], the corresponding and much clearer remark from [HERTZ, 1894] is quoted:
  "Wir machen uns innere Scheinbilder oder Symbole der äußeren Gegenstände, und zwar machen wir sie von solcher Art, daß die denknotwendigen Folgen der Bilder stets wieder die Bilder seien von naturnotwendigen Folgen der abgebildeten Gegenstände."

> "Using an expression of HERTZ (in the introduction to the *Prinzipien der Mechanik*), we can formulate our main question as follows: What are the necessary and sufficient and mutually independent conditions a system of things has to be subjected to, so that to each property of these things a geometric fact corresponds and conversely, thereby having these things provide a complete "image" of geometric reality."[24]

One can see in these remarks the shape of a certain logical or set-theoretic structuralism (that is *not* to be equated with contemporary philosophical forms of such a position) in the foundations of mathematics and of physics; geometry is viewed as the most perfect natural science.[25] How simple or how complex are these systems or structures supposed to be? What characteristics must they have in order to be adequate for both the systematic internal development of the corresponding theory and external applications? What are the things whose system is implicitly postulated? What is the mathematical connection, in particular, between the arithmetic and geometric structures?

DEDEKIND [1872] gives a precise notion of continuity for analysis; however, before doing so, he shifts his considerations to the geometric line. He observes that every point effects a cut, but — and that he views as central for the *continuity* or *completeness* of the line — every cut is effected by a point.[26] Transferring these reflections to cuts of rational numbers and, of course, having established that there are infinitely many cuts of rational numbers that are not effected by rational numbers, he "creates" a new, an irrational number α for any cut not effected by a rational:

> "We will say that this number α corresponds to that cut or that it effects this cut." [DEDEKIND, 1872, p.13]

DEDEKIND then introduces an appropriate order on cuts and establishes that the system of all cuts is indeed continuous. He shows that the continuity principle has to be accepted, if one wants to have the least-upper-bound principle for monotonically increasing, but bounded sequences of real numbers, as the two principles are indeed equivalent. He writes:

> "This theorem is equivalent to the principle of continuity, i.e., it loses its validity as soon as even just a single real number is viewed as not being contained in the domain R; ..."[27]

---

[24][TOEPELL, 1986, p. 204] and [HILBERT, 2004, p. 303]. The German text is:

> "Mit Benutzung eines Ausdrucks von HERTZ (in der Einleitung zu den *Prinzipien der Mechanik*) können wir unsere Hauptfrage so formulieren: Welches sind die nothwendigen und hinreichenden und unter sich unabhängigen Bedingungen, denen man ein System von Dingen unterwerfen muss, damit jeder Eigenschaft dieser Dinge eine geometrische Thatsache entspreche und umgekehrt, damit also diese Dinge ein vollständiges und einfaches "Bild" der geometrischen Wirklichkeit seien."

[25]There is a complementary discussion on pp. 103–104 and 119–122 of [HILBERT, 2004]. HILBERT's perspective is echoed in the views of ERNEST NAGEL and PATRICK SUPPES; cf. also [FRAASSEN, 1980].

[26]DEDEKIND uses *Stetigkeit* and *Vollständigkeit* interchangeably. The reader should notice the freedom with which DEDEKIND reinterprets in the following considerations an axiomatically characterized concept, namely a certain order relation: it is interpreted for the rational numbers, the geometric line, and the system of all cuts of rational numbers.

As it happens, HILBERT's first formulation of a geometric continuity principle in [HILBERT, 1894] (p. 92 of [HILBERT, 2004]) is a least-upper-bound principle for bounded sequences of geometric points.

DEDEKIND [1872] emphasizes the necessity of rigorous arithmetic concepts for analyzing the notion of space. He points out, reflecting the view of CARL FRIEDRICH GAUSZ concerning the fundamental difference between arithmetic and geometry, that space need not be continuous.[28]

The required arithmetization of analysis is established in DEDEKIND's thinking about the matter through three steps, namely,

(i) a logical characterization of the natural numbers,

(ii) the extension of that number concept to integer and rational numbers, and

(iii) the further extension to all real numbers we just discussed.

The first objective is accomplished in [DEDEKIND, 1888] via simply infinite systems, a notion that emerged in the 1870s after DEDEKIND had finished his essay [DEDEKIND, 1872].[29] DEDEKIND gives his final characterization in article 71 of [DEDEKIND, 1888] in the existential, algebraic form that is used later by HILBERT. But he avoids self-consciously the term axiom and speaks only of conditions (Bedingungen), which are easily seen to amount in this case to the well-known PEANO axioms.

DEDEKIND sought the ultimate basis for arithmetic in logic, broadly conceived. That view is formulated in the preface to [DEDEKIND, 1888]:

> "In calling arithmetic (algebra, analysis) merely a part of logic, I already claim that I consider the number concept as entirely independent of notions or intuitions of space and time, that I consider it rather as an immediate product of the pure laws of thought."[30]

Important for our considerations is the fact that DEDEKIND had developed towards the end of the 1870s not only the general concept of "system", but also that of a "mapping between systems". The latter notion corresponds for him to a crucial capacity of our mind [DEDEKIND, 1888, pp. III–IV of the preface]. He uses these concepts to formulate and prove that all simply infinite systems are isomorphic. He concludes from this result, in contemporary model-theoretic terms, that simply infinite systems are all elementarily equivalent and uses that fact to characterize the science of arithmetic:

> "The relations or laws, which are solely derived from the conditions $\alpha, \beta, \gamma, \delta$ in 71 and are therefore in all ordered systems always the same ..., form the next object of the *science of numbers* or of *arithmetic*."[31]

---

[27][DEDEKIND, 1872, p. 20]. The German text is: "Dieser Satz ist äquivalent mit dem Prinzip der Stetigkeit, d. h. er verliert seine Gültigkeit, sobald man auch nur eine einzige reelle Zahl in dem Gebiete R als nicht vorhanden ansieht; ..."

[28]In [DEDEKIND, 1888, p. VII], a particular countable model of real numbers is described in which all of EUCLID's construction can be interpreted.

[29]Cf. [SIEG & SCHLIMM, 2005]. In that paper one finds also a discussion of DEDEKIND's approach to the second issue based on unpublished manuscripts.

[30][DEDEKIND, 1888, p. III]. The German text is: "Indem ich die Arithmetik (Algebra, Analysis) nur einen Teil der Logik nenne, spreche ich schon aus, daß ich den Zahlbegriff für gänzlich unabhängig von den Vorstellungen oder Anschauungen des Raumes und der Zeit, daß ich ihn vielmehr für einen unmittelbaren Ausfluß der reinen Denkgesetze halte."

Natural numbers are not viewed as objects of a particular system, but are obtained by free creation or what WILLIAM W. TAIT has called "DEDEKIND abstraction"; that is clearly articulated in article 73 of [DEDEKIND, 1888].

Similar metamathematical investigations are adumbrated for real numbers in *Stetigkeit und irrationale Zahlen* [DEDEKIND, 1872]. DEDEKIND can be interpreted, in particular, as having characterized *fully continuous systems* as ordered fields that satisfy his principle of continuity, i.e., the principle that every cut of such a system is effected by an element. The system of all cuts of rational numbers constitutes a model; that is in modern terms the essential content of §§ 5 and 6 of [DEDEKIND, 1872], where DEDEKIND verifies that this system satisfies the continuity principle. In contemporaneous letters to RUDOLF LIPSCHITZ, DEDEKIND emphasizes that real numbers should not be identified with cuts, but rather viewed as newly created things corresponding to cuts.

The informal comparison of the geometric line with the system of cuts of rational numbers in §§ 2 and 3 of [DEDEKIND, 1872] contains almost all the ingredients for establishing rigorously that these two structures are isomorphic; missing is the concept of mapping. As I mentioned, that concept was available to DEDEKIND in 1879 and with it these considerations can be extended directly to show that arbitrary fully continuous systems are isomorphic. Finally, the methodological remarks about the arithmetic of natural numbers can now be extended to that of the real numbers. As a matter of fact, HILBERT articulates DEDEKIND's way of thinking of the system of real numbers in [HILBERT, 1922b], when describing the *axiomatische Begründungsmethode* for analysis:

> "The continuum of real numbers is a system of things that are connected to each other by certain relations, so-called axioms. In particular the definition of the real numbers by the DEDEKIND cut is replaced by two continuity axioms, namely, the Archimedian axiom and the so-called completeness axiom. In fact, DEDEKIND cuts can then serve to determine the individual real numbers, but they do not serve to define [the concept of] real number. On the contrary, conceptually a real number is just a thing of our system. ... The standpoint just described is altogether logically completely impeccable, and it only remains thereby undecided, whether a system of the required kind can be thought, i.e., whether the axioms do not lead to a contradiction."[32]

That is fully in DEDEKIND's spirit: HILBERT's critical remarks about the definition of real numbers as cuts do not apply to DEDEKIND, as should be clear from the above discussion, and the issue of consistency had been an explicit part of DEDEKIND's logicist program.

---

[31][DEDEKIND, 1888, p.17]. The German text is: "Die Beziehungen oder Gesetze, welche ganz allein aus den Bedingungen $\alpha, \beta, \gamma, \delta$ in 71 abgeleitet werden und deshalb in allen geordneten Systemen immer dieselben sind ... bilden den nächsten Gegenstand der *Wissenschaft von den Zahlen* oder der *Arithmetik*."

[32][HILBERT, 1922b, pp. 158–159]. The German text is: "Das Kontinuum der reellen Zahlen ist ein System von Dingen, die durch bestimmte Beziehungen, sogenannte Axiome, miteinander verknüpft sind. Insbesondere treten an Stelle der Definition der reellen Zahlen durch den DEDEKINDschen Schnitt die zwei Stetigkeitsaxiome, nämlich das Archimedische Axiom und das sogenannte Vollständigkeitsaxiom. Die DEDEKINDschen Schnitte können dann zwar zur Festlegung der einzelnen reellen Zahlen dienen, aber sie dienen nicht zur Definition der reellen Zahl. Vielmehr ist begrifflich eine reelle Zahl eben ein Ding unseres Systems. ... Der geschilderte Standpunkt ist vollends logisch vollkommen einwandfrei, und es bleibt nur dabei unentschieden, ob ein System der verlangten Art denkbar ist, d. h. ob die Axiome nicht etwa auf einen Widerspruch führen."

2.2 Consistency and logical models

Logicians in the 19th century viewed consistency of a notion from a semantic perspective. In that tradition, DEDEKIND addressed the problem for his characterization of natural numbers via simply infinite systems. The methodological need for doing that is implicit in [DEDEKIND, 1872], but is formulated most clearly a year after the publication of *Was sind und was sollen die Zahlen?* [DEDEKIND, 1888] in a letter to KEFERSTEIN [DEDEKIND, 1889]:

> "After the essential nature of the simply infinite system, whose abstract type is the number sequence N, had been recognized in my analysis (71, 73) the question arose: does such a system *exist* at all in the realm of our thoughts? Without a logical proof of existence, it would always remain doubtful whether the notion of such a system might not perhaps contain internal contradictions. Hence the need for such a proof (articles 66 and 72 of my essay)."[33]

DEDEKIND attempted in article 66 to prove the existence of an infinite system and to provide in article 72 an example of a simply infinite system within logic. In other words, he attempted to establish the existence of a *logical* model and to guarantee in this way that the notion of a simply infinite system indeed does not contain internal contradictions.

HILBERT turned to considerations for natural numbers only around 1904. Until then he had taken their proper foundation for granted and focused on the theory of real numbers: As I reported earlier from [BORN, 1904], HILBERT thought that the concept of irrational numbers created the greatest difficulties. These difficulties have been overcome, HILBERT claims in [BORN, 1904]. When the concept of natural number is secured, he continues there, the further steps towards the real numbers can be taken without a problem. This dramatic change of viewpoint helps us to understand more clearly some aspects of HILBERT's earlier considerations related to the consistency of the arithmetic of real numbers.

> What did HILBERT see as the "greatest difficulties" for the general concept of irrational numbers?

That question will be explored now.

In contrast to DEDEKIND, HILBERT had formulated a quasi-syntactic notion of consistency in [HILBERT, 1899; 1900a], namely, no finite number of logical steps leads from the axioms to a contradiction. This notion is "*quasi*-syntactic", as no deductive principles are explicitly provided. HILBERT did not seek to prove consistency by syntactic methods, however. On the contrary, the relative consistency proofs given in [HILBERT, 1899] are all straightforwardly semantic. To prove the consistency of the arithmetic of real numbers, one needs "a suitable modification of familiar methods of reasoning" [HILBERT, 1900a]. In the Paris talk, HILBERT suggests finding a *direct* proof and remarks more expansively:

> "I am convinced that it must be possible to find a direct proof for the consistency of the arithmetical axioms [as proposed in [HILBERT, 1900a] for the real numbers; WS], by means of a careful study and suitable modification of the known methods of reasoning in the theory of irrational numbers."[34]

---

[33]In [HEIJENOORT, 1971, p.101]. The German text is:
"Nachdem in meiner Analyse der wesentliche Charakter des einfach unendlichen Systems, deßen abstracter Typus die Zahlenreihe N ist, erkannt war (71, 73), fragte es sich: *existirt* überhaupt ein solches System in unserer Gedankenwelt? Ohne den logischen Existenz-Beweis würde es immer zweifelhaft bleiben, ob nicht der Begriff eines solchen Systems vielleicht innere Widersprüche enthält. Daher die Nothwendigkeit solcher Beweise (66, 72 meiner Schrift)."

HILBERT believed at this point, it seems to me, that the genetic build-up of the real numbers could *somehow* be exploited to yield the blueprint for a consistency proof in DEDEKIND's logicist style. There are difficulties with the genetic method that prevent it from providing directly a proper foundation for the general concept of irrational numbers. HILBERT's concerns are formulated most clearly in these retrospective remarks from [HELLINGER, 1905]:

> "It [the genetic method, WS] defines things by generative processes, not by properties what really must appear to be desirable. Even if there is no objection to defining fractions as systems of two integers, the definition of irrational numbers as a system of infinitely many numbers must appear to be dubious. Must this number sequence be subject to a law, and what is to be understood by a law? Is an irrational number being defined, if one determines a number sequence by throwing dice? These are the kinds of questions with which the genetic perspective has to be confronted."[35]

Precisely this issue was to be overcome (or to be sidestepped) by the axiomatic method. HILBERT [1900a] writes:

> "Under the conception described above, the doubts which have been raised against the existence of the totality of real numbers (and against the existence of infinite sets generally) lose all justification; for by the set of real numbers we do not have to imagine, say, the totality of all possible laws according to which the elements of a fundamental sequence can proceed, but rather — as just described — a system of things whose mutual relations are given by the finite and closed system of axioms I–IV, ..."[36]

---

[34][EWALD, 1996, p. 1104]. The German text from [HILBERT, 1901, p. 55] is:

"Ich bin nun überzeugt, daß es gelingen muß, einen direkten Beweis für die Widerspruchslosigkeit der arithmetischen Axiome zu finden, wenn man die bekannten Schlußmethoden in der Theorie der Irrationalzahlen im Hinblick auf das bezeichnete Ziel genau durcharbeitet und in geeigneter Weise modifiziert."

In [BERNAYS, 1935a, pp. 198–199], this idea is re-emphasized:

"Zur Durchführung des Nachweises gedachte HILBERT mit einer geeigneten Modifikation der in der Theorie der reellen Zahlen angewandten Methoden auszukommen."

[35][HELLINGER, 1905, pp. 10–11]. The German text is:

"Sie definiert Dinge durch Erzeugungsprozesse, nicht durch Eigenschaften, was doch eigentlich wünschenswert erscheinen muß. Ist nun auch gegen die Definition der Brüche als System zweier Zahlen nichts einzuwenden, so muß doch die Definition der Irrationalzahlen als ein System von unendlich vielen Zahlen bedenklich erscheinen. Muß diese Zahlenreihe einem Gesetz unterliegen, und was hat man unter einem Gesetz zu verstehen? Wird auch eine Irrationalzahl definiert, wenn man eine Zahlenreihe durch Würfeln feststellt? Dieser Art sind die Fragen, die man der genetischen Betrachtung entgegenhalten muß."

[36][EWALD, 1996, p. 1095]. Here is the German text from [HILBERT, 1900a, p. 261]:

"Die Bedenken, welche gegen die Existenz des Inbegriffs aller reellen Zahlen und unendlicher Mengen überhaupt geltend gemacht worden sind, verlieren bei der oben gekennzeichneten Auffassung jede Berechtigung: unter der Menge der reellen Zahlen haben wir uns hiernach nicht etwa die Gesamtheit aller möglichen Gesetze zu denken, nach denen die Elemente einer Fundamentalreihe fortschreiten können, sondern vielmehr — wie eben dargelegt ist — ein System von Dingen, deren gegenseitige Beziehungen durch das obige endliche und abgeschlossene System von Axiomen I–IV gegeben sind, ..."

In the Paris talk, HILBERT [1901] articulates exactly the same point by re-emphasizing "the continuum ... is not the totality of all possible series in decimal fractions, or of all possible laws according to which the elements of a fundamental sequence may proceed." Rather, it is a system of things whose mutual relations are governed by the axioms. The completeness axiom can be plausibly interpreted as guaranteeing the continuity of the system without depending on any particular method of generating real numbers.

Going back to the 1889/90 lectures [HILBERT, 1889/90], one sees that HILBERT then shared a KRONECKERian perspective on irrationals. He starts out with a brief summary of "Das Nothwendigste über den Zahlbegriff" and uses DEDEKIND cuts to define the notion of an irrational number. He shifts for the practical presentation of real numbers from cuts to fundamental sequences and asserts, making the problematic aspect fully explicit, "It is absolutely necessary to have a general method for the representation of irrational numbers."[37] The need to have a method of representation is not bound to fundamental sequences. Cuts are to be given by a *rule* that allows the partition of all rational numbers into smaller and greater ones: the real number effecting the cut is taken to be "the carrier of that rule". So we have a sense of the difficulties as HILBERT saw them. In order to support the claim that HILBERT thought, nevertheless, that the genetic build-up of the real numbers could *in some way* be exploited to yield a logicist consistency proof, we look closely at HILBERT's treatment of arithmetic in later lectures, in particular [HILBERT, 1894; 1897/98].

Let me begin with some deeply programmatic, logicist statements found in the *Introduction* of the notes of the lecture [HILBERT, 1898/99] by HILBERT's student HANS VON SCHAPER:

> "It is important to fix precisely the starting-point of our investigations: as given we consider the laws of pure logic and in particular all of arithmetic."[38]

Clearly, arithmetic is here taken in the narrower sense of the theory of natural numbers. He adds parenthetically, "On the relation between logic and arithmetic cf. DEDEKIND, *Was sind und was sollen die Zahlen?*" Clearly, for DEDEKIND, arithmetic is a part of logic; and that is not a novel insight for HILBERT in 1899. On the contrary: HILBERT opens the lectures [HILBERT, 1889/90] with the sentence, "The theoretical foundation for the mathematical sciences is the concept of positive integer, defined as *Anzahl*." This remark is deepened in the long introduction to the lectures on *Projektive Geometrie* [HILBERT, 1891], where HILBERT emphasizes that the results of pure mathematics are obtained by pure thought (reines Denken). He lists as parts of pure mathematics number theory, algebra, and complex analysis (Funktionentheorie). Mirroring DEDEKIND's perspective on the arithmetization of analysis from the preface to [DEDEKIND, 1888, p.VI], HILBERT asserts that the results of pure mathematics *can* and *have to be reduced* to relations between integers:

---

[37][HILBERT, 1889/90, p. 3]. The German text is:"Es ist unumgänglich nöthig, eine allgemeine Methode zur Darstellung der irrationalen Zahlen zu haben."

[38]The German text is found in [SCHAPER, 1898/99, p. 2], and also in [TOEPELL, 1986, pp. 203–204] and [HILBERT, 2004, p. 303]:
> "Es ist von Wichtigkeit, den Ausgangspunkt unserer Untersuchungen genau zu fixieren: *Als gegeben betrachten wir die Gesetze der reinen Logik und speciell die ganze Arithmetik.* (Ueber das Verhältnis zwischen Logik und Arithmetik vgl. DEDEKIND, Was sind und was sollen die Zahlen?) Unsere Frage wird dann sein: *Welche Sätze müssen wir zu dem eben definierten Bereich ‚adjungieren', um die Euklidische Geometrie zu erhalten?"*

In [TOEPELL, 1986] one finds additional remarks of interest for the issues discussed here, namely, on pp. 195, 206, and 227–228.

"*Nowadays* a proposition is being considered as proved only, when it *ultimately expresses* a relation between integers. Thus, the *integer is the element*. We can *obtain by pure thinking* the concept of integer, for example <by> *counting thoughts themselves*. Methods, foundations of pure mathematics belong to the sphere of pure thought."[39]

This very paragraph ends with the striking assertion, "I don't need anything but purely logical thought, when dealing with number theory or algebra."[40] HILBERT expresses here very clearly a thoroughgoing DEDEKINDian logicism with respect to natural numbers.[41]

The summer lecture *Die Grundlagen der Geometrie* [HILBERT, 1894] has a special section on "Die Einführung der Zahl". After all, HILBERT argues:

"In all of the exact sciences one obtains precise results only after the introduction of number. It is always of deep epistemological significance to see in detail, how this measuring is done."[42]

The number concept discussed now is that of real numbers; it is absolutely central for the problems surrounding the quadrature of the circle treated in the winter term 1894/95. The quadrature problem is the red thread for historically informed lectures entitled, not surprisingly, *Quadratur des Kreises*. Its solution requires the subtlest mathematical speculations with considerations from geometry as well as from analysis and number theory. In the winter term 1897/98, HILBERT presents these topics again, but this time in the *second* part of the course; its new *first* part is devoted entirely to a detailed discussion of the concept of number. That expansion is indicated by the new title of the lectures, *Zahlbegriff und Quadratur des Kreises* [HILBERT, 1897/98].

HILBERT remarks that the foundations of analysis have been examined through the work of WEIERSTRASZ, CANTOR, and DEDEKIND. He lists three important points:

"Today [we take as a] principle: Reduction to the laws for the whole rational numbers and purely logical operations.

1, 2, 3 ... = Anzahl = natural number

How one gets from this number [concept] to the most general concept of complex number, that is the foundation, i.e., the theory of the concept of number."[43]

---

[39][HILBERT, 2004, p. 22]. The German text is:
"*Es gilt heutzutage* ein Satz erst dann als bewiesen, wenn er eine Beziehung zwischen ganzen Zahlen *in letzter Instanz zum Ausdruck bringt*. Also die *ganze Zahl ist das Element*. Zum Begriff der ganzen Zahl können wir auch *durch reines Denken gelangen*, etwa <indem ich> die *Gedanken selber zähle*. Methoden, Grundlagen der reinen Mathematik gehören dem reinen Denken an." [HILBERT, 1891]

[40][HILBERT, 2004, p. 22]. The German sentence is:
"Ich brauche weiter nichts als *rein logisches* Denken, wenn ich mit Zahlentheorie oder Algebra mich beschäftige."

[41]In his extensive editorial note 6 of [HILBERT, 2004, p. 23], RALF HAUBRICH also points out the deep connection to DEDEKIND's views.

[42][HILBERT, 1894, p. 85]. The German text is:
"In allen exakten Wissenschaften gewinnt man erst dann präzise Resultate, wenn die Zahl eingeführt ist. Es ist stets von hoher erkenntnistheoretischer Bedeutung zu verfolgen, wie dies Messen geschieht."

In the lectures [HILBERT, 1897/98], the genetic build-up is to be captured axiomatically; that I take to be the gist of the discussion of the third point in the above list, namely, how one gets from the natural numbers to the most general concept of complex number. The motivating focus lies on the solution of problems by an extension of the axioms (Lösung von Problemen durch Erweiterung der Axiome); the discussion ends with the remark:

> "We have extended the number concept in such a way, or to put it in a better way, we have enlarged the stock of axioms in such a way, that we can solve all definite equations."[44]

At each step in this extending process, consistency of the axioms is required, but HILBERT does not give arithmetical or logical proofs, only geometric pictures.

In the winter term 1899/1900, HILBERT gave again the lectures *Zahlbegriff und Quadratur des Kreises* and refined the discussion of the axioms for the real numbers significantly; it is a transition to his paper *Über den Zahlbegriff* [HILBERT, 1900a]. On pages 38–43 added to the notes of the previous lectures of the same title [HILBERT, 1897/98], HILBERT extends the discussion of "komplexe Zahlensysteme" from *Grundlagen der Geometrie*.[45] He formulates the continuity principle as "BOLZANO's Axiom", i.e., every fundamental sequence has a limit; he shows also that this axiom has the Archimedian one as a consequence. Strikingly, he actually claims (on p. 42) that the full system is consistent! Before completing [HILBERT, 1900a], he obviously recognized

(i) that he could formulate the second continuity principle in such a way that it did not imply the Archimedian axiom, and

(ii) that the consistency of the system had *not* been established.

So we are back to the issue of consistency from the very beginning of this §2.2. The long digression on the "greatest difficulties" does not provide a clear answer, but a seemingly rhetorical question:

> How could HILBERT think about addressing the consistency problem by "a careful study and suitable modification of the known methods of reasoning in the theory of irrational numbers", if he did not have in mind a programmatically guided refinement of the steps presented here, i.e., of the foundation they constitute?

---

[43][HILBERT, 1897/98, p. 2]. The German text is:

> "Heute Grundsatz: Zurückführung auf die Gesetze der ganzen rationalen Zahlen und rein logische Operationen.
>
> 1, 2, 3 ... = Anzahl = natürliche Zahl
>
> Wie man von dieser Zahl zum allgemeinsten Begriff der komplexen Zahl kommt, das ist jenes Fundament, d. h. die Lehre vom Zahlbegriff."

[44][HILBERT, 1897/98, p. 23]. The German text is:

> "Wir haben den Zahlbegriff so erweitert, oder besser gesagt, den Bestand der Axiome so vermehrt, dass wir alle definiten Gleichungen lösen können."

HALLETT gives a complementary account of the completeness principle that emphasizes its rôle in HILBERT's work concerning geometry; that is presented in §5 of HALLETT's Introductory Note to [HILBERT, 1899]; [HILBERT, 2004, pp. 426–435].

[45]Indeed, pages 38 and 39 must be the page proofs of [HILBERT, 1899, pp. 26–28].

2.3 Consistency and proofs

HILBERT knew about the difficulties in set theory through his correspondence with GEORG CANTOR, but he did not — as far as I can see — move away from this programmatic position and the associated strategy for proving consistency until 1903 or, at the latest, 1904.[46]

One reason for making this conjecture rests on the fact that HILBERT gave his lectures on *Zahlbegriff und Quadratur des Kreises* [HILBERT, 1897/98] again in 1899/1900 and 1901/1902. We have only the "Disposition" of those lectures [HILBERT, 1901/02], but it is very detailed and does not contain any indication whatsoever of substantive changes.

In the summer term of 1904, HILBERT lectured again on *Zahlbegriff und Quadratur des Kreises* [HILBERT, 1904]. In the lecture notes written by MAX BORN one finds a dramatic change: HILBERT discusses the paradoxes in detail for the first time and sketches various foundational approaches. These discussions are taken up in HILBERT's talk at the Heidelberg Congress in August of 1904, where he presents a syntactic approach to the consistency problem. The goal is still to guarantee the existence of a suitable system, and the method seems inspired by one important aspect of the earlier investigations, as the crucial new step is the *simultaneous development of arithmetic and logic*!

HILBERT's view of the geometric axioms as characterizing a system of things that presents a "complete and simple image of geometric reality" is, after all, complemented by a traditional one: the axioms must allow the establishment, purely logically, of all geometric facts and laws. This rôle of the axioms, though clearly implicit in the formulation of the central question of [HILBERT, 1894], is emphasized in the notes [SCHAPER, 1898/99] of the lectures [HILBERT, 1898/99] and described in the Introduction of [HILBERT, 1899] in a methodologically refined way:

> "The present investigation is a new attempt of formulating for geometry a *simple* and *complete* system of mutually independent axioms; it is also an attempt of deriving from them the most important geometric propositions in such a way that the significance of the different groups of axioms and the import of the consequences of the individual axioms is brought to light as clearly as possible."[47]

This second important rôle of axioms is reflected also in the Paris talk, where HILBERT [1901] states that the totality of real numbers is "... a system of things whose mutual relations are governed by the axioms set up and for which all propositions, and only those, are true which can be derived from the axioms by a finite number of logical inferences." As in HILBERT's opinion "the concept of the continuum is strictly logically tenable in this sense only," the axiomatic method has to confront two fundamental problems that are formulated in [HILBERT, 1900a], at first for geometry and then also for arithmetic:

---

[46] PECKHAUS characterizes 1903 in his book [PECKHAUS, 1990] as a "Zäsurjahr" in Göttingen; see pp. 55–58. PECKHAUS also points out that DEDEKIND, in that very year, did not allow the re-publication of [DEDEKIND, 1888], because — as DEDEKIND put it in the preface to the third edition of 1911 — he had doubts concerning some of the important foundations of his view. Further evidence for this conjecture will be presented at the beginning of §3 below, in particular HILBERT's own recollections and remarks by BERNAYS.

[47] [HILBERT, 2004, p. 436]. The German text from [HILBERT, 1899, Einleitung, p. 3] is: "Die vorliegende Untersuchung ist ein neuer Versuch, für die Geometrie ein *einfaches* und *vollständiges* System von einander *unabhängiger* Axiome aufzustellen und aus denselben die wichtigsten geometrischen Sätze in der Weise abzuleiten, dass dabei die Bedeutung der verschiedenen Axiomgruppen und die Tragweite der aus den einzelnen Axiomen zu ziehenden Folgerungen möglichst klar zu Tage tritt."

> "The necessary task then arises of showing the consistency and the completeness of these axioms, i.e., it must be proved that the application of the given axioms can never lead to contradictions, and, further, that the system of axioms suffices to prove all geometric propositions."[48]

It is not clear, whether completeness of the respective axioms requires the proof of *all* true geometric or arithmetic statements, or whether provability of those that are part of the established corpora is sufficient; the latter would be a *quasi-empirical* notion of completeness. It should be mentioned that DEDEKIND [1888] attended consciously to the quasi-empirical completeness of the axiomatic conditions characterizing simply infinite systems. Through the proof principle of induction and the definition principle of recursion he provided the basic principles needed for the actual development of elementary number theory. The same observation can be made for the principle of continuity and DEDEKIND's treatment of analysis in [DEDEKIND, 1872].

But what are the logical steps that are admitted in proofs? FREGE criticized DEDEKIND on that point in the Preface of his *Grundgesetze der Arithmetik* [FREGE, 1893/1903], claiming that the brevity of DEDEKIND's development of arithmetic in [DEDEKIND, 1888] is only possible, "because much of it is not really proved at all". FREGE continues:

> "... nowhere is there a statement of the logical or other laws on which he builds, and, even if there were, we could not possibly find out whether really no others were used — for to make that possible the proof must be not merely indicated but completely carried out."[49]

Apart from making the logical principles explicit there is an additional aspect that is hinted at in FREGE's critique (and detailed in other writings). Though FREGE's critique applies as well to HILBERT's *Grundlagen der Geometrie*, POINCARÉ's review [POINCARÉ, 1902] of that book brings out this additional aspect in a dramatic way, namely, the idea of formalization as machine executability. POINCARÉ writes:

> "M. HILBERT has tried, so-to-speak, putting the axioms in such a form that they could be applied by someone who doesn't understand their meaning, because he has not ever seen either a point, or a line, or a plane. It must be possible, according to him [HILBERT, WS], to reduce reasoning to purely mechanical rules."

Indeed, POINCARÉ suggests giving the axioms to a reasoning machine, like JEVONS' logical piano, and observing whether all of geometry would be obtained. Such formalization might seem "artificial and childish", were it not for the important question of completeness:

> "Is the list of axioms complete, or have some of them escaped us, namely those we use unconsciously? ... One has to find out whether geometry is a logical consequence of the explicitly stated axioms, or in other words, whether the axioms, when given to the reasoning machine, will make it possible to obtain the sequence of all theorems as output [of the machine, WS]."[50]

---

[48][EWALD, 1996, pp. 1192–1193]. The German text from [HILBERT, 1900a] is: "Es entsteht dann die notwendige Aufgabe, die Widerspruchslosigkeit und Vollständigkeit dieser Axiome zu zeigen, d. h. es muß bewiesen werden, daß die Anwendung der aufgestellten Axiome nie zu Widersprüchen führen kann, und ferner, daß das System der Axiome zum Nachweis aller geometrischen Sätze ausreicht."

[49][FREGE, 1980, p.139]

[50][POINCARÉ, 1902, pp. 252–253].

At issue is, of course, what logical steps can the reasoning machine take? In HILBERT's lectures one finds increasingly references to, and investigations of, logical calculi. The calculi HILBERT considers remain very rudimentary until the lectures [HILBERT, 1917/18], where modern mathematical logic is all of a sudden being developed. Back in the Heidelberg talk, the novel syntactic approach to consistency proofs is suggested not for an axiom system concerning real numbers but rather concerning natural numbers; that will be the central topic of the next section. The logical calculus, if it can be called that, is extremely restricted — it is purely equational!

In late 1904, possibly even in early 1905, HILBERT sent a letter to his friend and colleague ADOLF HURWITZ. Clearly, he was not content with the proposal he had detailed in Heidelberg:

> "It seems that various parties started again to investigate the foundations of arithmetic. It has been my view for a long time that exactly the most important and most interesting questions have not been settled by DEDEKIND and CANTOR (and a fortiori not by WEIERSTRASZ and KRONECKER). In order to be forced into the position to reflect on these matters systematically, I announced a seminar on the 'logical foundations of mathematical thought' for next semester."[51]

The lectures from the summer term 1905 [HILBERT, 1905b] are as special as those of 1904 [HILBERT, 1904], but for a different reason: one finds in them a critical examination of logical principles and a realization that a broader logical calculus is needed that captures, in particular, universal statements and inferences. HILBERT discusses in some detail (on pp. 253–266) the shortcomings of the approach he pursued in his Heidelberg address; see §3.1 below. In his subsequent lectures on the foundations of mathematics, HILBERT does not really progress beyond the reflections presented in [HELLINGER, 1905]. In a letter from GERHARD HESSENBERG to LEONARD NELSON (dated 7 February 1906, which I received through PECKHAUS), one finds not only a discussion of RUSSELL's early logicism, but strikingly, DEDEKIND and HILBERT are both described as logicists. HESSENBERG remarks:

> "DEDEKIND's standpoint is also a logicist one, but altogether noble and nowhere brazenly dogmatic. One sees and feels everywhere the struggling, serious human being. But given all the superficialities I have found up to now in RUSSELL, I can hardly assume that he is going to provide more than something witty. ZERMELO believes that also HILBERT's logicism cannot be carried out, but considers POINCARÉ's objections as unfounded."[52]

---

[51] HILBERT mentions in this letter that [ZERMELO, 1904] had just been published. ZERMELO's paper is an excerpt of a letter to HILBERT that was written on 24 September 1904. The German text is from [DUGAC, 1976, p. 271]: "Die Beschäftigung mit den Grundlagen der Arithmetik wird jetzt, wie es scheint, wieder von den verschiedensten Seiten aufgenommen. Dass gerade die wichtigsten u. interessantesten Fragen von DEDEKIND und CANTOR noch nicht (und erst recht nicht von WEIERSTRASZ und KRONECKER) erledigt worden sind, ist eine Ansicht, die ich schon lange hege und, um einmal in die Notwendigkeit versetzt zu sein, darüber im Zusammenhang nachzudenken, habe ich für nächsten Sommer ein zweistündiges Colleg über die ‚logischen Grundlagen des math. Denkens' angezeigt."

[52] The German text is: "DEDEKINDs Standpunkt ist ja auch logizistisch, aber durchaus vornehm and nirgends unverfroren dogmatisch. Man sieht und fühlt überall den kämpfenden ernsten Menschen. Aber bei den Oberflächlichkeiten, die ich bis jetzt in RUSSELL gefunden habe, kann ich kaum annehmen, daß er mehr als geistreiches bringt. ZERMELO ist der Ansicht, daß auch HILBERTs Logizismus undurchführbar ist, hält aber POINCARÉs Einwendungen für unbegründet."

It is only in the Zürich talk *Axiomatisches Denken* [HILBERT, 1917a] that a new perspective emerges. In that pivotal essay, HILBERT remarks that the consistency of the axioms for the real numbers can be reduced, by employing set-theoretic concepts, to the very same question for integers; that result is attributed to the theory of irrational numbers developed by WEIERSTRASZ and DEDEKIND. HILBERT continues, insisting on a logicist approach:

> "In only two cases is this method of reduction to another more special domain of knowledge clearly not available, namely, when it is a matter of the axioms for the *integers* themselves, and when it is a matter of the foundation of *set theory*; for here there is no other discipline besides logic to which it were possible to appeal.
>
> But since the examination of consistency is a task that cannot be avoided, it appears necessary to axiomatize logic itself and to prove that number theory as well as set theory are only parts of logic."[53]

HILBERT remarks that RUSSELL and FREGE provided the basis for this approach. Given HILBERT's closeness to DEDEKIND's perspective, it is not surprising that he was attracted by the logicist considerations of *Principia Mathematica* [WHITEHEAD & RUSSELL, 1910–1913]. In 1920, however, HILBERT and BERNAYS reject the logicist program of the RUSSELLian kind. The reason is quite straightforward: for the development of mathematics the axiom of reducibility is needed, and that is in essence a return to existential axiomatics with all its associated problems. In particular, no reduction of epistemological interest has been achieved.

I want to end this section by pointing out that ZERMELO's axiomatization of set theory in [ZERMELO, 1908], inspired by both DEDEKIND and HILBERT, is another instance of existential axiomatics. Points 1 through 4 in the first section of [ZERMELO, 1908] are structured in the same way as the axiomatic formulations in [HILBERT, 1899; 1900a]. According to EBBINGHAUS,[54] ZERMELO had intended to give a consistency proof for his axiom system, but was encouraged by HILBERT to publish the axiomatization first and leave the consistency proof for another occasion. As to the nature of ZERMELO's considerations at that point I am not sure at all. ZERMELO [1908] just remarks:

> "I have not yet even been able to prove rigorously that my axioms are consistent, though this is certainly very essential; instead I have had to confine myself to pointing out now and then that the antinomies discovered so far vanish one and all if the principles here proposed are adopted as a basis."[55]

His later investigations on the foundations of set theory in [ZERMELO, 1930] are parallel to DEDEKIND's metamathematical reflections described above.

---

[53][EWALD, 1996, p.1113]. Here is the German text from [HILBERT, 1917a, p.153]:

"Nur in zwei Fällen nämlich, wenn es sich um die Axiome der *ganzen Zahlen* selbst und wenn es sich um die Begründung der *Mengenlehre* handelt, ist dieser Weg der Zurückführung auf ein anderes spezielleres Wissensgebiet offenbar nicht gangbar, weil es außer der Logik überhaupt keine Disziplin mehr gibt, auf die alsdann eine Berufung möglich wäre.

Da aber die Prüfung der Widerspruchslosigkeit eine unabweisbare Aufgabe ist, so scheint es nötig, die Logik selbst zu axiomatisieren und nachzuweisen, daß Zahlentheorie sowie Mengenlehre nur Teile der Logik sind."

3 Direct consistency proofs

It seems then that HILBERT modified his basic attitude as a DEDEKIND*ian logicist* only after the discovery of ZERMELO and RUSSELL's elementary contradiction. That paradox — which had a "catastrophic effect on the mathematical world when it became known" — convinced him that there was a *deep* problem, that difficulties appeared already at an earlier stage, and that the issue had to be addressed in different ways.[56]

The notes for the lectures *Zahlbegriff und Quadratur des Kreises* [HILBERT, 1904] by BORN [1904] contain, as we saw already, a detailed discussion of the paradoxes. They examine also DEDEKIND's contradictory system of all things, but both praise his epochal achievements and pointedly insist that fundamental difficulties remain.

> "He [DEDEKIND] arrived at the view that the standpoint of considering the integers as obvious [selbstverständlich] cannot be sustained; he recognized that the difficulties KRONECKER saw in the definition of irrationals arise already for integers; furthermore, if they are removed here, they disappear there. This work [[DEDEKIND, 1888]] was epochal, but it did not yet provide something definitive, certain difficulties remain. These difficulties are connected, as for the definition of the irrationals, above all to the concept of the infinite; ..."[57]

This set the stage for HILBERT's Heidelberg talk in August 1904. There he indicates a novel way of solving the consistency problem for arithmetic which is taken now in a much more restricted sense, namely, as dealing with just the natural numbers.

---

[54]EBBINGHAUS quotes in his book [EBBINGHAUS, 2007, p. 78] from a letter of ZERMELO to HILBERT, dated 25 March 1907 in which ZERMELO writes:

> "According to your wish I will finish my set theory as soon as possible, although actually I wanted to include a proof of consistency."

[55]All the remarks from [ZERMELO, 1908] are found in [HEIJENOORT, 1971, pp. 200–201].

[56][HILBERT, 1926, p.169], HILBERT writes, retrospectively:

> "Insbesondere war es ein von ZERMELO und RUSSELL gefundener Widerspruch, dessen Bekanntwerden in der mathematischen Welt geradezu von katastrophaler Wirkung war."

That this paradox played a special rôle is also confirmed by BERNAYS, who reported to CONSTANCE REID:

> "Under the influence of the discovery of the antinomies in set theory, HILBERT temporarily thought that KRONECKER had probably be right there. But soon he changed his mind. Now it became his goal, one might say, to do battle with KRONECKER with his own weapons of finiteness by means of a modified conception of mathematics."   [REID, 1970, p.173]

[57]Here is the German text from [BORN, 1904, p.166]:

> "Er drang zu der Ansicht durch, dass der Standpunkt mit der Selbstverständlichkeit der ganzen Zahlen nicht aufrecht zu erhalten ist; er erkannte, dass die Schwierigkeiten, die KRONECKER bei der Definition der irrationalen Zahlen sah, schon bei den ganzen Zahlen auftreten und dass, wenn sie hier beseitigt sind, sie auch dort wegfallen. Diese Arbeit war epochemachend, aber sie lieferte doch nichts definitives, es bleiben gewisse Schwierigkeiten übrig. Diese bestehen hier, wie bei der Definition der irrationalen Zahlen, vor allem im Begriff des Unendlichen; ..."

3.1 Heidelberg 1904 [HILBERT, 1905a]

The shift from arithmetic as pertaining not to the real but to the natural numbers is crucial for HILBERT's presentation in Heidelberg. The consistency of the system considered there would guarantee, as HILBERT puts it, "the consistent existence of the so-called smallest infinite." Its axioms are those for identity extended by DEDEKIND's requirements for a *simply infinite system*; the induction principle (which follows from DEDEKIND's minimality condition) is included, but neither formulated properly nor treated in the consistency proof. Using modern notation the axioms that are actually investigated can be given in this way:

1. $x = x$
2. $x = y \ \& \ W(x) \to W(y)$
3. $x' = y' \to x = y$
4. $x' \neq 1$

$W(x)$ stands for an arbitrary proposition containing the variable x. The rules, implicit in HILBERT's description of "consequence", are modus ponens and a rule of substitution that allows replacing variables by arbitrary sign combinations; other "modes of logical inferences" are only alluded to. Finally, HILBERT most strongly emphasized that the problem for the real numbers is solved once matters are settled for the natural numbers:

> "The existence of the totality of real numbers can be demonstrated in a way similar to that in which the existence of the smallest infinite can be proved; in fact, the axioms for real numbers as I have set them up ... can be expressed by precisely such formulas as the axioms hitherto assumed. ... and the axioms for the totality of real numbers do not differ qualitatively in any respect from, say, the axioms necessary for the definition of the integers. In the recognition of this fact lies, I believe, the real refutation of the conception of arithmetic associated with L. KRONECKER ...."[58]

In the consistency proof for the above system HILBERT uses the property of homogeneity; an equation $a = b$ is called *homogeneous* if and only if a and b have the same number of symbol occurrences. It is easily seen, by induction on derivations, that all equations derivable from axioms 1–3 are homogeneous. A contradiction can be obtained only by establishing an unnegated instance of 4 from 1–3, but such an instance is necessarily inhomogeneous and thus not provable. HILBERT comments:

> "The considerations just sketched constitute the first case in which a direct proof of consistency has been successfully carried out for axioms, whereas the method of a suitable specialization, or of the construction of examples, which is otherwise customary for such proofs — in geometry in particular — necessarily fails here."[59]

---

[58]In [HEIJENOORT, 1971, pp. 137–138]. Here is the German text from [HILBERT, 1905a, p. 185]: "Ähnlich wie die Existenz des kleinsten Unendlich bewiesen werden kann, folgt die Existenz des Inbegriffs der reellen Zahlen: in der Tat sind die Axiome, wie ich sie für die reellen Zahlen aufgestellt habe, genau durch solche Formeln ausdrückbar, wie die bisher aufgestellten Axiome. ... und die Axiome für den Inbegriff der reellen Zahlen unterscheiden sich qualitativ in keiner Hinsicht etwa von den zur Definition der ganzen Zahlen notwendigen Axiomen. In der Erkenntnis dieser Tatsache liegt, wie ich meine, die sachliche Widerlegung der von L. KRONECKER vertretenen ... Auffassung der Grundlagen der Arithmetik."

HILBERT stressed in his Heidelberg talk the programmatic goal of developing logic and mathematics simultaneously. The actual work has significant shortcomings, as there is no calculus for sentential logic, there is no proper treatment of quantification, and induction is not incorporated into the argument. In sum, there *is* an important shift from semantic arguments to syntactic ones, but the very set-up is inadequate as a formal framework for mathematics. It should also be noted very clearly that the ultimate goal of the consistency proof remains to guarantee the existence of a set, here of the smallest infinite one.

HILBERT was obviously not fully satisfied with the work he had presented in Heidelberg, and in the summer term of 1905 he did give lectures under the title *Logische Principien des mathematischen Denkens* [HILBERT, 1905b]. Towards the end of the term he reviewed the consistency proof sketched above and considers it "as the first *attempt* [my emphasis, WS] of a consistency proof without giving a special example". He articulates what he considers to be problematic and continues:

> "Carrying out this thought in a precise way would at any rate present considerable difficulties, in particular the exact justification of the statement that 'all' inferences from the first axioms lead only to homogeneous equations. One arrives with such considerations finally at the thought that proofs should be viewed as mathematical things; the logical calculus indicates that."[60]

But how is it possible to characterize "proofs" mathematically, if they constitute an infinite totality? And, even if one has succeeded to do that, how can one give arguments concerning them? Those questions were already on HILBERT's mind during the summer of 1904; in BORN's notes from that term one finds remarks concerning concepts that have particular defining characteristics. Such concepts are viewed as consistent just in case it is not possible to obtain a contradiction by drawing all possible inferences from their characteristics. HILBERT then concludes:

> "But in proceeding in this way one obviously uses arithmetic already; in a sense, one does so when forming the plural, but certainly when speaking of all inferences etc.[61]

These problematic aspects are sharpened, but not resolved in [HELLINGER, 1905].

---

[59][HEIJENOORT, 1971, p.135]. The German text from [HILBERT, 1905a, p.181] is:

> "Die eben skizzierte Betrachtung bildet den ersten Fall, in dem es gelingt, den direkten Nachweis für die Widerspruchslosigkeit von Axiomen zu führen, während die sonst — insbesondere in der Geometrie — für solche Nachweise übliche Methode der geeigneten Spezialisierung oder Bildung von Beispielen hier notwendig versagt."

[60][HELLINGER, 1905, pp. 264–265]. The German text is:

> "Eine genaue Durchführung dieses Gedankens würde immerhin noch beträchtliche Schwierigkeiten bieten, insbesondere die exakte Begründung der Aussage, daß ‚alle' Schlüsse aus den ersten Axiomen nur auf homogene Gleichungen führen. Man kommt bei solchen Überlegungen schließlich darauf, den Beweis selbst als ein mathematisches Ding anzusehen, worauf ja auch der Logikkalkül hinweist."

[61][BORN, 1904, p.170]. The German text is:

> "Dabei benutzt man aber offenbar schon die Arithmetik, eigentlich schon bei der Bildung des Plurals, noch mehr aber, wenn man von allen Schlüssen etc. spricht."

3.2 Obstacles

When HILBERT suggests, as I reported above, that proofs might be viewed as "mathematical things", he considers one possibility of doing so "that is more formal and is completely based on the logical calculus" [HELLINGER, 1905, p. 266]. As this formal approach requires a complete formulation of all axioms, in particular, of those that concern "Every", he encounters even there the very same issue. After all, replacing variables by arbitrary combinations is the principal operation and, on account of the great variety of potential combinations, it is difficult to provide a rigorous formulation. HILBERT concludes these reflections by stating:

> "Accordingly the main task would be to formulate exactly the axioms for 'Every' and the concept of combination that is closely connected with them."[62]

As is well known, POINCARÉ challenged the presumed foundational import of HILBERT's consistency proof, because it proceeds by induction; this is absolutely to the point. He also brought out other *methodological shortcomings* that are closely linked with the attempts HILBERT made in the Heidelberg talk to overcome the difficulties I just sketched.[63]

It becomes plain how penetrating POINCARÉ's considerations were when one reads the remarks of BERNAYS on HILBERT's Heidelberg paper that give essentially a précis of POINCARÉ's critique, but also of HILBERT's reflection on the issues; BERNAYS [1935a] writes:

> "In addition, the systematic standpoint of HILBERT's proof theory is not yet fully and clearly developed in the Heidelberg address. Some places indicate that HILBERT wants to avoid the intuitive conception of number and replace it by its axiomatic introduction. Such a procedure would lead to a circle in the proof-theoretic considerations".[64]

BERNAYS emphasizes that HILBERT did not express distinctions that would be central to the later HILBERT Program; for example, HILBERT did not bring out the restriction on the contentual application of the forms of existential and universal judgments.[65]

POINCARÉ's incisive analysis of the syntactic, proof-theoretic approach and his own insight into its shortcomings shifted HILBERT's attention from the stand he had advocated in the Heidelberg talk. HILBERT's lectures, given throughout the period from 1905 to 1917, document his continued foundational concerns.

---

[62][HELLINGER, 1905, p. 269]. The German text is:

"Hiernach wäre dann die exakte Formulierung der Axiome des 'Jedes' und des damit in naher Verbindung stehenden Begriffes der Kombination die Hauptaufgabe."

[63][POINCARÉ, 1905], translated on pp. 1026–1027 of [EWALD, 1996].

[64][BERNAYS, 1935a, p. 200]. The German text is:

"Außerdem ist auch der methodische Standpunkt der HILBERTschen Beweistheorie in dem Heidelberger Vortrag noch nicht zur vollen Deutlichkeit entwickelt. Einige Stellen deuten darauf hin, daß HILBERT die anschauliche Zahlvorstellung vermeiden und durch die axiomatische Einführung des Zahlbegriffs ersetzen will. Ein solches Verfahren würde in den beweistheoretischen Überlegungen einen Zirkel ergeben."

[65][BERNAYS, 1935a, p. 200]. This is part of the following text: "Auch wird der Gesichtspunkt der Beschränkung in der inhaltlichen Anwendung der Formen des existentialen und des allgemeinen Urteils noch nicht ausdrücklich und restlos zur Geltung gebracht."

Under the impact of a detailed study of *Principia Mathematica* [WHITEHEAD & RUSSELL, 1910–1913] that began in 1913, HILBERT flirted again with logicism.[66] What resulted from this study, most importantly, were the lectures *Prinzipien der Mathematik* [HILBERT, 1917/18] signaling the emergence of modern mathematical logic. In the following two years, the programmatic logicism that was underlying those lectures was abandoned for good reasons. A radical constructivism was adopted instead and subsequently also abandoned — for equally good, but obviously different reasons. The finitist consistency program as we think of it was formulated only in the lectures [HILBERT, 1921/22].[67]

The evolution towards the program started, however, in the summer term of 1920 when HILBERT came back to the proof-theoretic approach of [HILBERT, 1905a]. The notes from that term contain in § 7 a consistency proof for exactly the same fragment of arithmetic that was investigated in the Heidelberg talk. Its formulation is informed by the investigations of 1917/18: the language is more properly described; the basic vocabulary consists of variables $a, b, \ldots$, non-logical constants $1, +$, and all numerals; $=$ and $\neq$ are relation symbols, and $\rightarrow$ is the sole logical symbol. The axioms are:

$$1 = 1$$
$$a = b \;\rightarrow\; a+1 = b+1$$
$$a+1 = b+1 \;\rightarrow\; a = b$$
$$a = b \;\rightarrow\; (a = c \;\rightarrow\; b = c)$$
$$a+1 \neq 1$$

Modus ponens and substitution, now applicable only to numerals, are the sole inference rules. The combinatorial argument is different from that given in [HILBERT, 1905a], mainly by the introduction of the notion "kürzbar":

> "If one considers a proof with respect to a particular concrete property it has, then it is possible that the removal of some formulas in this proof still leaves us with a proof that has that particular property. In this case we are going to say that the proof is *kürzbar* with respect to the given property."[68]

HILBERT establishes three lemmata. The first claims that a theorem can contain at most two occurrences of $\rightarrow$, the second asserts that no statement of the form $(A \rightarrow B) \rightarrow C$ can be proved, and the third expresses that a formula $a = b$ is provable only if a and b are the same term. On the surface, HILBERT does not use an induction principle in these proofs.

---

[66] Around 1913, HILBERT became familiar with some of RUSSELL's writings and intended to invite RUSSELL to give lectures in Göttingen; the outbreak of World War I made that impossible. In the notes from the winter term 1914/15 one finds brief remarks about type theory and in serendipitously preserved notes of a student (found in the Institute for Advanced Study in Princeton) even more extended ones. Of greatest significance in indicating the degree to which the HILBERT group discussed the details of [WHITEHEAD & RUSSELL, 1910–1913] is BEHMANN's dissertation; that is described in [MANCOSU, 1999a]. Not only the details of [WHITEHEAD & RUSSELL, 1910–1913] were of significance; on the contrary, the broad logicist outlook had a real impact on HILBERT. The notes [GOEB, 1917] for HILBERT's course on *Set Theory* [HILBERT, 1917b] and HILBERT's Zürich talk *Axiomatisches Denken* [HILBERT, 1917a] reveal a logicist direction of his work.

[67] For details see [SIEG, 1999], but also for the relationship between [HILBERT, 1917/18] and [HILBERT & ACKERMANN, 1928].

[68] [HILBERT, 1920, p. 38]. The German text is: "Betrachtet man einen Beweis in Hinsicht auf eine bestimmte, konkret aufweisbare Eigenschaft, welche er besitzt, so kann es sein, dass, nach Wegstreichung einiger Formeln in diesem Beweise noch immer ein Beweis (...) übrig bleibt, welcher auch noch jene Eigenschaft besitzt. In diesem Fall wollen wir sagen, dass der Beweis sich in bezug auf die betreffende Eigenschaft kürzen lässt."

He compares, playing on the double meaning of "kürzbar", the structure of the proof to that for the irrationality of $\sqrt{2}$. In that classical indirect argument one starts out by assuming there are p and q, $q \neq 0$, such that $\sqrt{2} = p/q$; without loss of generality p and q are then taken to be relatively prime, i.e., the fraction p/q is not "kürzbar".

The presentation of the consistency proof is simplified in [HILBERT, 1922b].[69] Though "Kürzbarkeit" is only implicit in the argument, the strategic point of the modified argument is made quite explicit; HILBERT insists that POINCARÉ has been refuted.

> "POINCARÉ's objection, claiming that the principle of mathematical induction [vollständige Induktion] cannot be proved but by mathematical induction [vollständige Induktion], has been refuted by my theory."[70]

Is this to be taken in the strong sense that no induction principle is being used? Or does one use perhaps a special procedure that is based on the construction and deconstruction of numerals and that, by its very nature, is different from the full principle? From a mathematical point of view, HILBERT uses the "least number principle" that is equivalent to induction; it is used in an elementary form for just purely existential statements.

3.3 New directions

In the second part of [HILBERT, 1922b],[71] the theory is significantly expanded. The basic system is modified by replacing the axiom $1 = 1$ with $a = a$, but also by including the axioms $1 + (a+1) = (1+a) + 1$ and $a \neq 1 \rightarrow a = \delta(a) + 1$, where δ is the predecessor function. A schema for identity
$$a = b \ \rightarrow \ (A(a) \rightarrow A(b))$$

---

[69]This paper was based on talks given in the spring and summer of 1921 in Copenhagen, respectively Hamburg, where HILBERT presented his *new* investigations on the foundations of arithmetic. Reports on the three Copenhagen talks were published in Berlingske Tidende. HILBERT gave these lectures on March 14, 15, and 17. The first talk had the title *Natur und mathematisches Erkennen*; the second and third were entitled *Axiomenlehre und Widerspruchsfreiheit*. As to the Hamburg talks: BEHNKE [1976] gives a brief description of the impact of the Hamburg lecture:

> "Im Sommer 1922 [sic!] trug HILBERT hier seinen viel zitierten ersten Aufsatz zur Neubegründung der Mathematik vor. Es war eine Entgegnung auf die Schriften von HERMANN WEYL und L. E. J. BROUWER, die Unruhe unter die Mathematiker gebracht hatten, weil sie das tertium non datur bestritten und damit schon die Existenz der reellen Zahlen in Frage stellten. In den Diskussionen zu HILBERTS Vorträgen [sic!] sprachen ERNST CASSIRER, damals Professor der Philosophie in Hamburg, HEINRICH SCHOLZ aus Kiel, der spätere Begründer der mathematischen Logik in Deutschland, [ARNOLD] KOWALEWSKI – Königsberg, der für die Philosophie des "Als ob" von VAIHINGER warb, und mancher andere. Die Erörterungen waren viel reger und grundsätzlicher, als es sonst nach mathematischen Vorträgen der Fall zu sein pflegt. Man begriff, daß HILBERT sich um die entscheidende Sicherung der Mathematik gegen Angriffe der Intuitionisten wandte."

The detailed information concerning the talks in Copenhagen was conveyed to me by VINCENT HENDRICKS, that concerning the Hamburg talks by ALI BEHBOUD.

[70][HILBERT, 1922b, p.161]. The German text is:

> "Sein [POINCARÉs] Einwand, dieses Prinzip [der vollständigen Induktion] könnte nicht anders als selbst durch vollständige Induktion bewiesen werden, ist durch meine Theorie widerlegt."

[71]The editors of [HILBERT, 1932ff., Vol. 3] remark in note 2 on page 168 that the consistency arguments stem from an earlier stage of proof theory, and that a new direction of HILBERT's considerations is indicated on the following pages.

is also added. (Upper case Roman letters are viewed as formula variables.) The system includes rules for universal quantifiers; but as negation is available only for identities, existential quantifiers cannot be introduced. The induction principle and certain logical principles are adjoined to the basic theory. The former principle is given in the form

$$(a)(A(a) \to A(a+1)) \to \{A(1) \to (Z(b) \to A(b))\}.$$

The logical axioms are to reflect the usual inference procedures and are divided into two groups; those in the first group concern the conditional $\to$, and those of the second group deal with negation. Here are the axioms of the first group:

$$A \to (B \to A)$$
$$\{A \to (A \to B)\} \to (A \to B)$$
$$\{A \to (B \to C)\} \to \{B \to (A \to C)\}$$
$$(B \to C) \to \{(A \to B) \to (A \to C)\}.$$

Negation is available only for equations. It is thus treated in a restricted way:

$$a \neq a \to A,$$
$$(a = b \to A) \to ((a \neq b \to A) \to A).$$

As to this restricted use of negation, HILBERT remarks that the system is to be kept constructive; thus, the proper metamathematical direction of the program has not yet been taken.[72] Eight arithmetical axioms concerning Z, addition, and subtraction round out the system.

HILBERT [1922b] emphasizes that the axioms ensure that all statements of arithmetic can be obtained in a formal way. The editors of HILBERT's *Gesammelte Abhandlungen* mention [HILBERT, 1932ff., Vol. 3, p. 176, Note 1] that "a schema for the introduction of functions by recursion equations" has to be added if this last goal is to be reached. As to the extent of the claimed consistency result, they assert [HILBERT, 1932ff., Vol. 3, p. 176,

---

[72]HILBERT emphasizes:

"Wir haben bisher außer dem Zeichen $\to$ und dem Allzeichen kein anderes logisches Zeichen eingeführt und insbesondere für die logische Operation ‚nicht' die Formalisierung vermieden. Dieses Verhalten gegenüber der Negation ist für unsere Beweistheorie charakteristisch: ein formales Äquivalent für die fehlende Negation liegt lediglich in dem Zeichen $\neq$, durch dessen Einführung die Ungleichheit gewissermaßen ebenso positiv ausgedrückt und behandelt wird, wie die Gleichheit, deren Gegenstück sie ist. Inhaltlich kommt die Negation nur im Nachweise der Widerspruchsfreiheit zur Anwendung, und zwar nur, insoweit es unserer Grundeinstellung entspricht. Mit Rücksicht auf diesen Umstand bringt uns, wie ich glaube, unsere Beweistheorie zugleich auch eine erkenntnistheoretische wichtige Einsicht in die Bedeutung und das Wesen der Negation." [HILBERT, 1922b, p.173]

When discussing that the existential quantifier can be defined in "formal logic" with the universal quantifier and negation, HILBERT points out:

"Da aber in unserer Beweistheorie die Negation keine direkte Darstellung haben darf, so wird hier die Formalisierung von 'es gibt' dadurch erreicht, daß man individuelle Funktionszeichen mittels einer Art impliziter Definition einführt, indem gewissermaßen das, 'was es gibt', durch eine Funktion wirklich hergestellt wird." [HILBERT, 1922b, p.174]

HILBERT discusses then as an example the predecessor function. In [HILBERT, 1923a, p.179, note 1], HILBERT acknowledges very briefly:

"In meiner vorhin zitierten Abhandlung [[HILBERT, 1922b]; WS] hatte ich dieses Zeichen noch vermieden; es hat sich herausgestellt, daß bei der gegenwärtigen, ein wenig veränderten Darstellung meiner Theorie der Gebrauch des Zeichens 'nicht' ohne Gefahr geschehen kann."

This is not a minor shift in perspective, but a methodologically dramatic one!

Note 2] that it holds only if the universal quantifier is excluded and the induction axiom is replaced by the induction rule. Taking into account these modifications, consistency is claimed for a theory that includes primitive recursive arithmetic. There is no indication of a proof, but the result can obviously not be obtained by extending the combinatorial argument I described above.[73]

[HILBERT, 1922b] was based, as described in a previous note, on talks he had given in the spring and summer of 1921 in Copenhagen, respectively Hamburg. The first Copenhagen talk has been preserved as a manuscript in HILBERT's own hand. It is worthwhile to quote its last paragraph in order to re-emphasize HILBERT's broad vision for mathematics.

> "We went rapidly through those chapters of theoretical physics that are currently most important. If we ask, what kind of mathematics is being considered by physicists, then we see that it is *analysis* that serves physicists in its complete *content* and *extension*. Indeed, it does so in two different ways: first it serves to *clarify* and *formulate* their ideas, and second — as an instrument of calculation — it serves to obtain quickly and reliably numerical results, which help to check the correctness of their ideas. Apart from this face seen by physicists, there is a completely different face that is directed towards philosophy; the features of that face deserve no less our interest. That topic will be discussed in my subsequent talks."[74]

HILBERT did not only expound his particular perspective, but argued also against the stand of BROUWER and WEYL. Summarizing his position in the published essay, HILBERT [1922b] contrasts their constructive tendencies with his own, claiming that WEYL has "failed to see the path to the fulfillment of these [constructive, WS] tendencies" and that "only the path taken here in pursuit of axiomatics will do full justice to the constructive tendencies":

> "The goal of securing a firm foundation for mathematics is also my goal. I should like to regain for mathematics the old reputation for incontestable truth, which it appears to have lost as the result of the paradoxes of set theory; but I believe that this can be done while fully preserving its accomplishments. The method I follow in pursuit of this goal is none other than the axiomatic method; its essence is as follows."[75]

---

[73]After having made the claim that the consistency of that system can be obtained, HILBERT goes on to say:
"Aber der wesentlichste Schritt bleibt noch zu tun übrig, nämlich der Nachweis der Anwendbarkeit des logischen Prinzips *"tertium non datur"* in dem Sinne der Erlaubnis, auch bei unendlich vielen Zahlen, Funktionen oder Funktionenfunktionen schließen zu dürfen, daß eine Aussage entweder für alle diese Zahlen, Funktionen bzw. Funtionenfunktionen gilt oder daß notwendig unter ihnen eine Zahl, Funktion bzw. Funtionenfunktion vorkommt, für die die Aussage nicht gilt." [HILBERT, 1922b, p.176]

It is then claimed that by introducing appropriate functionals the foundation of classical analysis can be provided and that the consistency of the extended system can be established.

[74][HILBERT, 1921, pp. 28–29]. The German text is: "Wir haben so im Fluge, die gegenwärtig wichtigsten Kapitel der theoretischen Physik durcheilt. Fragen wir, welcher Art Mathematik dabei für den Physiker in Betracht kommt, so sehen wir, dass es die *Analysis* ist, [die] in ihrem gesamten *Inhalt und Umfang* dem Physiker dient und zwar auf zweierlei Weisen: erstens zur *Klärung und Formulierung* seiner Ideen und zweitens als Instrument der Rechnung zur raschen und sicheren Gewinnung numerischer Resultate, durch die er die Richtigkeit seiner Ideen prüft. Ausser diesem Antlitz, in das der Physiker schaut, hat die Mathematik aber noch ein ganz anderes Gesicht, das auf die Philosophie gerichtet ist, [und] dessen Züge nicht minder unser Interesse verdienen. Darüber will ich in meinen folgenden Vorträgen handeln."

Having described the nature of the axiomatic method, he points to the task of recognizing the consistency of the arithmetical axioms including at this point axioms for number theory, analysis, and set theory. This task leads naturally to the metamathematical investigation of suitable formalisms, in which parts of mathematics can be carried out and proofs are given. The concepts of proof and provability are thus "relativized" to the underlying axiom system. HILBERT emphasizes,

> "This relativism is natural and necessary; it causes no harm, since the axiom system is constantly being extended, and the formal structure, in keeping with our constructive tendency, is becoming more and more complete."[76]

HILBERT's constructivism comes in at this point in yet a different way, namely, through constructing ever more complete formalisms for the development of mathematics.

The terms "HILBERTsche Beweistheorie" and "finite Mathematik" appear for the first time in the notes [HILBERT, 1921/22]. "HILBERTsche Beweistheorie" is of course self-explanatory, but "finite Mathematik" is not, unless one interprets it simply as having been chosen in contrast to "transfinite Mathematik". HILBERT and BERNAYS give no philosophical explication. Rather, they insist that an intuitive foundation ("anschauliche Begründung") is to be given and develop paradigmatically elementary number theory on such a foundation; their development is a marvelously detailed and thorough presentation of the beginnings of recursive number theory. The opposition between "finite" and "transfinite Mathematik" is founded on that between "finite" and "transfinite Logik", as the step from "finite" to "transfinite Mathematik" is taken when the classical logical equivalences between the quantifiers are taken to be valid even when we are dealing with infinite totalities. But that cannot be done in "finite Mathematik": "Assuming the unrestricted validity of those equivalences amounts to thinking of totalities with infinitely many individuals as completed ones."[77]

Elementary number theory as HILBERT and BERNAYS develop it should not be understood as all of their "finite Mathematik". On the contrary, they envision a dramatic expansion; an expansion that is needed, on the one hand, to fully develop analysis and set theory and, on the other hand, to recognize "on the basis of finitist logic" why and to what extent "the application of transfinite inferences [in analysis and set theory] always leads to correct results". The strategy that is suggested reflects what HILBERT had done already, to some extent, around the turn of the century, namely, to shift the KRONECKERian constructivity requirements from mathematics itself to metamathematics. We have to expand, so they demand, the domain of objects that are being considered:

---

[75][EWALD, 1996, p.1119]. Here is the German text from [HILBERT, 1922b, p.160]:
"Das Ziel, die Mathematik sicher zu begründen, ist auch das meinige; ich möchte der Mathematik den alten Ruf der unanfechtbaren Wahrheit, der ihr durch die Paradoxien der Mengenlehre verloren zu gehen scheint, wiederherstellen; aber ich glaube, daß dies bei voller Erhaltung ihres Besitzstandes möglich ist. Die Methode, die ich dazu einschlage, ist keine andere als die axiomatische; ihr Wesen ist dieses."

[76][EWALD, 1996, p.1127]. Here is the German text from [HILBERT, 1922b, p.169]:
"Dieser Relativismus ist naturgemäß und notwendig; aus ihm entspringt auch keinerlei Schaden, da das Axiomensystem beständig erweitert und der formale Aufbau, unserer konstruktiven Tendenz entsprechend, immer vollständiger wird."

[77][HILBERT, 1921/22, p. 2a of Part II]. The German text is:
"Denn die Annahme der unbeschränkten Gültigkeit jener Äquivalenzen kommt darauf hinaus, dass wir uns Gesamtheiten von unendlich vielen Individuen als etwas fertig Vorhandenes denken."

> "... i.e., we have to apply our intuitive considerations also to figures that are not number signs. Thus we have good reasons to distance ourselves from the earlier dominant principle according to which each theorem of pure mathematics is ultimately a statement concerning integers. This principle was viewed as expressing a fundamental methodological insight, but it has to be given up as a prejudice.
>
> We have to adhere firmly to one demand, namely, that the figures we take as objects must be completely surveyable and that only discrete determinations are to be considered for them. It is only under these conditions that our claims and considerations have the same reliability and evidence as in intuitive number theory."[78]

So at this point, HILBERT and BERNAYS find, simultaneously, a resolution for two problems, namely, to fix the standpoint on the basis of which consistency proofs are to be carried out ("finite Mathematik") and on the basis of which the formal development of mathematics can be pushed along in a systematic way. This resolution addresses in particular, so it was thought, the problem that had plagued HILBERT in [HILBERT, 1905a], but also before and after, namely, to capture the content of a statement concerning all natural numbers or all proofs of a formal system without needing the "axiomatic definition of the system (Gesamtbegriffes) of natural numbers [or of proofs, WS] as one thing". A systematic formal development had already been required in [HELLINGER, 1905, p. 250], but was viewed as extremely difficult. It was hoped then, that it would lead to a complete description of mathematical and thus of all theoretical thinking; in addition, the effectiveness requirements for the metamathematical notions would achieve an almost LEIBNIZian end:

> "... and in this way, by writing down certain symbols in limited number, every question shall be decidable by an 'experiment with chalk or pen'."[79]

To recognize early and deep roots of principled aspects of HILBERT's thinking seems to me to be important; it is equally important to see how the study of *Principia Mathematica* [WHITEHEAD & RUSSELL, 1910–1913] opened a new way to the formalization of mathematics that had not been available to HILBERT in [HILBERT, 1905a], in spite of FREGE's work.

---

[78][HILBERT, 1921/22, pp. 4a, 5a of Part III]. The German text is:

> "... d. h. wir müssen unsere anschaulichen Überlegungen auch auf andere Figuren als auf Zahlzeichen anwenden. Wir sehen uns somit veranlasst, von dem früher herrschenden Grundsatz abzugehen, wonach jeder Satz der reinen Mathematik letzten Endes in einer Aussage über ganze Zahlen bestehen sollte. Dieses Prinzip, in welchem man eine grundlegende methodische Erkenntnis erblickt hat, müssen wir als Vorurteil preisgeben.
>
> An einer Forderung aber müssen wir festhalten, dass nämlich die Figuren, die wir als Gegenstände nehmen, vollständig überblickbar sind, und dass an ihnen nur *diskrete* Bestimmungen in Betracht kommen. Denn nur unter diesen Bedingungen können unsere Behauptungen und Überlegungen die gleiche Sicherheit und Handgreiflichkeit haben wie in der anschaulichen Zahlentheorie."

[79][HELLINGER, 1905, p. 275]. The quotation is part of the following German text:

> "Jeder mathematische Satz und Beweis soll auf eine hinschreibbare Form gebracht werden, und jede Frage soll damit durch ein Hinschreiben gewisser Zeichen in begrenzter Zahl, durch ein ‚Experiment mit Kreide oder Feder' entschieden werden können."

What is the status of "finite" in "finite Mathematik" in that historical regard? Does it indicate a new philosophical perspective with a particular foundational claim that emerged in the early 1920s? The way in which the concept is introduced in [HILBERT, 1921/22], very matter of fact and without great fanfare, almost leads one to suspect that HILBERT and BERNAYS employ a familiar one.[80] That suspicion is hardened by *aspects of the past* and an *attitude* that is pervasive until 1932: as to the attitude, finitism and intuitionism were considered as co-extensional until GÖDEL and GENTZEN proved in 1932 the consistency of classical arithmetic relative to its intuitionistic version; as to aspects of the past, there are HILBERT's many and oft-repeated remarks about constructivist demands, mostly connected to KRONECKER. HILBERT himself remarked in 1931 that KRONECKER's conception of mathematics "essentially coincides with our finitist mode of thought".[81]

There is also a very informative paper by FELIX BERNSTEIN, entitled "Die Mengenlehre GEORG CANTORs und der Finitismus" [BERNSTEIN, 1919]; it is one of the few papers listed at the end of the Introduction to the second edition of *Principia Mathematica* in 1927. BERNSTEIN characterizes "Finitismus" as a familiar foundational stance that opposed set theory since its very beginnings in CANTOR's and DEDEKIND's work. Somewhat indiscriminately, KRONECKER and HERMITE, BOREL and POINCARÉ, RICHARD and LINDELÖF, but also BROUWER and WEYL (on account of [WEYL, 1918]) are viewed as members of the finitist movement. BERNSTEIN's paper was published before HILBERT and BERNAYS viewed "finite Mathematik" as foundationally significant for proof theory. (The characteristic features of finitism BERNSTEIN formulated are actually close to those of "finite Mathematik" HILBERT and BERNAYS emphasized in [HILBERT, 1921/22].)

That concrete background of the term "finitism" could of course be a topic of detailed historical analysis, but it won't be discussed any further here. I just state as a fact that in the lecture notes from the 1920s no detailed discussion of "finite Mathematik" is found. The most penetrating analysis is given in [BERNAYS, 1930/31], still emphasizing the co-extensionality of finitism and intuitionism. Indeed, BERNAYS interprets BROUWER's mathematical work as showing that considerable parts of analysis and set theory can be "given a finitist foundation".[82] What is found in the lecture notes is innovative metamathematical work that tries to overcome in leaps and bounds the obstinate difficulties of giving consistency proofs for stronger formal theories, but in the end unsuccessfully. That's what I am going to describe in some detail now.

---

[80]It is most informative to look at the "Disposition" (outline) for the lectures [HILBERT, 1921/22], proposed by BERNAYS in a letter to HILBERT dated 17 October 1921. The actual lectures followed that "Disposition". But what is in the lectures called "finitist" is here still simply called "constructive". The "Disposition" is reprinted in full in [SIEG, 1999, pp. 35–36].

[81][EWALD, 1996, p.1151].

[82][BERNAYS, 1930/31, pp. 41–42]. The quoted remark is part of this longer passage:

"BROUWER hat den Standpunkt KRONECKERs nach zwei Richtungen weitergeführt: einerseits in Hinsicht auf die philosophische Motivierung, durch die Aufstellung seiner Theorie des ‚Intuitionismus', andererseits dadurch, daß er gezeigt hat, wie man im Gebiete der Analysis und der Mengenlehre den finiten Standpunkt zur Anwendung bringen und, durch eine Umgestaltung der Begriffsbildung und der Schlussweisen von Grund auf, diese Theorien wenigstens zu einem beträchtlichen Teil auf finitem Wege begründen kann."

BERNAYS attached to "Intuitionismus" in the above passage a footnote, saying:

"Es scheint mir im Interesse der Klärung der Diskussion angezeigt, den Ausdruck ‚Intuitionismus' als Bezeichnung einer philosophischen Ansicht zu gebrauchen, im Unterschied von dem Terminus ‚finit', der eine bestimmte Art des Schließens und der Begriffsbildung bezeichnet."

4 Finitist proof theory

In the winter term 1921/22, HILBERT and BERNAYS not only introduced the terms "HILBERTsche Beweistheorie" and "finite Mathematik", but they also used novel techniques to give a finitist proof of the restricted consistency result formulated in [HILBERT, 1922b]. The proof is presented in KNESER's *Mitschrift* of these lectures [KNESER, 1921/22]. Indeed, the proof-theoretic considerations start on 2 February and end on 23 February 1922; they include a treatment of definition by (primitive) recursion and proof by induction. The latter principle is formulated as a rule and restricted to quantifier-free formulas. The part of the official notes [HILBERT, 1921/22] that presents the consistency proof was written by BERNAYS after the term had ended and contains a different argument; that argument pertains only to the basic system and is sketched in [HILBERT, 1923a]. I am going to describe a third modification found in KNESER's *Mitschrift* of the 1922/23 lectures [KNESER, 1922/23]. Apart from this important beginning of proof-theoretic investigations, I will discuss HILBERT's *Ansatz* for the treatment of quantifiers, some relevant further developments in the 1920s due to WILHELM ACKERMANN and JOHN VON NEUMANN, and the refined and systematic presentation of this work in the first and second volume of *Grundlagen der Mathematik* [HILBERT & BERNAYS, 1934; 1939].

4.1 Proof transformations

The work in 1921/22 gave a new direction to proof-theoretic investigations and required the partial resolution of some hard technical, as well as deep methodological issues. In the first group of issues we find in particular the treatment of quantifiers; that is absolutely crucial, if consistency is to be established for the kind of formalisms that had been used in the lectures [HILBERT, 1917/18] to represent classical mathematical practice. The second group of issues concerns the extent of metamathematical means (is contentual induction admitted?) and the character of the formal systems to be investigated (are they to be quasi-constructive?).

The last issue has a direct resolution: it is addressed in [HILBERT, 1921/22] by a modification of the logical calculus, i.e., by replacing the (in-) equalities in the axioms for negation by (negations $\overline{B}$ of) arbitrary formulas B. The axioms are given in the form:

$$A \rightarrow \left(\overline{A} \rightarrow B \right)$$
$$(A \rightarrow B) \rightarrow \left\{ \left(\overline{A} \rightarrow B \right) \rightarrow B \right\}$$

With this emendation we have the system that is used in [HILBERT, 1923a, p. 153]; the first axiom is called the *law of contradiction* and the second the *principle of tertium non datur*.

Though the technical resolution is straightforward, it required a new philosophical perspective that is, perhaps first and best, articulated in [BERNAYS, 1922] written during the early fall of 1921. The paper starts out with a discussion of *existential axiomatics*, which presupposes, as we saw in § 2, a system of objects satisfying the structural conditions formulated by the axioms. The assumption of such a system contains, BERNAYS writes, "something so-to-speak transcendent for mathematics". Thus, the question arises, "which principled position with respect to it should be taken". BERNAYS remarks that it might be perfectly coherent to appeal to an intuitive grasp of the natural number sequence or even of the manifold of real numbers. However, that could obviously not be an intuition in any primitive sense, and one should take very seriously the tendency in the exact sciences to use, as far as possible, only restricted means for acquiring knowledge. BERNAYS is thus led to a programmatic demand:

"Under this perspective we are going to try, whether it is not possible to give a foundation to these transcendent assumptions in such a way that only *primitive intuitive knowledge is used*."[83]

Clearly, contentual mathematics is to be based on primitive intuitive knowledge and that includes for BERNAYS already at this stage induction.

BERNAYS' perspective is clearly reflected in the notes [HILBERT, 1921/22]. The notes contain a substantial development of finitist arithmetic that involves from the very outset contentual induction ("anschauliche Induktion"). This step was not a small one for HILBERT. In 1904 he had described KRONECKER as a dogmatist because of his view "that integers — and in fact integers as a general notion (parameter value) — are directly and immediately given". In the intervening years he scolded KRONECKER for just that reason and accused him of not being radical enough. BERNAYS reports that HILBERT viewed this expansion of contentual arithmetic as a *compromise*, and it is indeed a compromise between a radical philosophical position and intrinsic demands arising from metamathematical investigations. But it is also a compromise in that it gives up the direct parallelism with the case of physics, when statements that can be verified by observation are viewed as corresponding to statements that can be checked by calculation.

For the central steps of the consistency argument we need a definition: a formula is called *numeric* if it is built up solely from $=$, $\neq$, numerals, and sentential logical connectives. The first step transforms formal proofs with a numeric end formula into proofs that contain only closed formulas, without changing the end formula. In the second step these modified proofs are turned into configurations that may no longer be proofs, but consist only of numeric formulas; that is achieved by reducing the closed terms to numerals. Finally, all formulas in these configurations are brought into disjunctive normal form and syntactically recognized as "true". Assume now that there is a formal proof of $0 \neq 0$; the last observation allows us to infer that the end formula of this proof is true. However, $0 \neq 0$ is not true and hence not provable.

This proof is given with many details in [HILBERT, 1921/22]. It is sketched in [HILBERT, 1923a, p.184] and presented in the winter term of 1922/23 according to KNESER's *Mitschrift* [KNESER, 1922/23] in a slightly modified form. The papers [BERNAYS, 1922], [HILBERT, 1922b; 1923a] were all available for the winter term 1922/23, as indicated explicitly by KNESER. The system in [KNESER, 1922/23] uses a calculus for sentential logic that incorporates the connectives $\&$ and $\vee$. The main point of this expanded language and calculus is to provide a more convenient formal framework for representing informal arguments. The axioms for the additional connectives are given as follows:

$$\begin{aligned} A \,\&\, B &\to A \\ A \,\&\, B &\to B \\ A &\to (B \to A \,\&\, B) \\ A &\to A \vee B \\ B &\to A \vee B \\ (A \to C) &\to ((B \to C) \to (A \vee B \to C)) \end{aligned}$$

Clearly, the axioms correspond directly to "natural" rules for these connectives, and one finds here the origin of GENTZEN's natural deduction calculi.

---

[83][BERNAYS, 1922, p.11]. The German text is:
"Unter diesem Gesichtspunkt werden wir versuchen, ob es nicht möglich ist, jene transzendenten Annahmen in einer solchen Weise zu begründen, daß nur *primitive anschauliche Erkenntnisse zur Anwendung kommen.*"

In [BERNAYS, 1927], the completeness of this calculus is asserted and its methodological advantages are discussed. BERNAYS writes:

> "The initial formulas can be chosen in quite different ways. In particular one has tried to get by with the smallest number of axioms, and in this respect the limit of what is possible has indeed been reached. One supports the purpose of logical investigations better, if one separates, as it is done in the axiomatics of geometry, different groups of axioms in such a way that each of these groups expresses the rôle of one particular logical operation."[84]

This formal system is used in [ACKERMANN, 1925a], and also in [HILBERT, 1928], where, however, (as already in [HILBERT, 1926]) an interestingly different formulation of the axioms for negation is given, called the *principle of contradiction* and the *principle of double negation*:

$$\left(A \rightarrow B \ \& \ \overline{B} \right) \rightarrow \overline{A}$$
$$\overline{\overline{A}} \rightarrow A$$

These are axiomatic formulations of GENTZEN's natural deduction rules for classical negation; the earlier axioms for negation can be derived easily. (This new formulation can be found in [HEIJENOORT, 1971] on p. 382 and more fully on pp. 465–466.)

The consistency argument for the system of arithmetic described above, but with this logical calculus, is given in [KNESER, 1922/23, pp. 20–22].

The form of this consistency argument is used also in the first volume of *Grundlagen der Mathematik* [HILBERT & BERNAYS, 1934, §6]. It begins by separating linear proofs into proof threads ("Auflösung in Beweisfäden"). Then variables are eliminated in the resulting proof trees ("Ausschaltung der Variablen"), and the numerical value of functional terms is determined by calculation ("Reduktion der Funktionale"). That makes it possible to determine the truth or falsity of formulas in the very last step. Here that is not done by turning formulas into disjunctive normal form, but rather simply by truth-table computations ("Wahrheitsbewertung"). Incorporating definition by recursion and proof by induction into these considerations is fairly direct.

From a contemporary perspective this argument shows something very important. If a formal theory contains a certain class of finitist functions, then it is necessary to appeal to a wider class of functions in the consistency proof: an *evaluation function* is needed to determine uniformly the numerical value of terms, and such a function is no longer in the given class. The formal system considered in the above consistency proof includes primitive recursive arithmetic, and the consistency proof goes beyond the means available in primitive recursive arithmetic. At this early stage of proof theory, finitist mathematics is consequently stronger than primitive recursive arithmetic. This assessment of the relative strength is sustained, as we will see, throughout the development reported in this essay.

---

[84][BERNAYS, 1927, p.10]. The German text is:

> "Die Wahl der Ausgangsformeln kann auf sehr verschiedene Weise getroffen werden. Man hat sich besonders darum bemüht, mit einer möglichst geringen Zahl von Axiomen auszukommen, und hat hierin in der Tat die Grenze des Möglichen erreicht. Den Zwecken der logischen Untersuchung wird aber besser gedient, wenn wir, entsprechend wie in der Axiomatik der Geometrie, verschiedene *Axiomengruppen* voneinander sondern, derart, daß jede von ihnen die Rolle einer logischen Operation zum Ausdruck bringt."

It is informative to compare these remarks with HILBERT's in the Introduction to [HILBERT, 1899], quoted in §2.2.

4.2 HILBERT's *Ansatz*: the ε-substitution method

ACKERMANN reviewed in §II of [ACKERMANN, 1925a] the above consistency proof; the section is entitled *The consistency proof before the addition of the transfinite axioms* ("Der Widerspruchsfreiheitsbeweis vor Hinzunahme der transfiniten Axiome"). The very title reveals the restricted programmatic significance of this result, as it concerns a theory that is part of finitist mathematics and need not be secured by a consistency proof. The truly expanding step that goes beyond the finitist methods just described involves the treatment of quantifiers. Already in [HILBERT, 1922b] we find a brief indication of his *Ansatz*, which is elaborated in [KNESER, 1921/22]. Indeed, the treatment of quantifiers is just the first of three steps that have to be taken: the second concerns the general induction principle for number theory, and the third an expansion of proof-theoretic considerations to analysis.

HILBERT sketched in his *Leipzig* talk of September 1922, which was published as [HILBERT, 1923a], the consistency proof from § 4.1. However, the really dramatic aspect of his proof-theoretic discussion is the treatment of quantifiers with the τ-function, the dual of the later ε-operator. The logical function τ associates with every predicate $A(a)$ a particular object $\tau_a(A(a))$ or simply τA. It satisfies the *transfinite axiom* $A(\tau A) \to A(a)$, which expresses, according to HILBERT, "if a predicate A holds for the object τA, then it holds for all objects a". The τ-operator allows the definition of the quantifiers:

$$(a)\,A(a) \quad \sim \quad A(\tau A)$$
$$(Ea)\,A(a) \quad \sim \quad A(\tau(\overline{A}))$$

HILBERT extends the consistency argument then to the "first and simplest case" that goes beyond the finitist system: this "Ansatz" will evolve into the ε-substitution method. It is only in the Leipzig talk that HILBERT gave proof theory its principled formulation and discussed its technical tools — for the first time outside of Göttingen. As far as analysis is concerned, HILBERT had taken as the formal frame for its development ramified type theory with the axiom of reducibility in [HILBERT, 1917/18]. In his Leipzig talk he considers a third-order formulation; appropriate functionals (Funktionenfunktionen) allow him to prove (on pp. 189–191)

(i) the least-upper-bound principle for sequences and sets of real numbers and

(ii) ZERMELO's choice principle for sets of sets of real numbers.

He conjectures that the consistency of the additional transfinite axioms can be patterned after that for τ. He ends the paper with the remark:

> "Now the task remains of precisely carrying out the basic ideas I just sketched; its solution completes the founding of analysis and prepares the ground for the founding of set theory."[85]

In Supplement IV of the second volume of *Grundlagen der Mathematik*, HILBERT & BERNAYS [1939] give a beautiful exposition of analysis and emphasize, correctly, that the formalisms for their "deductive treatment of analysis" were used in HILBERT's early lectures on proof theory and that they were described in [ACKERMANN, 1925a].

---

[85][HILBERT, 1923a, p.191]. The German text is: "Es bleibt nun noch die Aufgabe einer genauen Ausführung der soeben skizzierten Grundgedanken; mit ihrer Lösung wird die Begründung der Analysis vollendet und die der Mengenlehre angebahnt sein."

The above concrete proof-theoretic work was directly continued in ACKERMANN's thesis, on which [ACKERMANN, 1925a] is based. The paper was submitted to *Mathematische Annalen* on 30 March 1924 and published in early 1925.[86] It starts out, as I mentioned, with a concise review of the earlier considerations for quantifier-free systems. In 1923, the ε-symbol had replaced the τ-symbol (as is clear from [KNESER, 1922/23]), and in ACKERMANN's dissertation the ε-calculus as we know it replaced the τ-calculus from 1922. The transfinite axiom of the ε-calculus is obviously dual to the axiom for τ and is formulated in [ACKERMANN, 1925a] in §III entitled, "Die Widerspruchsfreiheit bei Hinzunahme der transfiniten Axiome und der höheren Funktionstypen." For number variables the crucial axiom is formulated as $A(a) \to A(\varepsilon A)$ and allows of course the definition of quantifiers:

$$(Ea)\, A(a) \quad \sim \quad A(\varepsilon A)$$
$$(a)\, A(a) \quad \sim \quad A(\varepsilon(\overline{A}))$$

The remaining transfinite axioms are adopted from [HILBERT, 1923a]. However, the ε-symbol is actually characterized as the least-number operator, and the recursion schema with just number variables is extended to a schema that permits also function variables.

The connection to the mathematical development in HILBERT's paper is finally established in §IV, where ACKERMANN explores the "Tragweite" of the axioms. At first it was believed that ACKERMANN had completed the task indicated by HILBERT at the end of the Leipzig talk and had established not only the consistency of full classical number theory but even of analysis. However, a note was added in proof (p. 9) that significantly restricted the result; NEUMANN noted a mistake in ACKERMANN's arguments. In his own paper *Zur Hilbertschen Beweistheorie* [NEUMANN, 1927], submitted on 29 July 1925, a consistency proof was given for a system that covers, NEUMANN asserts (p. 46), RUSSELL's mathematics without the axiom of reducibility or WEYL's system in [WEYL, 1918]. ACKERMANN corrected his arguments and obtained an equivalent result; that corrected proof is discussed in [HILBERT, 1928] and [BERNAYS, 1928].[87] (It is also the basis for the corresponding presentation in [HILBERT & BERNAYS, 1939]; see §4.4 below.) In his *Bologna* talk of 1928, HILBERT [1930b] stated, in line with NEUMANN's observation, that the consistency of full number theory had been secured by the proofs of ACKERMANN and NEUMANN; according to BERNAYS in his preface to [HILBERT & BERNAYS, 1939], that belief was sustained until 1930.

A number of precise metamathematical problems were formulated in HILBERT's Bologna talk, among them the completeness question for first-order logic, which is presented in a new formulation suggested by the LÖWENHEIM–SKOLEM Theorem. The latter theorem also suggested, so it seems, that higher mathematical theories might not be syntactically complete. In any event, HILBERT's Bologna address shows most dramatically through its broad perspective and clear formulation of open problems, how far mathematical logic had been moved in roughly a decade. This is a remarkable achievement with impact not just on proof theory; the (in-) completeness theorems, after all, are to be seen in this broader foundational enterprise.[88]

---

[86] In his lecture *Über das Unendliche* given on 4 June 1925 in Münster [HILBERT, 1926], he presented an optimistic summary of the state of proof-theoretic affairs together with bold new considerations. Even before this talk, HILBERT had turned his attention to issues beyond the consistency of number theory and analysis, namely, the solution of the continuum problem.

[87] See [HEIJENOORT, 1971], pp. 477, 489, but also note 3 on p. 489 and the introductions to these papers.

[88] A remark indicating the depth of DEDEKIND's continued influence: HILBERT formulated as Problem I of his Bologna talk the consistency of the ε-axioms for function variables and commented, "The solution of Problem I justifies also DEDEKIND's ingenious considerations in his essay *Was sind und was sollen die Zahlen?*"

In terms of specifically proof-theoretic results no striking progress was made, in spite of the fact that most talented people had been working on the consistency program. That state of affairs had intrinsic reasons. BERNAYS asserts in his preface to [HILBERT & BERNAYS, 1934] that a presentation of proof-theoretic work had almost been completed in 1931. But, he continues, the publication of papers by HERBRAND and GÖDEL produced a deeply changed situation that resulted in an extension of the scope of the work and its division into two volumes. The eight sections of the first volume can be partitioned into three groups: §§ 6–8 give refined versions of results that had been obtained in the 1920s, investigating the consistency problem and other metamathematical questions for a variety of (sub-) systems of number theory; §§ 3–5 develop systematically the logical framework of first-order logic (with identity); §§ 1 and 2 discuss the central foundational issues and the finitist standpoint in most informative and enlightening ways. Let me start with a discussion of the central issues as viewed in 1934.

4.3 *Grundlagen der Mathematik* [HILBERT & BERNAYS, 1934; 1939]

[HILBERT & BERNAYS, 1934, § 1] begins with a general discussion of axiomatics, at the center of which is a distinction between *contentual* and *formal axiomatic theories*. Contentual axiomatic theories like EUCLID's geometry, NEWTON's mechanics, and CLAUSIUS' thermodynamics draw on experience for the introduction of their fundamental concepts and principles, all of which are understood contentually. Formal axiomatic theories like HILBERT's axiomatization of geometry, by contrast, abstract away such intuitive content. They begin with the assumption of "a fixed system of things (or several such systems)", which is delimited from the outset and constitutes a domain of individuals for all predicates from which the statements of the theory are built up. The existence of such a domain of individuals constitutes "an idealizing assumption that properly augments the assumptions formulated in the axioms".[89] HILBERT and BERNAYS refer to this approach as *existential axiomatics*. They clearly consider formal axiomatics to be a sharpening of contentual axiomatics, but are also quite explicit that the two types of axiomatics complement each other and are both necessary.

The consistency of a formal axiomatic theory with a finite domain can be established by exhibiting a model satisfying the axioms. However, for theories F with infinite domains consistency proofs present a special problem, because "the question of the existence of an infinite manifold cannot be decided by appealing to extra-mathematical objects. Rather, the question must be solved within mathematics itself."[90] One must treat the consistency problem for such a theory F as a *logical problem* — from a proof-theoretic perspective. This involves formalizing the principles of logical reasoning and proving that from F one cannot derive (with the logical principles) both a formula and its negation. Proof theory need not prove the consistency of every F; it suffices to find a proof for a system F with a structure that is

(i) surveyable to make a consistency proof for the system plausible and

(ii) rich enough so that the existence of a model S for F guarantees the satisfiability of axiom systems for branches of physics and geometry.

Arithmetic (including number theory and analysis) is considered to be such an F.

---

[89][HILBERT & BERNAYS, 1934, p. 2, lines 2–10]. Also [HILBERT & BERNAYS, 1968, p. 2, lines 2–10], see this volume.

[90][HILBERT & BERNAYS, 1934, p. 17, lines 15–17]. Also [HILBERT & BERNAYS, 1968, p. 17, lines 15–17], see this volume.

For a consistency argument to be foundationally significant it must avoid the idealizing existence assumptions made by formal axiomatic theories. But if a proof-theoretic justification of arithmetic by elementary means were possible, might it not also be possible to develop arithmetic directly, free from non-elementary assumptions and thus not requiring any additional foundational justification? The answer to this question involves elementary presentations of parts of number theory and formal algebra; these presentations serve at the same time to illuminate the finitist standpoint. Finitist deliberations take here their purest form, i.e., the form of *"thought experiments* with things that are assumed to be *concretely present"*.[91] The word "finitist" is intended to convey the idea that a consideration, a claim or definition respects

(i) that *objects are representable*, in principle, and

(ii) that *processes are executable*, in principle.[92]

A direct, elementary justification for all of mathematics is not possible, HILBERT and BERNAYS argue, because already in number theory and analysis one uses non-finitist principles. While one might circumvent the use of such principles in number theory (as only the existence of the domain of integers is assumed), the case for analysis is different. There the objects of the domain are themselves infinite sets of integers, and the principle of the excluded middle is applied in this extended domain. Thus one is led back to using finitist proof theory as a tool to secure the consistency of mathematics. (This restriction is relaxed only at the end of the second volume when "extensions of the methodological framework of proof theory" are considered.) The first stage of this project, the proof-theoretic investigation of appropriate logical formalisms, occupies §§ 3–5. These systems are close to contemporary ones; they can be traced back to [HILBERT, 1917/18], and were presented also in [HILBERT & ACKERMANN, 1928]. The second stage is carried out in §§ 6–8 and involves the investigation of sub-systems of first-order number theory; there are three different groups of such systems.

The first group consists of weak fragments of number-theory that contain few, if any, function symbols and extend predicate logic with equality by axioms for 0, successor and $<$; some of the fragments involve quantifier-free induction. Metamathematical relations between them are explored, and independence as well as consistency results are established. The main technique for giving consistency proofs is that discussed in § 4.1 above. However, since the formalisms contain quantifiers, an additional procedure is required, namely a procedure that assigns *reducts*, quantifier-free formulas acting as witnesses, to formulas containing quantifiers. The method underlying this procedure is due to HERBRAND and PRESBURGER. Consistency is inferred from results that involve the notion of *verifiability*, extending the notion of truth to formulas containing free variables, bound variables, and recursively defined function signs.[93] To establish the consistency of a formalism F, one

---

[91][HILBERT & BERNAYS, 1934, p. 20, lines -9--8]. Also [HILBERT & BERNAYS, 1968, p. 20, lines -8--7], see this volume.

[92][HILBERT & BERNAYS, 1934, p. 32, lines 21–25]. Also [HILBERT & BERNAYS, 1968, p. 32, lines 21–25], see this volume.

[93]More precisely, letting A be a formula of the formalism F:

(i) if A is a numeric formula, it is verifiable if it is true;

(ii) if A contains free numeric variables (but no formula variables or bound variables), it is verifiable if one can show by finitist means that the substitution of arbitrary numerals for variables (followed by the evaluation of all function-expressions and their replacement through their numerical values) yields a true numeric formula;

(iii) if A contains bound variables but no formula variables, it is verifiable if its reduct is verifiable (according to (i) and (ii)).

proves that every formula not containing formula variables is verifiable, if it is derivable in F. Since $0 \neq 0$ is not verifiable, it is not derivable in F; thus, F is consistent.

The second group of sub-systems of number theory contains formalisms arising from the elementary calculus with free variables (the quantifier free fragment of the predicate calculus) through the addition of functions defined by recursion. HILBERT and BERNAYS discuss at the beginning of §7 the formalization of the definition principle for recursion, reminding the reader how they viewed recursive definition of a function in their discussion of finitist number theory in §1. There they emphasize that by "function" they understand an intuitive instruction ("anschauliche Anweisung") on account of which a numeral is associated with a given numeral, or a pair of numerals, or a triple ... of numerals. The proof method of mathematical induction (vollständige Induktion) is not considered as an "independent principle" but as a consequence that is gathered from the concrete build-up of numerals.[94] Similarly, definition by recursion is not viewed as an independent definition principle but rather as "an agreement concerning an abbreviated description of certain construction processes by means of which one obtains a numeral from one or several given numerals."[95]

> "These processes or procedures can be mimicked formally by the introduction of function symbols together with recursion equations."[96]

They treat *in a first step* ("zunächst") what they consider as the simplest schema of recursion (and "restrict *for the time being* the notion of recursion to it"). That simplest schema of recursion is the schema of what we (and they later on p. 326) call *primitive recursion*, namely,
$$f(a,\ldots,k,0) = \mathfrak{a}(a,\ldots,k),$$
$$f(a,\ldots,k,n') = \mathfrak{b}(a,\ldots,k,n,f(a,\ldots,k,n)).$$
$\mathfrak{a}$ and $\mathfrak{b}$ denote previously defined functions and $a,\ldots,k,n$ are numerical variables. What then is the formalization of the intuitive procedure associated with a recursive definition that is formally reflected by the recursion schema for f? It is given by a derivation — via the recursion equations and the axioms for identity — of the equation $f(a,\ldots,k,z) = t(a,\ldots,k)$ for each numeral z, where t is a term not containing f. Then they argue that the value of $f(a,\ldots,k,z)$ can be determined for each sequence of numerals $a,\ldots,k$ and conclude their considerations with:

> "Thus we can completely mimick the recursive calculation procedure of finitist number theory in our formalism by the deductive application of the recursion equations."[97]

---

[94][HILBERT & BERNAYS, 1934, p. 24, lines -9--7]. Also [HILBERT & BERNAYS, 1968, p. 24, lines -6--4], see this volume.

[95][HILBERT & BERNAYS, 1934, p. 24, lines 18–23]. Also [HILBERT & BERNAYS, 1968, p. 24, lines 18–23], see this volume.

[96][HILBERT & BERNAYS, 1934, p. 287]. This is a translation of part of the following German text:
 "Dieses Verfahren [der Berechnung von Funktionswerten in der finiten Zahlentheorie] können wir auch im Formalismus nachbilden, indem wir allgemein die Einführung von Funktionszeichen in Verbindung mit Rekursionsgleichungen zulassen."

[97][HILBERT & BERNAYS, 1934, p. 292]. The German text is:
 "Somit können wir das rekursive Berechnungsverfahren der finiten Zahlentheorie in unserem Formalismus durch die deduktive Anwendung der Rekursionsgleichungen vollkommen nachbilden."

They then prove on pp. 298–300 a general *Consistency Theorem*:

> "Let F be a formalism extending the elementary calculus with free variables by verifiable axioms (that may contain recursively defined functions whose defining equations are taken as axioms) and the schema of quantifier-free induction, then every derivable formula of F (without formula variables) is verifiable."

This theorem is explicitly taken to establish the consistency of a number of formalisms including in particular that of recursive number theory. Having formally developed the latter theory to indicate the strength of this recursive treatment of number theory, they state:

> "This recursive number theory is closely related to intuitive number theory, as we have considered it in § 2, because all of its formulas can be given *a finitist contentual interpretation*. This contentual interpretability follows from the verifiability of all derivable formulas of recursive number theory, as we established it already. Indeed, in this area verifiability has the character of a direct contentual interpretation, and thus the proof of consistency could be obtained here so easily."

As we have a finitist consistency proof for recursive arithmetic, finitist mathematics here goes beyond recursive arithmetic. Their notion of recursive number theory was at first, as I emphasized above, to involve only primitive recursions; so finitist mathematics is stronger than primitive recursive arithmetic. Indeed, after the above remark they continue the discussion of the relationship between intuitive and recursive number theory as follows:

> "Recursive number theory is differentiated from intuitive number theory by its formal restrictions; its only method for forming concepts, apart from explicit definitions, is the recursion schema, and its methods of derivation are also firmly delimited.
>
> However, without taking away from recursive number theory what is the characteristic feature of its method, we can admit certain *extensions of the recursion schema* as well as of the induction schema. We'll discuss these [extensions] now briefly."

The extending formalisms involve the schema of nested recursion ("verschränkte Rekursion") that allows in particular the definition of the ACKERMANN function; they remark that their previous consistency results are extended to these proper extensions of recursive number theory as well; thus, the bounds of finitist mathematics are pushed still further.[98]

The third group of formalisms that are actually equivalent to full number theory is investigated towards the end of § 7 and in § 8. The first of these is the formalism of the axiom system (Z); call this formalism (Z). HILBERT and BERNAYS comment that the techniques used in the consistency proofs for fragments of number theory cannot be generalized to (Z), as any reduction procedure for (Z) would provide a decision procedure for (Z) and, thus, allow solving all number-theoretic problems. They leave the possibility of such a procedure as an open problem (whose solution, if it exists, they view as a long way off) and focus on examining whether (Z) provides the means for formalizing informal number theory.

With this end in mind, it is proved in §8 that all recursive functions are representable in (Z). For this proof they establish three separate claims:[99]

(1) the least number operator μ can be explicitly defined in terms of the ι-symbol;

(2) any recursive definition can be explicitly defined in (Z_μ), i.e. (Z) extended by defining axioms for the μ-operator;

(3) the addition of the ι-rule to (Z) is a conservative extension of (Z).

Volume I concludes with the remark that the above results entail the consistency of (Z_μ) relative to that of (Z), but that none of the results or methods considered so far suffice to show that (Z) is consistent.

4.4 Limited results for quantifiers

The consistency proofs in §7(a) of [HILBERT & BERNAYS, 1934] are given for quantifier-free systems. In the [HILBERT & BERNAYS, 1939], these theories are embedded in the system of full predicate logic together with the ε-axioms in the form $A(a) \to A(\varepsilon_x(A(x)))$. The first crucial task is to eliminate all references to bound variables from proofs of theorems that do not contain them; axioms used in these proofs must not contain bound variables either. In the formulation of HILBERT and BERNAYS, the consistency of a system of proper axioms relative to the predicate calculus together with the ε-axioms is to be reduced to the consistency of the system relative to the elementary calculus (with free variables).[100] The consistency of the latter system is recognized on account of a suitable finitist interpretation.

---

[98] The two long quotations above are both found on p. 325 of [HILBERT & BERNAYS, 1934]. The remark concerning the finitist provability of the consistency of the extended formalisms is on p. 325; more precisely, it is asserted there, all formulas (not containing formula variables) that are provable in the extended formalisms are verifiable. Here is the first German text:

"Diese rekursive Zahlentheorie steht insofern der anschaulichen Zahlentheorie, wie wir sie im §2 betrachtet haben, nahe, als ihre Formeln sämtlich *einer finiten inhaltlichen Deutung fähig* sind. Diese inhaltliche Deutbarkeit ergibt sich aus der bereits festgestellten Verifizierbarkeit aller ableitbaren Formeln der rekursiven Zahlentheorie. In der Tat hat in diesem Gebiet die Verifizierbarkeit den Charakter einer direkten inhaltlichen Interpretation, und der Nachweis der Widerspruchsfreiheit war daher auch hier so leicht zu erbringen."

And here is the second passage:

"Der Unterschied der rekursiven Zahlentheorie gegenüber der anschaulichen Zahlentheorie besteht in ihrer formalen Gebundenheit; sie hat als einzige Methode der Begriffsbildung, außer der expliziten Definition, das Rekursionsschema zur Verfügung, und auch die Methoden der Ableitung sind fest umgrenzt.

Allerdings können wir, ohne der rekursiven Zahlentheorie das Charakteristische ihrer Methode zu nehmen, gewisse *Erweiterungen des Schemas der Rekursion* sowie auch des Induktionsschemas zulassen. Auf diese wollen wir noch kurz zu sprechen kommen."

[99] The second claim is first established for primitive recursively defined functions, but then extended to the case of nested recursion; cf. [HILBERT & BERNAYS, 1934, p. 421]. The latter result is attributed to NEUMANN and GÖDEL. As to the general issues compare NEUMANN's letter to GÖDEL of 29 November 1930; the letter is reprinted and translated in [GÖDEL, 1986ff., Vol. V].

[100] [HILBERT & BERNAYS, 1939, p. 33]. Referring to the first ε-theorem they write there: "Die Bedeutung dieses Theorems besteht ja darin, daß es die Frage der Widerspruchsfreiheit eines Systems von eigentlichen Axiomen ohne gebundene Variablen bei Zugrundelegung des Prädikatenkalkuls und des ε-Axioms zurückführt auf die seiner Widerspruchsfreiheit bei Zugrundelegung des elementaren Kalkuls mit freien Variablen."

Thus, it is emphasized, operating with the ε-symbol can be viewed as "merely an auxiliary calculus, which is of considerable advantage for many metamathematical considerations."[101]

In the framework of the extended calculus bound variables are associated with just ε-terms, as the quantifiers can be defined. The initial elimination result is the *First ε-Theorem*:

> If the axioms $A_1, \ldots, A_k$ and the conclusion of a proof do not contain bound individual variables or (free) formula variables, then all bound variables can be eliminated from the proof.

The argument can be extended to cover proofs of purely existential formulas, but the proofs yield then as their conclusion suitable disjunctions of instances of the end formula. Based on this extension a *Consistency Theorem* of a quite general character is proved:

> If the axioms $A_1, \ldots, A_k$ are verifiable, then
> (i) any provable formula containing no bound variables is verifiable, and
> (ii) for any provable, purely existential formula $(Ex_1)\cdots(Ex_n)\,A(x_1,\ldots,x_n)$ (with only the variables shown) there are variable-free terms $t_1,\ldots,t_n$ such that $A(t_1,\ldots,t_n)$ is true.

This theorem is applied to establish the consistency (i) of Euclidean and Non-Euclidean geometry without continuity assumptions in § 1.4, and (ii) of arithmetic with recursive definitions, but only quantifier-free induction in §§ 2.1 and 2.2. In essence then, the consistency theorem from [Herbrand, 1932] (cf. § 5.1) has been reestablished in a subtly more general way: Hilbert and Bernays allow the introduction of a larger class of recursive functions.[102] Putting the result in the appropriate systematic context, we see that the consistency proof of [Hilbert, 1922b] for the quantifier-free system of primitive recursive arithmetic has been extended to cover that system's expansion by full classical quantification theory. Indeed, the extensions from [Hilbert & Bernays, 1934] are also covered.

The remainder of § 2 discusses the difficulty of extending the elimination procedure in the proof of the first ε-theorem to a system with full induction and examines Hilbert's original *Ansatz* for eliminating ε-symbols.[103] §§ 3 and 4 investigate the formalism for predicate logic and begin in § 3 with a proof of the *Second ε-Theorem*:

> If the axioms and the conclusion of a proof (in predicate logic with identity) do not contain ε-symbols, then all ε-symbols can be eliminated from the proof.

---

[101][Hilbert & Bernays, 1939, p. 13]. The quotation is part of this German text: "Es besteht ja keineswegs die Notwendigkeit, das ε-Symbol in den endgültigen deduktiven Aufbau des logisch-mathematischen Formalismus einzubeziehen. Vielmehr kann das Operieren mit dem ε-Symbol als ein bloßer Hilfskalkul angesehen werden, der für viele metamathematische Überlegungen von erheblichem Vorteil ist."

[102][Hilbert & Bernays, 1939, p. 52]. That is made explicit in note 1 on page 52. Though this larger class of functions (all general recursive ones?) allows a generalized notion of verifiability, the latter would no longer guarantee a finitist consistency proof. They write there:
> "Diese [allgemeine Anweisung zur Einführung von Funktionszeichen in [Herbrand, 1932]] ist insofern etwas enger als die hier formulierte Anweisung, als Herbrand verlangt, daß das Verfahren der Wertbestimmung sich, so wie bei den Schematen der rekursiven Definition, durch eine finite Deutung der Axiome ergeben soll."

[103] As to the character of the original and the later version of the elimination method and Ackermann's work see pp. 21, 29-30, 92ff., 121ff., the note on p.121, as well as Bernays' preface.

HERBRAND's theorem is obtained as well as a variety of criteria for the refutability of formulas in predicate logic; proofs of the LÖWENHEIM–SKOLEM Theorem and of GÖDEL's completeness theorem are also given. These considerations are used to establish results concerning the decision problem; solvable cases as well as reduction classes are discussed. In § 4, GÖDEL's "arithmetization of metamathematics" is presented in great detail and applied to obtain a fully formalized proof of the completeness theorem.

The completeness theorem can be taken as stating that the consistency of an axiom system relative to the calculus of predicate logic coincides with satisfiability of the system by an arithmetic model. The formalized proof is intended to establish a kind of finitist equivalent to a consequence of this formulation, namely, that the consistency relative to the predicate calculus guarantees consistency, as it is formulated on p. 205, in an open contentual sense ("im unbegrenzten inhaltlichen Sinne"). The finitist equivalent is formulated roughly as follows: if a formula is irrefutable in predicate logic, then it remains irrefutable in "every consistent number-theoretic formalism," i.e., in every formalism that is consistent and remains consistent when the axioms of (Z_μ) and verifiable formulas are added. That result expresses a strong deductive closure of the predicate calculus, but only if (Z_μ) is consistent.[104] Thus, there is another reason for establishing finitistically the consistency of this number-theoretic formalism.

5 Incompleteness

The second volume picks up where the first left off. As we saw, it presents in §§ 1 and 2 HILBERT's proof-theoretic "Ansatz" based on the ε-symbol as well as related consistency proofs; this is the first main topic. The methods used there open a simple approach to HERBRAND's theorem, which is at the center of § 3. The discussion of the decision problem at the end of § 3 leads, after a thorough discussion of the "method of the arithmetization of metamathematics", in § 4 to a proof-theoretic sharpening of GÖDEL's completeness theorem. The remainder of the second volume of *Grundlagen der Mathematik* is devoted to the second main topic, the examination of the fact, "which is the basis for the necessity to expand the frame of the contentual inference methods, which are admitted for proof theory, beyond the earlier delimitation of the 'finitist standpoint'."[105] Of course, GÖDEL's incompleteness theorems are at the center of that discussion, and they were also at the center for the changed situation that forced a re-thinking of the proof-theoretic enterprise. This re-thinking took two directions,

(i) exploring the extent of finitist mathematics, and

(ii) demarcating the appropriate methodological standpoint for proof theory.

---

[104][HILBERT & BERNAYS, 1939, p. 253]. Here is the German text of which the quoted remark is a part: "Das hiermit bewiesene Theorem hat allerdings seine Bedeutung als ein Vollständigkeitssatz, d. h. als Ausdruck einer Art von deduktiver Abgeschlossenheit des Prädikatenkalkuls, nur unter der Voraussetzung der Widerspruchsfreiheit des zahlentheoretischen Formalismus."

[105]BERNAYS wrote in "Zur Einführung" of [HILBERT & BERNAYS, 1939]: "Das zweite Hauptthema bildet die Auseinandersetzung des Sachverhalts, auf Grund dessen sich die Notwendigkeit ergeben hat, den Rahmen der für die Beweistheorie zugelassenen inhaltlichen Schlußweisen gegenüber der vorherigen Abgrenzung des ‚finiten Standpunktes' zu erweitern."

5.1 Changed situation

The consistency result in [HERBRAND, 1932] took an additional step beyond ACKERMANN's, as far as the very claim is concerned; at the time, it was the strongest proof-theoretic result established by finitist means. Comparing ACKERMANN's result with his own, HERBRAND wrote to BERNAYS in a letter dated 7 April 1931:

> "In my arithmetic the axiom of induction [vollständige Induktion] is restricted, but one may use a variety of functions other than those that are defined by simple recursion: in this direction, it seems to me, that my theorem goes a little farther than yours [i.e., ACKERMANN's]."

GÖDEL formulated HERBRAND's central result in this beautiful and penetrating way:

> "If we take a theory which is constructive in the sense that each existence assertion made in the axioms is covered by a construction, and if we add to this theory the non-constructive notion of existence and all the logical rules concerning it, e.g., the law of excluded middle, we shall never get into any contradiction."
> [GÖDEL, 1933a, p. 52]

Obviously, that formulation covers equally well the above result obtained by HILBERT & BERNAYS [1939]. When HERBRAND wrote his last paper [HERBRAND, 1932], he knew already of GÖDEL's incompleteness theorems and NEUMANN's related conjecture. The latter had drawn the consequences of GÖDEL's results most sharply and dramatically:

> "If there is a finitist consistency proof at all, then it can be formalized. Therefore, GÖDEL's proof implies the impossibility of any [such] consistency proof."[106]

Was this the end of the proof-theoretic program? An answer to this question required an answer to another question: What is finitist mathematics by means of which the proof-theoretic program was supposed to be carried out? There is, as I described above, no sharp and clear answer to be found in HILBERT and BERNAYS' work from the 1920s as to the upper limits. At the very start of HILBERT's new proof-theoretic investigations, finitist considerations are extended beyond primitive recursive arithmetic, and it is clear that finitist arguments make use only of proof by induction and definition by recursion within a quantifier-free framework. That is clear also from the perspective of "outsiders" in the early 1930s, cf. [HERBRAND, 1932] and [GÖDEL, 1933a]. At issue was, and to some extent still is, what are the recursion schemata that are finitistically allowed? Again, there is no sharp and clear answer to be extracted (or to be expected) from the early work of HILBERT and BERNAYS. However, one fact can be stated: NEUMANN, HERBRAND, GÖDEL [1933b], and even HILBERT & BERNAYS [1934; 1939] consider the ACKERMANN function as a finitist one. I.e., the lower bound on the strength of finitist methods is given by a proper extension of primitive recursive arithmetic. The issue of an upper bound was much discussed in 1931 and subsequent years, when people tried to assess the impact of the incompleteness theorems. I mentioned already NEUMANN's immediate and decidedly pessimistic judgment.

---

[106][ANON, 1931]. The German text is:

> "Wenn es einen finiten Widerspruchsbeweis überhaupt gibt, dann lässt er sich auch formalisieren. Also involviert der GÖDELsche Beweis die Unmöglichkeit eines Widerspruchsbeweises überhaupt."

GÖDEL was much more cautious in [GÖDEL, 1931] and also in his letter to HERBRAND dated 25 July 1931; in the former he asserted that the second incompleteness theorem does not contradict HILBERT's formalist viewpoint:

> "... this viewpoint presupposes only the existence of a consistency proof in which nothing but finitist means of proof are used, and it is conceivable that there exist finitist proofs that cannot be expressed in the formalism of P (or of M or A)."[107]

Both NEUMANN and HERBRAND argued in correspondence with GÖDEL against that position and conjectured that finitist arguments may be formalizable already in elementary number theory (and if not there, then undoubtedly in analysis). That would establish the limited reach of HILBERT's finitist consistency program.[108] What is the perspective HILBERT and BERNAYS took on these deliberations in volume II of *Grundlagen der Mathematik*?

5.2 GÖDEL's theorems

HILBERT and BERNAYS' discussion of the incompleteness theorems begins with a thorough investigation of semantic paradoxes. This investigation does not try to "solve" the paradoxes in the case of natural languages, but focuses on the question under what conditions analogous situations can occur in the case of *formalized languages*. These conditions are formulated quasi-axiomatically for deductive formalisms F taking for granted that there is a bijection between the expressions of F and natural numbers, a "GÖDEL-numbering." The formalism F and the numbering are required to satisfy roughly two *representability conditions*:

(R1) primitive recursive arithmetic is "contained in" F, and

(R2) the syntactic properties and relations of F's expressions, as well as the processes that can be carried out on them, are given by primitive recursive predicates and functions.

For the consideration of the first incompleteness theorem the second representability condition is made more specific. It is now required that the *substitution function* (yielding the number of the expression obtained from an expression with number **k**, when every occurrence of the number variable a is replaced by a numeral **l**) is given primitive recursively by a binary function $s(k, l)$ and the *proof predicate* by a binary relation $B(m, n)$ (holding when **m** is the number of a sequence of formulas that constitutes an F-derivation of the formula

---

[107][GÖDEL, 1931, p.194]. The German text is:

> "Es sei ausdrücklich bemerkt, daß Satz XI (und die entsprechenden Resultate über M, A) in keinem Widerspruch zum HILBERTschen formalistischen Standpunkt stehen. Denn dieser setzt nur die Existenz eines mit finiten Mitteln geführten Widerspruchsfreiheitsbeweises voraus und es wäre denkbar, daß es finite Beweise gibt, die sich in P (bzw. M, A) nicht darstellen lassen."

The formalism P is that of *Principia Mathematica*, M is that of NEUMANN's set theory, and A that of analysis. The letter to HERBRAND is found in [GÖDEL, 1986ff., Vol.V].

[108]That discussion in their correspondence with GÖDEL is described in [SIEG, 2005]. There is also an informative letter of HERBRAND to his friend CLAUDE CHEVALLEY; the letter was written on 3 December 1930, when HERBRAND was in Berlin working on proof theory and discussing GÖDEL's theorems with NEUMANN. (The letter is presented and analyzed in [SIEG, 1994, appendix].)

with number **n**). Consider, as GÖDEL did, the formula $\overline{B(m,s(a,a))}$; according to the first representability condition this is a formula of the formalism F and has a number, say **p**. Because of the defining property of $s(k,l)$, the value of $s(\mathbf{p},\mathbf{p})$ is then the number **q** of the formula $\overline{B(m,s(\mathbf{p},\mathbf{p}))}$. The equation $s(\mathbf{p},\mathbf{p}) = \mathbf{q}$ is provable in F; thus, $\overline{B(m,s(\mathbf{p},\mathbf{p}))}$ is actually equivalent to $\overline{B(m,\mathbf{q})}$ and expresses that "the formula with number **q** is not provable in F". As **q** is the number of $\overline{B(m,s(\mathbf{p},\mathbf{p}))}$, this formula consequently expresses (via the equivalence) its own underivability. The argument adapted from that for the liar paradox leads, assuming that this formula is provable, directly to a contradiction in F. But instead of encountering a paradox one infers that the formula is not provable, if the formalism F is consistent.

HILBERT and BERNAYS discuss — following GÖDEL and assuming the ω-consistency of F — the unprovability of the sentence $\overline{(x)}\ \overline{B(x,\mathbf{q})}$. Then they establish the ROSSER version of the first incompleteness theorem; i.e., the independence of a formula R from F, assuming just F's consistency. Thus, a "sharpened version" of the theorem can be formulated for deductive formalisms satisfying certain conditions:

> "One can always determine a primitive recursive function f, such that the equation $f(m) = 0$ is not provable in F, while for each numeral **l** the equation $f(\mathbf{l}) = 0$ is true and provable in F; neither the formula $(x)(f(x) = 0)$ nor its negation is provable in F."
> [HILBERT & BERNAYS, 1939, p. 279], [HILBERT & BERNAYS, 1970, p. 288]

This sharpened version of the theorem asserts that every sufficiently expressive, sharply delimited, and consistent formalism is deductively incomplete.

For a formalism F that is consistent and satisfies the restrictive conditions, the proof of the first incompleteness theorem shows the formula $\overline{B(m,\mathbf{q})}$ to be unprovable. However, it also shows that the sentence $\overline{B(\mathbf{m},\mathbf{q})}$ holds and is provable in F, for each numeral **m**. The second incompleteness theorem is obtained by formalizing these considerations, i.e. by proving in F the formula $\overline{B(m,\mathbf{q})}$ from the formal expression C of F's consistency. That is possible, however, only if F satisfies additional conditions, the so-called *derivability conditions*. The formalized argument makes use of the representability conditions (R1) and (R2). The second condition now requires also that there is a primitive recursive function e, which when applied to the number **n** of a formula yields as its value the number of the negation of that formula. The derivability conditions are formulated as follows:

(D1) If there is a derivation of a formula with number **l** from a formula with number **k**, then the formula $(Ex)\,B(x,\mathbf{k}) \to (Ex)\,B(x,\mathbf{l})$ is provable in F;

(D2) The formula $(Ex)\,B(x,e(k)) \to (Ex)\,B(x,e(s(k,l)))$ is provable in F;

(D3) If $f(m)$ is a primitive recursive term with m as its only variable and if **r** is the number of the equation $f(a) = 0$, then the formula $f(m) = 0 \to (Ex)\,B(x,s(\mathbf{r},m))$ is provable in F.

Consistency is formally expressed by $(Ex)\,B(x,n) \to \overline{(Ex)}\,B(x,e(n))$; starting with that formula C as an assumption, the formula $\overline{B(m,\mathbf{q})}$ is obtained in F by a rather direct argument. So, in case the formalism F is consistent, no formalized proof of consistency, i.e., no derivation of the formula C, can exist in F.[109]

---

[109][HILBERT & BERNAYS, 1939, pp. 286–288].

There are two brief remarks with which I want to complement this metamathematical discussion of the incompleteness theorems. The first simply states that verifying the representability conditions and the derivability conditions is the central mathematical work that has to be done; HILBERT and BERNAYS accomplish this for the formalisms (Z_μ) and (Z).[110] Thus, the second volume of *Grundlagen der Mathematik* contains the first full argument for the second incompleteness theorem; after all, GÖDEL's paper contains only a minimal sketch of a proof. However, it has to be added — and that is the second remark — that the considerations are not fully satisfactory for a *general* formulation of the theorems, as there is no argument given why deductive formalisms should satisfy the particular restrictive conditions on their syntax. (This added observation points to one of the general methodological issues discussed briefly at the end of § 5.3.)

Existential formal axiomatics emerged in the second half of the 19th century and found its remarkable expression in 1899 through HILBERT's *Grundlagen der Geometrie*; its existential assumption constituted the really pressing issue for the various HILBERT programs during the period from 1899 to 1934, the date of the publication of the first volume of *Grundlagen der Mathematik*. The finitist consistency program began to be pursued in 1922 and is the intellectual thread holding the investigations in both volumes together. The ultimate goal of proof-theoretic investigations, as HILBERT formulated it in the preface to volume I, is to recognize the usual methods of mathematics, without exception, as consistent. HILBERT continued:

> "Regarding this goal, I would like to emphasize that an opinion, which had emerged intermittently — namely that some more recent results of GÖDEL would imply the infeasibility of my proof theory — has turned out to be erroneous."[111]

How is the program affected by those results? Is it indeed the case, as HILBERT expressed it also in that preface of [HILBERT & BERNAYS, 1934], that GÖDEL's theorems just force proof theorists to exploit the finitist standpoint in a sharper way?

5.3 Completeness and bounds on finitism

The second question is raised prima facie through the second incompleteness theorem. However, HILBERT and BERNAYS discuss also the effect of the first incompleteness theorem and ask explicitly, whether the deductive completeness of formalisms is a necessary feature for the consistency program to make sense. They touched on this very issue already in pre-GÖDEL publications: HILBERT in his Bologna talk of 1928, and BERNAYS in his penetrating article [BERNAYS, 1930/31]. HILBERT formulated in his talk the question of the syntactic completeness for number theory and analysis as Problem IV; he concluded the discussion by suggesting that "in höheren Gebieten" (higher than number theory) it is thinkable that a system of axioms could be consistently extended by a statement S, but also by its negation $\overline{S}$; the acceptance of one of the statements is then to be justified by "systematic advantages (principle of the permanence of laws, possibilities of further developments etc.)."[112]

---

[110]The considerations for the former systems start on p. 293, for the latter on p. 324.

[111][HILBERT & BERNAYS, 1934, p.VII]. Also [HILBERT & BERNAYS, 1968, p.VII], see this volume.

[112][HILBERT, 1930b, p. 6]. The German text is:
> "In höheren Gebieten wäre der Fall der Widerspruchsfreiheit von S und der von $\overline{S}$ denkbar; alsdann ist die Annahme einer der beiden Aussagen $S, \overline{S}$ als Axiom durch systematische Vorzüge (Prinzip der Permanenz von Gesetzen, weitere Aufbaumöglichkeiten usw.) zu rechtfertigen."

This paragraph does not appear in the original proceedings of the Bologna Congress in 1928.

HILBERT conjectured that number theory is deductively complete. That is reiterated in [BERNAYS, 1930/31] and followed by the remark that "the problem of a real proof for this is completely unresolved." The problem becomes even more difficult, BERNAYS continues, when we consider systems for analysis or set theory. However, this "Problematik" is not to be taken as an objection against the standpoint presented:

> "We only have to realize that the [syntactic] formalism of statements and proofs we use to represent our conceptions does not coincide with the [mathematical] formalism of the structure we intend in our thinking. The [syntactic] formalism suffices to formulate our ideas of infinite manifolds and to draw the logical consequences from them, but in general it [the syntactic formalism] cannot combinatorially generate the manifold as it were out of itself."[113]

That is also the central point in the general discussion of the first incompleteness theorem. Indeed, HILBERT and BERNAYS emphasize that in formulating the problems and goals of proof theory they avoided from the beginning "to introduce the idea of a total system for mathematics with a philosophically principled significance". It suffices to characterize the actual systematic structure of analysis and set theory in such a way that it provides an appropriate frame for (the reducibility of) the geometric and physical disciplines.[114] That point was already emphasized in §3.3.

From these reflective remarks it follows that the first incompleteness theorem for the central formalisms F (of number theory, analysis, and set theory) does not directly undermine HILBERT's program. It raises nevertheless in its sharpened form a peculiar issue: any finitist consistency proof for F would yield a finitist proof of a statement in recursive number theory — that is not provable in F. Finitist methods would thus go beyond those of analysis and set theory, even for the proof of number-theoretic statements. (That is of course the situation GÖDEL contemplated in his remarks, when claiming that his results don't contradict HILBERT's standpoint; cf. §5.1.) This is a "paradoxical" situation, in particular, as HILBERT and BERNAYS quite unambiguously state in the first volume (on p. 42), "finitist methods are included in the usual arithmetic". Consequently, even the first theorem forces us to address two general tasks, namely,

(i) exploring the extent of finitist methods, and

(ii) demarcating appropriately the methodological standpoint for proof theory.

---

[113][BERNAYS, 1930/31, p. 59]. The German text is:

> "Wir müssen uns nur gegenwärtig halten, daß der Formalismus der Sätze und Beweise, mit denen wir unsere Ideenbildung zur Darstellung bringen, nicht zusammenfällt mit dem Formalismus derjenigen Struktur, die wir in der Gedankenbildung intendieren. Der Formalismus reicht aus, um unsere Ideen von unendlichen Mannigfaltigkeiten zu formulieren und aus diesen die logischen Konsequenzen zu ziehen, aber er vermag im allgemeinen nicht, die Mannigfaltigkeit gleichsam aus sich kombinatorisch zu erzeugen."

[114][HILBERT & BERNAYS, 1939, p. 280]. The extended German text is:

> "Wir haben in unserer Darstellung der Ausgangsproblematik und der Zielsetzung der Beweistheorie von vornherein vermieden, den Gedanken eines Totalsystems der Mathematik in einer philosophisch prinzipielllen Bedeutung einzuführen, vielmehr uns begnügt, die tatsächlich vorhandene Systematik der Analysis und Mengenlehre als eine solche zu charakterisieren, die einen geeigneten Rahmen für die Einordnung der geometrischen und physikalischen Disziplinen bildet. Diesem Zweck kann ein Formalismus auch entsprechen, ohne die Eigenschaft der vollen deduktiven Abgeschlossenheit zu besitzen."

Tasks (i) and (ii) are usually associated with the second incompleteness theorem, which, as emphasized in § 5.2, allows us to infer directly and sharply that a finitist consistency proof for a formalism F (satisfying the representability and derivability conditions) cannot be carried out in F. HILBERT and BERNAYS explore the extent of finitist methods in § 5(a) by first trying to answer the question, in which formalism their various finitist investigations can actually be carried out. The immediate claim is that most considerations can be formalized, perhaps with a great deal of effort, already in primitive recursive arithmetic. But then they assert: "At various places this formalism is admittedly no longer sufficient for the desired formalization. However, in each of these cases the formalization is possible in (Z_μ)."[115] They point to the more general recursion principles from § 7 of the first volume as an example of "procedures of finitist mathematics" that cannot be captured in primitive recursive arithmetic, but can be formalized in (Z_μ).

In the remainder of § 5.3(a) they discuss "certain other typical cases", in which the boundaries of primitive recursive arithmetic are too narrow to allow a formalization of their prior finitist investigations. There is, first of all, the issue of an evaluation function that is needed for the consistency proof of primitive recursive arithmetic (already in volume I), but cannot be defined by primitive recursion (p. 341). Secondly, there is the general concept of a calculable function (p. 341). That concept is used (p. 189) to formulate a sharpened notion of satisfiability, i.e. *effective satisfiability*, in their treatment of solvable cases of the decision problem. Thirdly, they discuss (p. 344) the principle of induction for universally quantified formulas used in consistency proofs. The issue surrounding this principle is settled metamathematically, as we know, by later proof-theoretic work: the system of elementary number theory with this induction principle is conservative over recursive arithmetic, whether in the narrow or wider sense.[116]

As to the second issue (effective satisfiability), some remarks concerning Supplement II are relevant here, because the notion of a calculable function has to be sharpened in such a way as to be formalizable. The presentation of the negative solution of the decision problem in Supplement II is preceded by an analysis of the concept "reckonable function", i.e., of a function whose values can be calculated according to elementary rules. The latter rather vague notion is sharpened in a way that is methodologically very similar to their analysis of the incompleteness theorems, namely, by formulating *recursiveness conditions* for deductive formalisms that allow equational reasoning. The central condition requires the proof predicate to be primitive recursive. It is then shown that the functions calculable in formalisms satisfying these conditions are exactly the general recursive ones. Though their analysis is not fully satisfactory for the reason mentioned in § 5.2, it is nevertheless a major and concluding step for the analysis of effectively calculable functions as pursued in the mid-1930s by GÖDEL, CHURCH, KLEENE, and others. It should be emphasized that HILBERT and BERNAYS are perfectly clear about one fact, namely, that this class of

---

[115][HILBERT & BERNAYS, 1939, p. 340]. The extended German text is:

"An verschiedenen Stellen ist freilich dieser Formalismus [der rekursiven Zahlentheorie mit nur primitiver Rekursion, WS] nicht mehr für die gewünschte Formalisierung ausreichend. Doch zeigt sich dann jedesmal die Möglichkeit der Formalisierung in (Z_μ). Gewisse über die rekursive Zahlentheorie (im ursprünglichen Sinne) hinausgehende Verfahren der finiten Mathematik haben wir bereits im §7 besprochen, nämlich die Einführung von Funktionen durch verschränkte Rekursionen und die allgemeineren Induktionsschemata. Dabei erwähnten wir auch die Formalisierbarkeit dieser Rekursions- und Induktionsschemata im vollen zahlentheoretischen Formalismus."

[116]See [SIEG, 1991], § 2.1 and references there to work by CHARLES PARSONS.

functions goes beyond the class of functions introduced in [HERBRAND, 1932] by the axioms of his "Groupe C". Indeed, they write:

> "General instructions for the introduction of new function symbols were given by HERBRAND in his paper [[HERBRAND, 1932], WS] ... ("Groupe C"). These are somewhat narrower than the ones formulated here inasmuch as HERBRAND requires that the procedure for determining values [of functions] is to follow from a finitist interpretation of the axioms, as is the case for the schemata of recursive definitions."[117]

So it is also here quite clear that HILBERT and BERNAYS view HERBRAND's procedure of introducing function symbols as a definitely finitist one; recall that HERBRAND mentions explicitly that a symbol for the ACKERMANN function can be introduced in this way.

5.4 Methodological frame

The careful re-examination of their own proof-theoretic practice leads HILBERT and BERNAYS to the conclusion that some finitist considerations go beyond primitive recursive arithmetic, but can be formally captured in (Z_μ); most of this was pointed out already in volume I. It is at exactly this point that the second incompleteness theorem provides, as the title of §5 states, the "reason for extending the methodological frame for proof theory." Already in the transition from §4 to §5, HILBERT and BERNAYS state specifically that consequences of the theorem force us to view the domain of the contentual inference methods used for the investigations of proof theory more broadly "than it corresponds to our development of the finitist standpoint so far."[118]

The question is, whether there are any methods that can still be called properly "finitist" and yet go beyond (Z_μ). HILBERT and BERNAYS argue that this is not a precise question, as "finitist" is not a sharply delimited notion, but rather indicates methodological guidelines that enable us to recognize some considerations as definitely finitist and others as definitely non-finitist. The limits of finitist considerations are to be "loosened"; two possibilities for such loosenings are considered and quickly seen to be conservative.[119] Which further loosening is "admissible, if we want to adhere to the fundamental tendencies of proof theory?" Against this background two results, then quite recent, are examined: the reduction of classical arithmetic (Z) to the system **Z** of arithmetic with just minimal logic,

---

[117]Here is the German text:

> "Eine allgemeine Anweisung zur Einführung neuer Funktionszeichen durch Axiome wurde von HERBRAND in seiner Abhandlung [[HERBRAND, 1932], WS] ... gegeben ("Groupe C"). Diese ist insofern etwas enger als die hier formulierte Anweisung, als HERBRAND verlangt, daß das Verfahren der Wertbestimmung sich, so wie bei den Schematen der rekursiven Definition, durch eine finite Deutung der Axiome ergeben soll."   [HILBERT & BERNAYS, 1939, p. 52, Note 2]

[118][HILBERT & BERNAYS, 1939, p. 253]. The German text is: "Dieser Umstand [daß die bisherigen Methoden nicht zum Nachweis der Konsistenz des vollen zahlentheoretischen Formalismus ausreichen, WS] ... findet nun eine grundsätzliche Erklärung durch ein Theorem von GÖDEL über deduktive Formalismen, für welches der zahlentheoretische Formalismus einen ersten Anwendungsfall bildet und dessen Konsequenzen uns dazu nötigen, den Bereich der inhaltlichen Schlußweisen, die wir für die Überlegungen der Beweistheorie verwenden, weiter zu fassen, als es unserer bisherigen Durchführung des finiten Standpunktes entspricht."

[119]In the context of this discussion, on p. 348, a very concise explication of "Begriff der finiten Aussage" is given that reflects faithfully the informal considerations on which I have been reporting.

and GENTZEN's consistency proof for a version of **Z** (and thus of (Z)) using a special form of transfinite induction.

The reductive result HILBERT and BERNAYS formulate is a slightly stronger one than the one obtained by GÖDEL and, independently, by GENTZEN. The proof showing that (Z) is consistent relative to **Z** is an elementary finitist one. Thus, the obstacle for obtaining a finitist consistency proof for (Z) does not lie in the fact that it contains the typically non-finitist logical principles like tertium non datur! The obstacle appears already when one tries to give a finitist consistency proof for **Z**. The consistency of (Z) would be established on the basis of any assumptions, "which suffice to give a verifying interpretation of the restricted formalism".[120] Such a contentual verification, based on interpretations of KOLMOGOROV and HEYTING, is then examined with the conclusion that it involves the intuitionistic understanding of negation as absurdity. In using the underlying contentual concept of consequence, it is claimed, "we are totally turning away from HILBERT's methodological ideas for proof theory".[121] That is consonant with the view expressed in the first volume (p. 43) that intuitionism is a proper extension of finitist mathematics (in sharp contrast to the earlier perspective that was discussed in the second half of § 3.3). BERNAYS expressed that view also in contemporaneous papers and in many later comments, perhaps most dramatically in his article [BERNAYS, 1967] on HILBERT, where (on p. 502) the above relative consistency proof for (Z) is seen as the reason for the recognition "that intuitionistic reasoning is not identical with finitist reasoning, contrary to the prevailing views at the time."

The question is raised, whether — in a proof of the consistency of (Z) — the use of absurdity can be avoided, as well as the appeal to an interpretation of the formalism (viewed in contrast to its direct proof-theoretic examination). It is claimed that GENTZEN's consistency proof addresses both these issues. After a thorough discussion of the details of the system of ordinal notation (for ordinals less than the first epsilon number) and the justification of the principle of transfinite induction, but only the briefest indication of the structure of GENTZEN's proof, the main body of the first edition of the book concludes with some extremely general remarks about the significance of that proof: it provides a perspective for the proof-theoretic investigation also of stronger formalisms, when one clearly has to countenance the use of larger and larger ordinals. The volume concludes with the sentence: "If this perspective should prove its value, then GENTZEN's consistency proof would open a new phase of proof theory." In this way, it seems, BERNAYS sees GENTZEN's approach as overcoming "the temporary fiasco of proof theory" he discussed in the introduction to volume II and attributed to "... exaggerated methodological demands put on the theory". No explicit final and definitive judgment on the methodologically appropriate character of GENTZEN's consistency proof is articulated in the first edition of the book. However, in the introduction to the second edition BERNAYS states that the transfinite induction principle used in it is "a non-finitist tool".

---

[120][HILBERT & BERNAYS, 1939, p. 357f.]. That point is re-emphasized after the reductive argument has been completed. The German text there is:

"Stellen wir uns andererseits auf einen inhaltlichen Standpunkt, von dem aus die formalen Ableitungen in (ℨ) als Darstellung richtiger inhaltlicher Überlegungen deutbar sind, so ist für diesen auf Grund der festgestellten Beziehung zwischen den Formalismen (Z) und (ℨ) die Widerspruchsfreiheit des Systems (Z) ersichtlich."

[121][HILBERT & BERNAYS, 1939, p. 358]. The fuller German text is:

"Jedoch entfernen wir uns mit dem inhaltlichen Folgerungsbegriff total von HILBERTs methodischen Gedanken der Beweistheorie, ..."

In the introduction of the first edition and the detailed discussion sketched here, some see an ambiguity in HILBERT and BERNAYS' view as to whether the extension of the finitist standpoint necessitated by the incompleteness theorems is essentially still the finitist standpoint as articulated in §§ 1 and 2 of volume I or whether it is a proper extension compatible with the broader strategic considerations underlying proof theory. I think the ambiguity, if it is there at all, should be resolved in the latter sense; after all, the considerations in § 5.5 come under the heading "Transcending the former methodological standpoint of proof theory. Consistency proofs for the full number-theoretic formalism". One just has to distinguish very carefully, as BERNAYS does, between the two different tasks I described at the very beginning of this section:

(i) exploring the extent of finitist mathematics, and

(ii) demarcating the appropriate methodological standpoint for proof theory.

However, there is not even a broad demarcation of a new, wider methodological standpoint for proof theory; a reason for this lack is perhaps implicit in the remarks connecting the consistency proof for (Z) relative to intuitionistic arithmetic with GENTZEN's consistency proof (p. 359). It is claimed, first of all, that it is "unsatisfactory from the standpoint of proof theory" to have a consistency proof for (Z) that "rests mainly on an interpretation of a formalism". It is observed, secondly, that the only method of going beyond the formalism (Z) has been the formulation of truth definitions: a classical truth definition was given for (Z), and the formalization of the consistency proof based on an intuitionistic interpretation would amount to using a truth definition. Thirdly and finally, it is argued that a consistency proof is desirable that rests on "the direct treatment of the formalism itself"; that is seen in analogy to obtaining the consistency of (primitive) recursive arithmetic, where HILBERT and BERNAYS were not satisfied with the possibility of a finitist interpretation, but rather convinced themselves of the consistency by specific proof-theoretic methods. Where in this discussion is even an opening for a broader demarcation?

BERNAYS, in the "Nachtrag" to [BERNAYS, 1930/31], reflects on these issues and indicates, in particular, that the epistemological perspective that was underlying proof theory became problematic. Referring to his own essay he writes:

> "... the sharp distinction between what is intuitive and what is not, as it is used in the treatment of the problem of the infinite, apparently cannot be drawn so strictly, and the reflections on the formation of mathematical ideas still need to be worked out in more detail in this respect. Various considerations for this are contained in the following essays."[122]

Some indications of a general direction for philosophical reflections are indeed contained in the essays reprinted in [BERNAYS, 1976], but also in an essay that is not reprinted there, namely, [BERNAYS, 1954b]. There he envisions the appeal to what he calls *sharpened axiomatics* (verschärfte Axiomatik) and, opposing it to existential axiomatics, formulates as a

---

[122][BERNAYS, 1976, p. 61]. The German text is:

> "... die scharfe Unterscheidung des Anschaulichen und des Nicht-Anschaulichen, wie sie bei der Behandlung des Problems des Unendlichen angewandt wird, ist anscheined nicht so strikt durchführbar, und die Betrachtung der mathematischen Ideenbildung bedarf wohl in dieser Hinsicht noch der näheren Ausarbeitung. Für eine solche sind in den folgenden Abhandlungen verschiedene Überlegungen enthalten."

The "folgenden Abhandlungen" are referring, obviously, to the essays in [BERNAYS, 1976].

minimal requirement that "the objects [making up the intended model of the theory] are not taken from a domain that is thought as being already given, but are rather constituted by generative processes."[123] There is no indication in that paper or in other writings what kind of generative processes should be considered, and why that particular feature of domains should play a distinctive foundational rôle. Contemporary proof-theoretic investigations give such indications.

6 Outlook beyond 1939

What ACKERMANN [1934] formulated in his review of just the first volume of *Grundlagen der Mathematik*, holds even more for the complete two-volume work, namely, that it "is to be viewed in line with the great publications of FREGE, PEANO, and RUSSELL and WHITEHEAD." In contrast to the other works, these two volumes synthesize contemporaneous research and continue to have a deep effect on research in mathematical logic and proof theory.[124]

Ever since the GÖDEL–GENTZEN reduction of classical arithmetic to its intuitionistic version was obtained in 1932, *foundational reductions* have been pursued. We have been able to establish the consistency of *strong* sub-systems of analysis, respectively of *weak* sub-systems of set theory. These consistency results have been achieved relative to constructive theories (like theories of constructive number classes and MARTIN-LÖF type theory) or relative to intuitionistic number theory together with transfinite induction for long initial segments of the second number class. Classical analysis as presented in [HILBERT & BERNAYS, 1939, Supplement IV] can easily be carried out in the theory $(\Pi_1^1 - CA)\restriction$, which is consistent relative to the intuitionistic theory of finite number classes.[125]

---

[123][BERNAYS, 1954b, pp. 11–12]. The German text is:

"Die Mindest-Anforderung an eine verschärfte Axiomatik ist die, dass die Gegenstände nicht einem als vorgängig gedachten Bereich entnommen werden, sondern durch Erzeugungsprozesse konstituiert werden."

BERNAYS continues with a methodologically important remark:

"Es kann aber dabei die Meinung sein, dass durch diese Erzeugungsprozesse der Umkreis der Gegenstände determiniert ist; bei dieser Auffassung erhält das *tertium non datur* seine Motivierung. In der Tat kann Offenheit eines Bereiches in zweierlei Sinn verstanden werden, einmal nur so, dass die Konstruktionsprozesse über jeden einzelnen Gegenstand hinausführen, und andererseits in dem Sinne, dass der resultierende Bereich überhaupt nicht eine mathematisch bestimmte Mannigfaltigkeit darstellt. Je nachdem die Zahlenreihe in dem erstgenannten oder in dem zweiten Sinne aufgefasst wird, hat man die Anerkennung des *tertium non datur* in bezug auf die Zahlen oder den intuitionistischen Standpunkt. Bei dem finiten Standpunkt kommt noch die Anforderung hinzu, dass die Überlegungen an Hand der Betrachtung von endlichen Konfigurationen verlaufen, somit insbesondere Annahmen in der Form allgemeiner Sätze ausgeschlossen werden."

[124]KREISEL's *Survey of proof theory* [KREISEL, 1968] gives a most interesting, albeit somewhat idiosyncratic perspective. Proof-theoretic investigations in GENTZEN-style are presented by SCHWICHTENBERG [1977] and FEFERMAN [1981], and of course in the textbooks by SCHÜTTE [1977], TAKEUTI [1987], and POHLERS [1997]. The constructive understanding of formalisms via (GÖDEL's Dialectica) interpretation is presented in [TROELSTRA, 1973], and surveyed in [AVIGAD & FEFERMAN, 1998]. Other constructive approaches were initiated through MARTIN-LÖF's type theory [MARTIN-LÖF, 1984] and MYHILL's constructive set theory [MYHILL, 1975].

[125]$(\Pi_1^1 - CA)\restriction$ is the sub-system of second-order arithmetic with the comprehension principle for formulas of the form $(X)\phi(X)$, where the matrix ϕ is purely arithmetic, and with the induction principle formulated as a second-order axiom, not as a schema for all formulas of the theory.

While such foundational reductions were being pursued, it was also becoming clear through refined mathematical investigations that most of classical analysis can be carried out in $(\Pi_1^0 - CA)\upharpoonright$ or, what KREISEL called, the theory of arithmetic properties. This theory is like $(\Pi_1^1 - CA)\upharpoonright$, except that the comprehension principle is restricted essentially to arithmetic formulas; it is a conservative extension of (Z) and, thus, actually consistent relative to the system **Z**. The program of Reverse Mathematics, initiated by FRIEDMAN & SIMPSON [2000], has grown a sort of KRONECKERian offshoot that allows developing substantial parts of analysis and algebra already in conservative extensions of primitive recursive arithmetic and even weaker theories, like FERNANDO FERREIRA's feasible analysis [FERREIRA, 1994], [FERREIRA & FERREIRA, 2006]. As these conservation results mostly hold also for Π_2^0-sentences one achieves *computational reductions*: proofs of such statements yield SKOLEM functions that lie in particular classes of calculable functions, and proof-theoretic tools are used to extract computational information. KOHLENBACH has been very successful in obtaining mathematically significant results by this "proof mining" [KOHLENBACH & OLIVA, 2003].

All of this is in a certain sense internal to proof theory. However, we should consider the stimulus HILBERT's approach and questions provided to contemporaries outside the HILBERT School, for example, to NEUMANN, HERBRAND, GÖDEL, CHURCH, and TURING. There is indeed no foundational enterprise with a more profound or more far-reaching effect on the emergence and development of modern mathematical logic. The foundational concerns have led to a dramatic expansion of the reach of mathematical thought. HILBERT would be delighted with the development of mathematical logic in general, and proof theory in particular — in spite of the fact that his Second Problem hasn't been solved yet. He might just note optimistically "Wir müssen wissen, wir werden wissen." and follow this by "Es gibt kein Ignorabimus!" for added emphasis.

HILBERT's specific proposal of mediating between a restricted constructivist and a strong classical position does not work out, when finitism is equated with the methodological stance associated with proof theory. The reductive program that emerged from it provides, however, an important perspective on aspects of mathematical experience and helps us to gain a better understanding of the distinctive character of modern mathematics. It is centrally connected to the starting point of HILBERT's considerations, namely, the foundational opposition of DEDEKIND and KRONECKER. HILBERT's principal goal was to secure mathematics from contradictions — and thus to safeguard creative freedom within mathematics and for contexts of applications. In the Heidelberg talk of 1904 he formulated the creative principle that "in its freest use justifies us in forming ever new notions, with the sole restriction that we avoid a contradiction." As philosophers and mathematicians we should exploit HILBERT's complex insights into the workings of mathematics, instead of keeping him shackled to a narrow foundational position that was taken for exploratory, programmatic reasons.

Bilingual Part

On the following page, we start with the bilingual part.

It has the German original of the second edition [HILBERT & BERNAYS, 1968] on the left-hand side of each double page and the commented English translation on the right-hand side.

The deleted multi-page parts of the first edition [HILBERT & BERNAYS, 1968] would be found at the end of this bilingual part, but there are no such parts in HILBERT's and BERNAYS' prefaces and §§ 1 and 2, which constitute the bilingual part of this volume of our edition.

Die Grundlehren der mathematischen Wissenschaften

in Einzeldarstellungen
mit besonderer Berücksichtigung
der Anwendungsgebiete

Band 40

Herausgegeben von

J. L. Doob · E. Heinz · F. Hirzebruch · E. Hopf · H. Hopf
W. Maak · S. MacLane · W. Magnus · D. Mumford
M. M. Postnikov · F. K. Schmidt · D. S. Scott · K. Stein

Geschäftsführende Herausgeber

B. Eckmann und B. L. van der Waerden

Fundamental Knowledge of the Mathematical Sciences

in monographs
with special regard to
the areas of application

Volume 40

Edited by

Joseph Leo Doob · E. Heinz · Friedrich Hirzebruch
Eberhard Hopf · Heinz Hopf · W. Maak
Saunders MacLane · Wilhelm Magnus
David Mumford · Mikhail Mikhailovich Postnikov
Friedrich Karl Schmidt
Dana S. Scott · Karl Stein

Editors-in-Chief

Beno Eckmann
and
Bartel Leendert van der Waerden

II

D. Hilbert und P. Bernays

Grundlagen der Mathematik I

Zweite Auflage

Springer-Verlag Berlin Heidelberg New York 1968

DAVID HILBERT and PAUL BERNAYS

Foundations of Mathematics I

Second edition

Springer-Verlag Berlin Heidelberg New York 1968

Prof. Dr. Paul Bernays
CH-8002 Zürich, Bodmerstr. 11

Geschäftsführende Herausgeber:

Prof. Dr. B. Eckmann
Eidgenössische Technische Hochschule Zürich

Prof. Dr. B. L. van der Waerden
Mathematisches Institut der Universität Zürich

Alle Rechte vorbehalten. Kein Teil dieses Buches darf ohne schriftliche Genehmigung
des Springer-Verlages übersetzt oder in irgendeiner Form vervielfältigt werden.

© by Springer-Verlag Berlin · Heidelberg 1934 and 1968

Library of Congress Catalog Card Number 68-55369.

Printed in Germany

Titel-Nr. 5023

Prof. Dr. PAUL BERNAYS
CH–8002 Zürich, Bodmerstr. 11

Editors-in-Chief:

Prof. Dr. BENO ECKMANN
Swiss Federal Institute of Technology Zurich (ETH)

Prof. Dr. BARTEL LEENDERT VAN DER WAERDEN
Mathematical Institute of the University of Zurich (UZH)

All rights reserved. No part of this book may be translated or reproduced
in any form with out written permission of Springer-Verlag.
© by Springer-Verlag Berlin · Heidelberg 1934 and 1968
Library of Congress Catalog Card Number 68–55369.
Printed in Germany
Title-No. 5023

Vorwort zur zweiten Auflage

Schon vor etlichen Jahren haben der verstorbene Heinrich Scholz und Herr F. K. Schmidt mir vorgeschlagen, eine zweite Auflage der „Grundlagen der Mathematik" vorzunehmen, und Herr G. Hasenjaeger war auch zu meiner Unterstützung bei dieser Arbeit auf einige Zeit nach Zürich gekommen. Es zeigte sich jedoch bereits damals, daß eine Einarbeitung der vielen im Gebiet der Beweistheorie hinzugekommenen Ergebnisse eine völlige Umgestaltung des Buches erfordert hätte. Erst recht kann bei der jetzt vorliegenden zweiten Auflage, zu der wiederum Herr F. K. Schmidt den Anstoß gab, nicht davon die Rede sein, den Inhalt dessen, was seither in der Beweistheorie erreicht worden ist, zur Darstellung zu bringen. Das ist auch um so weniger erforderlich, als in der Zwischenzeit verschiedene namhafte Lehrbücher erschienen sind, welche die Beweistheorie und die an sie grenzenden Fragengebiete behandeln.

Andererseits sind doch etliche Dinge in den „Grundlagen der Mathematik" eingehender auseinandergesetzt, als man sie anderwärts findet, was sich auch an der Nachfrage nach dem seit längerem vergriffenen Buche geltend macht. Unter diesen Umständen erschien es als tunlich, das Buch im wesentlichen in seiner bisherigen Form zu belassen und die Änderungen und Ergänzungen auf solche Punkte zu beschränken, die in engem Zusammenhang mit dem Inhalt der ersten Auflage stehen.

Es wurde auch darauf verzichtet, Änderungen in der Symbolik und in der Terminologie vorzunehmen. Was insbesondere die logische Symbolik betrifft, so sind ohnehin deren mehrere in Gebrauch, und es macht keine Schwierigkeit von der einen zu einer andern überzugehen. Die einführenden Paragraphen, die die Problemstellung entwickeln, wurden fast unverändert übernommen.

Für den vorliegenden ersten Band seien als inhaltliche Änderungen und Hinzufügungen (abgesehen von etlichen Korrekturen und Verbesserungen im einzelnen) die folgenden erwähnt:

1. Im Aussagenkalkul eine eingehendere Behandlung der disjunktiven Normalform; 2. die Darstellung der von G. Hasenjaeger gegebenen Beantwortung einer bezüglich des Systems (B) seinerzeit offen gebliebenen Abhängigkeitsfrage; 3. die Einarbeitung einer Bemerkung von G. Kreisel, wonach bei der Behandlung der Theorie der $<$-Beziehung mittels der rekursiven δ-Funktion die Heranziehung der Summe nicht erfordert wird; 4. eine Ergänzung betreffend die rekursive Darstellung

Preface to the Second Edition

A number of years ago, the late HEINRICH SCHOLZ and Mr. FRIEDRICH KARL SCHMIDT suggested the undertaking of a second edition of the "Foundations of Mathematics"; and moreover, to assist me in this work, Mr. GISBERT HASENJAEGER came to Zürich for some time. Already back then, it became obvious that the integration of the many new results in the area of proof theory would have required a complete reorganization of the book. Furthermore, the present second edition (the impetus for which came again from Mr. FRIEDRICH KARL SCHMIDT) can by no means present the substance of the achievements in proof theory since the appearance of the first edition. Such a presentation of additional results, however, is now in demand to a lesser extent, because several notable textbooks covering proof theory and the related areas have been published in the meantime.

On the other hand, a number of subjects received deeper elaboration in the "Foundations of Mathematics" than can be found elsewhere. This was also confirmed by the demand for this book, which has been out of print for quite some time. In these circumstances, it seemed appropriate to maintain the original form of the book and to restrict the changes and extensions to points closely related to the content of the first edition.

Symbols and terminology were not changed either. There are many systems of logical symbols in use anyway, and there is no difficulty in switching from one to the other. Moreover, the introductory sections, which develop the ways of looking at the problems, were copied almost unchanged.

For the present first volume, the following substantial changes and extensions are to be noted (beside a number of corrections and improvements in detail):

1. in the propositional calculus, a more detailed treatment of the disjunctive normal form;[V.1]

2. the presentation of GISBERT HASENJAEGER's answer to the question of the dependence of System (B), which remained open in the first edition;[V.2]

3. the integration of a remark of GEORG KREISEL, according to which the treatment of the theory of the <-relation by means of the recursive δ-function does not require the introduction of the[V.3] addition;[V.4]

4. an extension regarding the recursive representation | of the maximum;[V.5]

[V.1] Actually, the new treatment of disjunctive normal form is not more detailed, but more concise. Cf. Note 55.2 on Page 55.

[V.2] This footnote is to be extended after translation of the respective section.

[V.3] The usage of the definite article is required for semantical reasons here, and the German expression is similarly curious as the definite article in the English translation. In the same way, we will have to make a difference, say, between "the implication" and "implication". The former refers more to the symbol than to the formal operation, but not just in the sense of syntax. It may refer also to symbolism in form of vague and philosophical ideas that do not necessarily require a definition.

[V.4] This footnote is to be extended after translation of the respective section.

[V.5] This footnote is to be extended after translation of the respective section.

des Maximums; 5. die formale Vorführung des Nachweises von TH. SKOLEM für die Entbehrlichkeit der erweiterten Induktionsschemata; 6. die Ersetzung des früheren sehr komplizierten Beweises für die Eliminierbarkeit der ι-Symbole durch einen einfacheren, auf einer Methode von B. ROSSER beruhenden Beweis von G. HASENJAEGER; 7. eine Verdeutlichung der Ausführung über die Vertretbarkeit der rekursiven Funktionen im System (Z).

An der Disposition des Buches wurde keine Änderung vorgenommen, im Hinblick darauf, daß das ausführliche Inhaltsverzeichnis eine hinlängliche Orientierung über den Inhalt und die Gedankengänge des Buches liefert. Auf dieses Inhaltsverzeichnis sei der Leser besonders hingewiesen.

Herr D. RÖDDING (Münster) hat ein Namenverzeichnis angelegt, das Sachverzeichnis erweitert und ein System von Rückverweisungen in der Form von Fußnoten zugefügt, durch welches vor allem eine verbesserte Möglichkeit gegeben werden soll, einzelne Partien des Buches gesondert zu lesen. Ich sage ihm hierfür meinen besten Dank.

Herrn GISBERT HASENJAEGER und Herrn GEORG KREISEL bin ich dankbar für das, was sie zum Inhalt der neuen Auflage beigetragen haben. In dem neuen Beweis für die Elimination der ι-Symbole ist die Arbeit zur Verwertung gekommen, die Herr HASENJAEGER seinerzeit hier für das Grundlagenbuch leistete.

Dankbar gedenke ich der stets wachen Teilnahme von HEINRICH SCHOLZ an den Arbeiten zu dieser Neuauflage und des Interesses, das Herr F. K. SCHMIDT der Durchführung ständig entgegenbrachte.

Herrn GERT MÜLLER danke ich von Herzen für das Vielseitige, das er zu der Herstellung der neuen Auflage beigetragen hat. Herrn DIRK SIEFKES (Heidelberg) danke ich herzlich für seine wertvolle Beteiligung bei den Korrekturen und speziell auch für die Ausführung der Ergänzungen im Sachverzeichnis. Herrn WALTER ZAUGG danke ich aufs beste für seine Hilfe bei der Niederschrift und Ausführung der Korrekturen.

Dem Verlage Springer bin ich für etliches freundliches Entgegenkommen dankbar, und im Hinblick auf das Vergangene besonders dafür, daß er auch in den schweren Zeiten die Verbindung mit mir aufrecht erhalten hat.

Zürich, im August 1968

P. BERNAY

5. the formal demonstration of the proof of THORALF SKOLEM of the dispensability of the extended induction schemata;[VI.1]

6. the replacement of a previously very complicated proof of the eliminability of the ι-symbols with a simpler proof by GISBERT HASENJAEGER, based on a method of BARKLEY ROSSER;[VI.2]

7. a clarification of the elaboration on the representability of the recursive functions in System (Z).[VI.3]

Having regard to the fact that the elaborate table of contents provides a sufficient overview of the content and the threads of the book, the disposition of the book has remained unchanged. We strongly advise the reader to make use of this table of contents.[VI.4]

Mr. DIETER RÖDDING (Münster) has contributed an index of persons, extended the subject index, and added a system of backward references in the form of footnotes, with the main intention of improving the possibility of reading different parts of the book separately. I would like to thank him very much for this.

I am grateful to Mr. GISBERT HASENJAEGER and Mr. GEORG KREISEL for their contributions to the content of the new edition. Our new proof of the eliminability of the ι-symbols exploits the work of Mr. HASENJAEGER for the Foundations book during his stay here in Zürich.

I gratefully acknowledge the always alert participation of HEINRICH SCHOLZ in the work toward this new edition, and the interest in its implementation that Mr. FRIEDRICH KARL SCHMIDT constantly displayed.

I sincerely thank Mr. GERT H. MÜLLER for his many-sided contribution to the creation of the new edition. I sincerely thank Mr. DIRK SIEFKES (Heidelberg) for his useful participation in the proof-reading and in particular also for carrying out the additions in the subject index. I thank Mr. WALTER ZAUGG very much for his support in writing down and carrying out the corrections.

I am grateful to the publisher Springer for a number of kind concessions and, regarding the past, I am especially grateful for their continued association with me during the hard times.[VI.5]

Zürich, August 1968

PAUL BERNAYS

[VI.1] This footnote is to be extended after translation of the respective section.
[VI.2] This footnote is to be extended after translation of the respective section.
[VI.3] This footnote is to be extended after translation of the respective section.
[VI.4] This table of contents provides additional numberings and subject labels to subsections that — in the text — have neither headers nor any annotation of their starting position at all.
[VI.5] The term "the hard times" refers to the time of the Nazi rule of injustice (Nazionalsozialistische Unrechtsherrschaft) in Germany (1933–1945). During this time, any contact the German publisher Springer had with BERNAYS to promote BERNAYS' work was an act of disobedience against the German State, just because of BERNAYS' Jewish parentage. See also Note 47.2 on Page 47.

Zur Einführung

Die Leitgedanken meiner Untersuchungen über die Grundlagen der Mathematik, die ich — anknüpfend an frühere Ansätze — seit 1917 in Besprechungen mit P. Bernays wieder aufgenommen habe, sind von mir an verschiedenen Stellen eingehend dargelegt worden.

Diesen Untersuchungen, an denen auch W. Ackermann beteiligt ist, haben sich seither noch verschiedene Mathematiker angeschlossen.

Der hier in seinem ersten Teil vorliegende, von Bernays abgefaßte und noch fortzusetzende Lehrgang bezweckt eine Darstellung der Theorie nach ihren heutigen Ergebnissen.

Dieser Ergebnisstand weist zugleich die Richtung für die weitere Forschung in der Beweistheorie auf das Endziel hin, unsere üblichen Methoden der Mathematik samt und sonders als widerspruchsfrei zu erkennen.

Im Hinblick auf dieses Ziel möchte ich hervorheben, daß die zeitweilig aufgekommene Meinung, aus gewissen neueren Ergebnissen von Gödel folge die Undurchführbarkeit meiner Beweistheorie, als irrtümlich erwiesen ist. Jenes Ergebnis zeigt in der Tat auch nur, daß man für die weitergehenden Widerspruchsfreiheitsbeweise den finiten Standpunkt in einer schärferen Weise ausnutzen muß, als dieses bei der Betrachtung der elementaren Formalismen erforderlich ist.

Göttingen, im März 1934

HILBERT

Vorwort zur ersten Auflage

Eine Darstellung der Beweistheorie, welche aus dem Hilbertschen Ansatz zur Behandlung der mathematisch-logischen Grundlagenprobleme erwachsen ist, wurde schon seit längerem von Hilbert angekündigt.

Die Ausführung dieses Vorhabens hat eine wesentliche Verzögerung dadurch erfahren, daß in einem Stadium, in dem die Darstellung schon ihrem Abschluß nahe war, durch das Erscheinen der Arbeiten von Herbrand und von Gödel eine veränderte Situation im Gebiet der Beweistheorie entstand, welche die Berücksichtigung neuer Einsichten

Introductory Note

Elsewhere I have exhaustively presented the guidelines of my investigations on the foundations of mathematics, which — on the basis of some earlier approaches — I picked up again in discussions with PAUL BERNAYS since 1917.

Also WILHELM ACKERMANN participated in these investigations; and from that time, other mathematicians joined in as well.

The advanced textbook of which this is the first volume was composed and is to be continued by BERNAYS, and aims at a presentation of the theory according to the current state of results.

Moreover, these results provide a direction for further research into proof theory, with the final goal of recognizing the consistency[VII.1] of all our customary methods of mathematics.

Regarding this goal, I would like to emphasize that an opinion, which had emerged intermittently — namely that some more recent results of GÖDEL[VII.2] would imply the infeasibility of my proof theory — has turned out to be erroneous. Indeed, that result shows only that — for more advanced consistency proofs — the finitistic[VII.3] standpoint has to be exploited in a manner that is sharper[VII.4] than the one required for the treatment of the elementary formalisms.[VII.5]

Göttingen, March 1934

HILBERT

[VII.1] We will use the English words "consistent" and "consistency" for the translation of the German words "widerspruchsfrei" and "Widerspruchsfreiheit". Literally, however, "widerspruchsfrei" would be "free of contradiction". Our translation is in accordance with [BERNAYS, 1930/31], more precisely with Line 1 and Note 11 on Page 46 of the reprint in [BERNAYS, 1976], where BERNAYS suggests to use "Konsistenz" ("consistency") instead of "Widerspruchsfreiheit".

[VII.2] Cf. [GÖDEL, 1931].

[VII.3] Until HILBERT's usage for his foundational standpoint, the German word "finit" was hardly used at all. When it was used, then in the grammatical term "finite Form eines Verbes" ("finite form of a verb"), the rarely occurring opposite of "Infinitiv" ("infinitive") (also called "infinite Form" ("infinite form")) and "Partizip" ("participle"). If we translated HILBERT's technical German term "finit" with the English word "finite", then there would be no way to distinguish this from the translation of the German word "endlich" ("finite"). As this distinction is crucial, however, we follow the standard of translating the technical German term "finit" as "finitistic". Note that also in German, the word "finitistisch" ("finitistic") is becoming standard for denoting HILBERT's special form of intuitionism. This change is caused by the dominance of the English language in science, and probably also by the fact that the German word "finit" has found additional usages in the meantime, such as "Finite-Elemente-Methode" ("finite-elements method"). Finally, our translation goes well together with the French word *"finitiste"*, which was chosen in the French translation [HILBERT & BERNAYS, 2001], and which was used already in 1912 by HENRI POINCARÉ (1854–1912) to denote this philosophical standpoint; cf. [HILBERT & BERNAYS, 2001, p. 32].

[VII.4] Note that the English verb "to sharpen" and the German verb "verschärfen" — in addition to their common origin and primary meaning related to cutting (*acer, acutus*) — share the secondary meaning of "to refine" (which seems to be the primary intention in the German original) or, more precisely, of "to give a keen edge or fine point to" [GOVE, 1993, p. 2088b, sharpen 1]. Thus, as a translation of the German comparative degree "schärfer", we prefer "sharper" to alternative translations; such as "more rigid" (because this misses the intended meaning); "more rigorous" (because we will use "rigorous" as the translation of "streng"); and "more refined" (because this is less literal). Moreover note that, according to the usage of the similar words "verschärft" ("sharpened") and "Verschärfung" ("sharpening") on Pages 1.a and 2, such a "manner that is sharper" may also include conceptual extensions, such as the "existential form". See also Note 1.6 on Page 1.b, and Note 6.7 on Page 6.b.

[VII.5] The question of what HILBERT exactly meant with this sentence is pertinent, and has been extensively discussed in the literature — without an emerging consensus. There is no publication of HILBERT's addressing the issue; his last proof-theoretical paper was published in 1931. The best information is still found in §6 of this first volume.

zur Aufgabe machte. Dabei ist der Umfang des Buches angewachsen, so daß eine Teilung in zwei Bände angezeigt erschien.

Über den Inhalt und Gedankengang des vorliegenden ersten Bandes gibt ein ausführliches Inhaltsverzeichnis Auskunft.

Hier sei besonders darauf hingewiesen, daß der logische Formalismus in den §§ 3—4 ganz von Anfang entwickelt wird. Die Behandlung unterscheidet sich gegenüber derjenigen in dem Buche von HILBERT und ACKERMANN: „Grundzüge der theoretischen Logik" (1928) vor allem in Hinsicht auf den Aussagenkalkul. Bei dem weiteren Kalkul hat insbesondere die Einsetzungsregel, deren bisherige Formulierung nicht genügend deutlich war[1], eine genauere Fassung erhalten.

Ebensowenig wie aus dem Gebiet der Logistik werden aus mathematischen Gebieten spezielle Vorkenntnisse vorausgesetzt.

In dieser Hinsicht möge sich ein Leser, der mit den Grundlagen der Geometrie oder vielleicht auch mit den Grundlagen der Analysis nicht näher vertraut ist, durch die im § 1 stehenden Hinweise auf HILBERTs „Grundlagen der Geometrie" und die im § 2 ausgeführte Betrachtung über die Methoden der Analysis nicht abschrecken lassen. Die beiden ersten Paragraphen dienen im wesentlichen nur der Einführung in die Problemstellung, während der eigentliche systematische Aufbau erst mit dem § 3 beginnt.

Für die §§ 7 und 8 ist allerdings eine gewisse Vertrautheit mit den Elementen der Zahlentheorie erwünscht.

Bei der Niederschrift der §§ 4—7 haben Herr ARNOLD SCHMIDT und Herr KURT SCHÜTTE durch Begutachtung und Vorschläge mitgewirkt. Ich spreche ihnen hierfür meinen herzlichen Dank aus. Herrn ARNOLD SCHMIDT danke ich noch ganz besonders für die sorgsame Mitarbeit an den Korrekturen, bei denen er mich durch mannigfache Ratschläge unterstützt hat.

Göttingen, im März 1934

P. BERNAYS

---

[1] Das Erfordernis einer deutlicheren Fassung dieser Regel ist besonders stark hervorgetreten durch die Kritik, welche H. SCHOLZ in seiner „Logistik" (Vorlesungen 1932—1933) an ihr geübt hat. Diese Kritik beruht auf einer von dem intendierten Sinn der Regel abweichenden Interpretation, welche durch die Ungenauigkeit der bisherigen Formulierung verursacht ist.

Preface to the First Edition

A presentation of proof theory as developed in the HILBERTian approach to the problems in the foundations of mathematics and logic was announced by HILBERT some time ago.

The execution of this enterprise received considerable delay because the whole field of proof theory was changed by the publication of the works of HERBRAND and GÖDEL when our work was already close to completion;[VIII.1] and this change put the consideration of new insights | into the agenda. As a consequence of this, the size of the book grew to the extent that a separation into two volumes seemed appropriate.

The content and the threads of the present first volume are displayed in an elaborate table of contents.

Note in particular that the logical formalism is developed in §§ 3–4 right from the scratch. The treatment here differs from the one in the book of HILBERT & ACKERMANN [1928] particularly with respect to the propositional calculus. Also the rule of substitution of the further calculus (the previous formulation of which was not sufficiently clear)[1] has received a more precise presentation.

Special knowledge of the reader in the field of formal logic or in the fields of mathematics is not presupposed.

Regarding this question, a reader who is not well acquainted with the foundations of geometry or the foundations of analysis should not get discouraged, neither by the references to HILBERT's "Foundations of Geometry" in § 1, nor by the treatment of the methods of analysis in § 2: Both sections essentially serve as introductions to our ways of looking at the problems, whereas the proper systematic development does not start until § 3.

For §§ 7 and 8, however, some acquaintance with the elements of number theory is expected.

In the writing of §§ 4–7, Mr. H. ARNOLD SCHMIDT and Mr. KURT SCHÜTTE have contributed with their reviews and suggestions. I would like to express my sincere thanks for this. Regarding the corrections, I am especially grateful to Mr. H. ARNOLD SCHMIDT for his careful assistance and for supporting me with his manifold advice.

Göttingen, March 1934

PAUL BERNAYS

---

[1] The demand for a clearer presentation of this rule has come to the fore especially in the critique that HEINRICH SCHOLZ has given in [SCHOLZ, 1932/33]. This critique is based on an interpretation that deviates from the intended meaning of the rules, and results from the imprecision of its previous formulation.[VIII.2]

[VIII.1] It is clear that the reference to GÖDEL refers primarily to [GÖDEL, 1931]. The reference to HERBRAND, however, seems to be more difficult to pin down: It seems that BERNAYS refers primarily to [HERBRAND, 1932], where [GÖDEL, 1931] is discussed. Closely related to the contemporary perception of GÖDEL's result and historically more important, however, is HERBRAND's letter to GÖDEL, cf. [GÖDEL, 1986ff., Vol. V, pp. 3–25], [SIEG, 2005], and [WIRTH &AL., 2009]. Finally, note that, on the other hand, with regard to the second volume of the "Foundations of Mathematics", BERNAYS' remark may also refer to HERBRAND's Fundamental Theorem [HERBRAND, 1930].

[VIII.2] QUINE [1982] goes into the matter of the rule of substitution with great care. [HILBERT & BERNAYS, 1934] gets the rule right and may have been the first publication to do so. By using different letters for free and bound variables, however, BERNAYS avoided one of the two difficulties mentioned in [QUINE, 1982, § 26]; namely the difficulty that (free) "variables of the substituted abstract must not be captured by quantifiers of the schema in which the substitution takes place", whereas the provision is still required that "quantifiers of the substituted abstract must not capture variables of the schema in which substitution takes place", cf. [QUINE, 1982, p.162] for both quotes.

Inhaltsverzeichnis

§ 1. Das Problem der Widerspruchsfreiheit in der Axiomatik als logisches Entscheidungsproblem 1

   a) Formale Axiomatik 1
      1. Verhältnis der formalen zur inhaltlichen Axiomatik; Frage der Widerspruchsfreiheit; Arithmetisierung 1
      2. Geometrische Axiome als Beispiel 3
      3. Rein logische Fassung der Axiomatik 7
   b) Das Entscheidungsproblem 8
      1. Allgemeingültigkeit und Erfüllbarkeit 8
      2. Entscheidung für endliche Individuenbereiche. 9
      3. Methode der Aufweisung 11
   c) Die Frage der Widerspruchsfreiheit bei unendlichem Individuenbereich 14
      1. Formeln, die nicht im Endlichen erfüllbar sind; die Zahlenreihe als Modell 14
      2. Problematik des Unendlichen 15
      3. Nachweis der Widerspruchsfreiheit als Unmöglichkeitsbeweis; Methode der Arithmetisierung 17

§ 2. Die elementare Zahlentheorie. — Das finite Schließen und seine Grenzen 20

   a) Die Methode der anschaulichen Überlegung und ihre Anwendung in der elementaren Zahlentheorie 20
      1. Begriff der Ziffer; Beziehung „kleiner"; Addition 20
      2. Rechengesetze; vollständige Induktion; Multiplikation; Teilbarkeit; Primzahl 23
      3. Rekursive Definitionen 25
      4. Unmöglichkeitsbeweis 27
   b) Weitere Anwendungen anschaulicher Überlegungen. 28
      1. Beziehung zwischen Zahlentheorie und Anzahlenlehre 28
      2. Standpunkt der formalen Algebra 29
   c) Der finite Standpunkt; Überschreitung dieses Standpunktes bereits in der Zahlentheorie 32
      1. Logische Charakterisierung des finiten Standpunktes. 32
      2. Das „tertium non datur" für ganze Zahlen; das Prinzip der kleinsten Zahl 34
   d) Nichtfinite Methoden in der Analysis 36
      1. Verschiedene Definitionen der reellen Zahl 36
      2. Obere Grenze einer Zahlenfolge; obere Grenze einer Zahlenmenge 38
      3. Das Auswahlprinzip 40

Table of Contents

§ 1 The Problem of Consistency in Axiomatics as a Logical Decision Problem 1

- (a) Formal axiomatics 1
  - 1. Relationship of formal to contentual axiomatics; issue of consistency; arithmetization 1
  - 2. Geometrical axioms as an example[IX.1] 4
  - 3. Abstract logical formulation of the axiomatics 7
- (b) The Entscheidungsproblem 8
  - 1. Validity and satisfiability 8
  - 2. Decision for finite domains of individuals 9
  - 3. Method of exhibition 11
- (c) The issue of consistency for infinite domains of individuals ... 14
  - 1. Formulas that are not satisfiable in the finite; the number series as a model 14
  - 2. Problems of the infinite[IX.1] 16
  - 3. Proof of consistency as impossibility proof; method of arithmetization .. 17

§ 2 Elementary Number Theory. — Finitistic Inference and its Limits 20

- (a) The method of intuitive consideration and its application in elementary number theory 20
  - 1. Notion of a numeral; relation "smaller than"; addition 20
  - 2. Laws of calculation; mathematical induction; multiplication; divisibility; prime number 23
  - 3. Recursive definitions[IX.1] 26
  - 4. Impossibility proof 27
- (b) Further applications of intuitive considerations 28
  - 1. Relation between number theory and the theory of cardinal numbers ... 28
  - 2. Standpoint of formal algebra 29
- (c) The finitistic standpoint; transgression of this standpoint already in number theory 32
  - 1. Logical characterization of the finitistic standpoint 32
  - 2. The "tertium non datur" for integers; the least-number principle ... 34
- (d) Non-finitistic methods in analysis 36
  - 1. Various representations of real numbers 36
  - 2. Least upper bound of a number sequence; least upper bound of a set of numbers 38
  - 3. The principle of choice 40
- (e) Investigations on the direct finitistic grounding of arithmetic; return to the previous way of posing the problem; proof theory 42

[IX.1]This section starts one page later in the translation.

e) Untersuchungen zur direkten finiten Begründung der Arithmetik; Rückkehr zur früheren Problemstellung; die Beweistheorie 42

§ 3. Die Formalisierung des logischen Schließens I: Der Aussagenkalkul 45
   a) Theorie der Wahrheitsfunktionen 45
     1. Die Wahrheitsfunktionen und ihre Schemata 45
     2. Ersetzbarkeit; Ersetzungsregeln 47
     3. Beispiele von Ersetzbarkeiten 50
     4. Dualität; konjunktive und disjunktive Normalform; identisch wahre Ausdrücke; Entscheidungsverfahren 52
     5. Ausgezeichnete Normalform; Entscheidung über Ersetzbarkeit; Beispiele 55
   b) Anwendung der Theorie der Wahrheitsfunktionen auf das logische Schließen; Formalisierung aussagenlogischer Schlüsse mittels der identisch wahren Ausdrücke, der Einsetzungsregel und des Schlußschemas 59
   c) Deduktive Aussagenlogik 63
     1. Problemstellung 63
     2. Ein System der deduktiven Aussagenlogik; Vollständigkeitseigenschaften dieses Systems 65
     3. Positive Logik; reguläre Implikationsformeln; positiv identische Implikationsformeln; Möglichkeiten der Kürzung 67
   d) Unabhängigkeitsbeweise nach der Methode der Wertung 71
     1. Die logische Interpretation als Wertung; das allgemeine Verfahren 71
     2. Unabhängigkeitsbeweise für das aufgestellte System; noch ein weiterer Unabhängigkeitsbeweis 74
     3. Anwendung der Wertungsmethode auf die Frage der Vertretbarkeit von Formeln durch Schemata 79
   e) Rückkehr zu der unter b) betrachteten Art der Formalisierung des Schließens; abkürzende Regeln; Bemerkung über den Fall eines Widerspruchs 84

§ 4. Die Formalisierung des Schließens II: Der Prädikatenkalkul 86
   a) Einführung der Individuenvariablen; Begriff der Formel; Einsetzungsregel; Beispiel; Vergleich mit dem inhaltlichen Schließen 86
   b) Die gebundenen Variablen und die Regeln für Allzeichen und Seinszeichen 91
     1. Unzulänglichkeit der freien Variablen 91
     2. Einführung der gebundenen Variablen; Allzeichen und Seinszeichen; Regel der Umbenennung; Vermeidung von Mehrdeutigkeiten; Erweiterung des Begriffs der Formel sowie der Einsetzungsregel . . . 93
     3. Heuristische Einführung der Regeln für die Allzeichen und Seinszeichen; inhaltliche Deutung der Formeln und Schemata 95
     4. Zusammenstellung der Regeln des Prädikatenkalkuls; Darstellung der Formen kategorischer Urteile; Ausschluß des leeren Individuenbereichs 100

§ 3 Formalization of Logical Inference I: The Propositional Calculus.   45

(a) Theory of truth functions 45
  1. Truth functions and their schemata 45
  2. Replaceability; replacement rules 47
  3. Examples of replaceabilities 50
  4. Duality; conjunctive and disjunctive normal form; identically true expressions; decision procedure 52
  5. Full normal form; decision on replaceability; examples 55

(b) Application of the theory of truth functions to logical inference; formalization of propositional inferences by means of the identically true expressions, the rule of substitution and the inference schema 59

(c) Deductive propositional logic 63
  1. Posing of the problem 63
  2. A system of deductive propositional logic; completeness properties of this system 65
  3. Positive logic; regular implication formulas; positive-identical implication formulas; possibilities of abbreviation 67

(d) Independence proofs by model construction 71
  1. Logical interpretation with models; the general approach 71
  2. Independence proofs for the system presented; yet another independence proof 74
  3. Application of model construction to the question of the representability of formulas by schemata 79

(e) Return to the mode of formalization of inference considered in (b); short-cutting rules; remark on the case of a contradiction[X.1] 82

§ 4 Formalization of Logical Inference II: The Predicate Calculus.   ... 86

(a) Introduction of the individual variables; notion of a formula; rule of substitution; example; comparison with contentual inference 86

(b) The bound variables and the rules for universal and existential quantifier symbols[X.2] 93
  1. Insufficiency of the free variables 93
  2. Introduction of the bound variables; universal and existential quantifier symbols; rule of renaming; avoidance of ambiguity; extension of the notion of a formula and extension of the rule of substitution 94
  3. Heuristic introduction of the rules for the universal and existential quantifier symbols; contentual interpretation of the formulas and schemata[X.3] 98
  4. Summary of the rules of the predicate calculus; representation of the forms of categorical judgment; exclusion of the empty domain of individuals 104

[X.1] Page number 84 given in the German original (instead of the correct 82) is a typo of the second edition.

[X.2] This section starts one page later in the translation.

[X.3] Page number 99 given in the German original (instead of the correct 98) is a typo of both editions. See also Note 82.1.

c) Ausführung von Ableitungen 106
  1. Einige abgeleitete Regeln 106
  2. Ableitung von Formeln 109
d) Systematische Fragen 117
  1. Begriff der f-zahlig identischen Formel und der im Endlichen identischen Formel; deduktive Abgeschlossenheit der Gesamtheit der f-zahlig identischen Formeln; Widerspruchsfreiheit des Prädikatenkalkuls; Vollständigkeitsfragen 118
  2. Exkurs über die mengentheoretische Prädikatenlogik; Vorläufiges zu den Fragen der Vollständigkeit; das Entscheidungsproblem und seine Verschärfung unter dem deduktiven Gesichtspunkt 125
e) Betrachtungen über den Formalismus des Prädikatenkalkuls 131
  1. Begriff der Überführbarkeit; abgeleitete Regeln 131
  2. Überführung von Formeln in pränexe Formeln; Beispiele; Abgrenzung von Fällen der Lösung des Entscheidungsproblems an Hand der pränexen Normalform 139
  3. Zerlegung einer Formel des einstelligen Prädikatenkalkuls in Primärformeln; Beispiel 145
f) Deduktionsgleichheit und Deduktionstheorem 148
  1. Begriff der Deduktionsgleichheit; zwei wesentliche Fälle von Deduktionsgleichheit; Überführbarkeit und Deduktionsgleichheit . . 148
  2. Das Deduktionstheorem 150
  3. Anwendungen des Deduktionstheorems: Zurückführung axiomatischer Fragen auf solche der Ableitbarkeit von Formeln des Prädikatenkalkuls; erleichterte Feststellung der Ableitbarkeit; Betrachtung einer gebräuchlichen Schlußweise 154
  4. Deduktionsgleichheit einer jeden Formel mit einer SKOLEMschen Normalform sowie auch mit einer Normaldisjunktion; Vereinfachung des Überganges 157

§ 5. Hinzunahme der Identität. Vollständigkeit des einstelligen Prädikatenkalkuls. 163

a) Erweiterung des Formalismus 163
  1. Das Gleichheitszeichen; Darstellung von Anzahlaussagen; die Gleichheitsaxiome und die formalen Eigenschaften der Identität 164
  2. Verwendung der Gleichheitsaxiome zu Umformungen, insbesondere solchen von Anzahlbedingungen; Anzahlformeln 168
  3. Zerlegung einer Formel des erweiterten einstelligen Prädikatenkalkuls in Primärformeln 178
  4. Ausdehnung des Begriffes der f-zahlig identischen Formel; deduktive Abgeschlossenheit der Gesamtheit der f-zahlig identischen Formeln; Eindeutigkeitsbetrachtung 183
  5. Hinzunahme von Funktionszeichen; Begriff des Terms; ableitbare Formeln 186
b) Lösung von Entscheidungsproblemen; Vollständigkeitssätze 189
  1. Entscheidung über die Ableitbarkeit solcher pränexer Formeln des Prädikatenkalkuls, bei denen jedes Allzeichen jedem Seinszeichen vorhergeht; Entscheidbarkeit im Endlichen 190

(c) Execution of derivations 106
    1. Some derived rules 106
    2. Derivation of formulas 109

(d) Systematic issues 117
    1. Notion of the k-identical formula and the finitely identical formula; deductive closedness of the totality of the k-identical formulas; consistency of the predicate calculus; issues of completeness 118
    2. Excursion on set-theoretic predicate logic; preliminary notes on the issues of completeness; the Entscheidungsproblem and its sharpening under the deductive aspect 125

(e) Discussion of the formalism of the predicate calculus 131
    1. Notion of convertibility; derived rules 131
    2. Conversion of formulas into prenex formulas; examples; classification of solved cases of the Entscheidungsproblem by means of the prenex normal form[XI.1] 140
    3. Decomposition of a formula of the monadic predicate calculus into prime formulas; example 145

(f) Deductive equivalence and deduction theorem 148
    1. Notion of deductive equivalence; two essential cases of deductive equivalence; convertibility and deductive equivalence 148
    2. The deduction theorem 150
    3. Applications of the deduction theorem; Reduction of axiomatic problems to problems of derivability of formulas of the predicate calculus; simplified determination of derivability; treatment of a customary mode of inference 154
    4. Deductive equivalence of every formula with some SKOLEM normal form and with some normal disjunction as well; simplification of the equivalence step[XI.1] 158

§ 5 Adding the Identity. Completeness of the Monadic Predicate Calculus 163

[XI.1] This section starts one page later in the translation.

2. Ableitbarkeit einer jeden im Endlichen identischen Formel des einstelligen Prädikatenkalkuls, Nachweis mit Hilfe des vorherigen Entscheidungsverfahrens; ein mengentheoretischer Nachweis und seine finite Verschärfung 192
3. Deduktionsgleiche Normalform für eine Formel des erweiterten einstelligen Prädikatenkalkuls 198
4. Vollständigkeitssätze für den erweiterten einstelligen Prädikatenkalkul 205

§ 6. **Widerspruchsfreiheit unendlicher Individuenbereiche. Anfänge der Zahlentheorie** 208

a) Überleitung von der Frage der Unableitbarkeit gewisser im Endlichen identischer Formeln des Prädikatenkalkuls zur Frage der Widerspruchsfreiheit eines zahlentheoretischen Axiomensystems 208
  1. Ersetzung der Formelvariablen durch Prädikatensymbole; eine Abhängigkeit zwischen den betrachteten Formeln 209
  2. Einbeziehung der Gleichheitsaxiome; die DEDEKINDsche Unendlichkeitsdefinition; Einführung des Strichsymbols 212
  3. Übergang zu Axiomen ohne gebundene Variablen unter Verschärfung der Existenzaxiome; das Symbol 0; Ziffern im neuen Sinne; PEANOsche Axiome; Zusammenstellung der erhaltenen Axiome 215

b) Allgemein logischer Teil des Nachweises der Widerspruchsfreiheit . . 219
  1. Spezialisierung der Endformel; Ausschluß gebundener Variablen; Auflösung in Beweisfäden 219
  2. Rückverlegung der Einsetzungen; Ausschaltung der freien Variablen; numerische Formeln; Definition von „wahr" und „falsch"; „Wahrheit" einer jeden ohne Benutzung gebundener Variablen ableitbaren Formel 224
  3. Einbeziehung der gebundenen Variablen; Maßregel zur Erhaltung der Schemata bei der Rückverlegung der Einsetzungen; Unzulänglichkeit des bisherigen Verfahrens 230

c) Durchführung des Nachweises der Widerspruchsfreiheit mittels eines Reduktionsverfahrens 233
  1. Ausschaltung der Allzeichen; die Reduktionsschritte; Begriff der Reduzierten 233
  2. Verifizierbare Formeln; Eindeutigkeitssatz; Hilfssätze 237
  3. Verifizierbarkeit einer jeden ableitbaren, von Formelvariablen freien Formel; Wiedereinbeziehung der Allzeichen; Ersetzbarkeit von Axiomen durch Axiomenschemata 243

d) Übergang zu einem (im Bereich der Formeln ohne Formelvariablen) deduktiv abgeschlossenen Axiomensystem 248
  1. Unableitbarkeit gewisser verifizierbarer Formeln durch das betrachtete Axiomensystem; Nachweis mit Hilfe von „Ziffern zweiter Art" 248
  2. Ansatz zur Vervollständigung des Axiomensystems; Ableitbarkeit einer Reihe von Äquivalenzen als hinreichende Bedingung 252
  3. Deduktive Zurückführung der Äquivalenzen auf fünf zu den Axiomen hinzuzufügende Formeln; Vereinfachungen; das System (A) 255
  4. Vollständigkeitseigenschaften des Systems (A) 262

§ 6 Consistency of Infinite Domains of Individuals. Beginnings of Number
     Theory 208

e) Einbeziehung der vollständigen Induktion 264
  1. Formalisierung des Prinzips der vollständigen Induktion durch eine Formel oder durch ein Schema; Gleichwertigkeit der beiden Formalisierungen; Unveränderteit des Bereiches der ableitbaren Formeln ohne Formelvariablen bei der Hinzunahme des Induktionsschemas zu dem System (A) 264
  2. Vereinfachung des Axiomensystems bei Hinzunahme des Induktionsaxioms; das System (B) 269
f) Unabhängigkeitsbeweise 274
  1. Unableitbarkeit des Induktionsaxioms aus dem System (A) 274
  2. Unabhängigkeitsbeweise mittels eines Substitutionsverfahrens . . 277
  3. Feststellung der übrigen Unabhängigkeiten durch Modifikationen des Reduktionsverfahrens 279
g) Darstellung des Prinzips der kleinsten Zahl durch eine Formel; Gleichwertigkeit dieser Formel mit dem Induktionsaxiom bei Zugrundelegung der übrigen Axiome des Systems (B) 283

§ 7. Die rekursiven Definitionen 286
a) Grundsätzliche Erörterungen 286
  1. Das einfachste Schema der Rekursion; Formalisierung des anschaulichen Berechnungsverfahrens; Gegenüberstellung von expliziter und rekursiver Definition 287
  2. Nachweis der Widerspruchsfreiheit der Hinzunahme rekursiver Definitionen im Rahmen des elementaren Kalkuls mit freien Variablen; Einbeziehung des Induktionsschemas 294
  3. Unmöglichkeit, die Widerspruchsfreiheit der rekursiven Definitionen schon aus der Widerspruchsfreiheit des vorherigen Axiomensystems zu folgern; Ersetzbarkeit arithmetischer Axiome durch rekursive Definitionen; explizite Definition von „$<$" durch eine rekursive Funktion; Ableitung der Grundeigenschaften von „$<$" 300
b) Die rekursive Zahlentheorie 309
  1. Ableitung der Gesetze für die Addition, Subtraktion, Multiplikation und die Beziehung „$<$" 309
  2. Darstellung von Aussagen durch Gleichungen der Form $t = 0$; Summen und Produkte mit variabler Gliederzahl; Darstellung von allgemeinen und existentialen Aussagen über Zahlen $\leq n$, sowie von Maximum- und Minimumausdrücken 313
  3. Teilbarkeit; Division mit Rest; kleinster von 1 verschiedener Teiler; Reihe der Primzahlen; Zerlegung der Zahlen in Primfaktoren; umkehrbar eindeutige Beziehung zwischen den Zahlen > 1 und den endlichen Folgen von Zahlen; Numerierung der Zahlenpaare; größter gemeinsamer Teiler; kleinstes gemeinsames Vielfaches . . 321
c) Erweiterungen des Schemas der Rekursion und des Induktionsschemas 330
  1. Rekursionen, die sich auf das einfachste Rekursionsschema (die primitive Rekursion) zurückführen lassen: Wertverlaufsrekursionen, simultane Rekursionen 331
  2. Verschränkte Rekursionen; Unzurückführbarkeit gewisser verschränkter Rekursionen auf primitive Rekursionen 335
  3. Erweiterte Induktionsschemata; ihre Entbehrlichkeit 348

§ 7 The Recursive Definitions 286

d) Vertretbarkeit rekursiver Funktionen; Übergang zu einem für die Zahlentheorie ausreichenden Axiomensystem 355
  1. Rückkehr zum vollen Formalismus; das System (C); Begriff einer wesentlichen Erweiterung eines Formalismus; Beispiele von nicht wesentlichen Erweiterungen; Vertretbarkeit einer Funktion 355
  2. Nachweis, daß die Summe und die Differenz im Formalismus des Systems (B) nicht vertretbare Funktionen sind; die Rekursionsgleichungen für die Summe als Axiome; das System (D) 362
  3. Nachweis der Widerspruchsfreiheit und Vollständigkeit des Systems (D) nach der Methode der Reduktion; Unvertretbarkeit des Produktes im Formalismus des Systems (D) 368
  4. Veränderte Situation bei Hinzunahme der Rekursionsgleichungen für das Produkt; das System (Z) 379
e) Ergänzende Betrachtungen über die Gleichheitsaxiome 382
  1. Ersetzung des zweiten Gleichheitsaxioms durch speziellere Axiome 382
  2. Anwendung auf die Axiomensysteme (A), (B), (Z) 384
  3. Anwendung auf das Entscheidungsproblem; Eliminierbarkeit der Gleichheitsaxiome aus einer Ableitung einer Formel des Prädikatenkalkuls. 388

§ 8. Der Begriff ,,derjenige, welcher'' und seine Eliminierbarkeit 392
a) Die ι-Regel und ihre Handhabung 392
  1. Inhaltliche Erörterung; Einführung der ι-Regel; Vermeidung von Kollisionen; Darstellung von Funktionen durch ι-Terme 392
  2. Einlagerung und Überordnung; abkürzende Symbole 397
  3. Die Funktion $\omega(A)$; Formalisierung des Begriffs der kleinsten Zahl durch die Funktion $\mu_x A(x)$; Eindeutigkeitsformeln 401
b) Deduktive Entwicklung der Zahlentheorie auf Grund des Axiomensystems (Z) unter Hinzunahme des formalisierten Begriffs der kleinsten Zahl 410
  1. Begriff ,,kleiner''; Kongruenz; Division mit Rest; Teilbarkeit; zueinander prime Zahlen 410
  2. Kleinstes gemeinsames Vielfaches von zwei Zahlen und von einer endlichen Folge von Zahlen; Maximum einer endlichen Folge von Zahlen 415
c) Zurückführung primitiver Rekursionen auf explizite Definitionen mittels der Funktion $\mu_x A(x)$ bei Zugrundelegung des Systems (Z) . . 421
  1. Heuristischer Ansatz 421
  2. Formale Durchführung; Möglichkeit der Verallgemeinerung des Verfahrens 425
d) Die Eliminierbarkeit der Kennzeichnungen (der ι-Symbole) 431
  1. Erweiterung der ι-Regel, Verhältnis zur ursprünglichen ι-Regel, die Terme $\iota_x^{(d)} A(x)$ 432
  2. Der Ansatz von ROSSER und seine Vereinfachung durch HASENJAEGER. Einsetzung von ι-Termen, das Axiom $\{\iota\}$, Beschaffenheit der in Frage stehenden formalen Systeme 435
  3. Erklärung der ,,Reduzierten'' einer Formel und Zurückführung des zu erbringenden Nachweises auf den Beweis der ι-freien Herleitbarkeit der nach einem gewissen Schema gebildeten Formeln . . 439

§ 8 The Notion "that which" and its Eliminability 392

4. Durchführung dieses Beweises 441
5. Formulierung des Eliminationstheorems, Überführbarkeit jeder Formel in ihre Reduzierte, Vergleich verschiedener Eliminationsverfahren 448

e) Folgerungen aus der Eliminierbarkeit der Kennzeichnungen 451
1. Die Vertretbarkeit der rekursiven Funktionen im System (Z) . . . 451
2. Allgemeines Verfahren der Ausschaltung von Funktionszeichen durch Einführung von Prädikatensymbolen; Ausschaltung von Individuensymbolen 455
3. Durchführung des Verfahrens an dem System (Z); Ausblick auf weitere Fragestellungen 458

f) Nachtrag: Ausdehnung des Satzes über die Vertretbarkeit des Gleichheitsaxioms (J_2) bei Hinzunahme der ι-Regel 464

Namenverzeichnis 467

Sachverzeichnis 468

§ 1. Das Problem der Widerspruchsfreiheit in der Axiomatik als logisches Entscheidungsproblem.

Der Stand der Forschungen im Gebiete der Grundlagen der Mathematik, an den unsere Ausführungen anknüpfen, wird durch die Ergebnisse von dreierlei Untersuchungen gekennzeichnet:

1. der Ausbildung der axiomatischen Methode, insbesondere an Hand der Grundlagen der Geometrie,

2. der Begründung der Analysis nach der heutigen strengen Methode durch die Zurückführung der Größenlehre auf die Lehre von Zahlen und Zahlenmengen,

3. der Untersuchungen zur Grundlegung der Zahlen- und Mengenlehre.

An den hierdurch erreichten Standpunkt knüpft sich auf Grund einer verschärften methodischen Anforderung eine weitergehende Aufgabestellung, bei der es sich um eine neue Art der Auseinandersetzung mit dem Problem des Unendlichen handelt. Wir wollen auf diese Problemstellung von der Betrachtung der Axiomatik aus hinführen.

Der Terminus „axiomatisch" wird teils in weiterem, teils in engerem Sinne gebraucht. In der weitesten Bedeutung des Wortes nennen wir die Entwicklung einer Theorie axiomatisch, wenn die Grundbegriffe und Grundvoraussetzungen als solche an die Spitze gestellt werden und aus ihnen der weitere Inhalt der Theorie mit Hilfe von Definitionen und Beweisen logisch abgeleitet wird. In diesem Sinne ist die Geometrie von EUKLID, die Mechanik von NEWTON, die Thermodynamik von CLAUSIUS axiomatisch begründet worden.

Eine Verschärfung, welche der axiomatische Standpunkt in HILBERTS „Grundlagen der Geometrie" erhalten hat, besteht darin, daß man von dem sachlichen Vorstellungsmaterial, aus dem die Grundbegriffe einer Theorie gebildet sind, in dem axiomatischen Aufbau der Theorie nur dasjenige beibehält, was als Extrakt in den Axiomen formuliert ist, von allem sonstigen Inhalt aber abstrahiert. Bei der Axiomatik in der engsten Bedeutung kommt noch als weiteres Moment die *existentiale Form* hinzu. Durch diese unterscheidet sich *die axiomatische Methode* von der *konstruktiven* oder *genetischen* Methode der Begründung einer Theorie[1]. Während bei der konstruktiven Methode die

---

[1] Vgl. zu dieser Gegenüberstellung den Anhang VI von HILBERTS Grundlagen der Geometrie: Über den Zahlenbegriff, 1900.

§ 1. The Problem of Consistency in Axiomatics as a Logical Decision Problem

We start from the state of research in the foundations of mathematics that is characterized by the results of three kinds of investigations:

1. the development of the axiomatic method, in particular in the foundations of geometry,[1.1]

2. the grounding of analysis with the current rigorous methods,[1.2] through the reduction of the theory of magnitudes to the theory of numbers and sets of numbers,

3. the investigations into the foundations of number theory and set theory.

Because of new sharpened methodological[1.3] requirements, the standpoint reached through these investigations is augmented with a further task: to deal with the problem of the infinite in a new way. We will approach this task on the basis of our treatment of axiomatics.

The technical term "axiomatic" will be used partly in a broader and partly in a narrower sense. We will call the development of a theory axiomatic in the broadest sense if the basic notions and presuppositions are stated first, and then the further content of the theory is logically derived with the help of definitions and proofs. In this sense, EUCLID provided an axiomatic grounding for geometry, NEWTON for mechanics, and CLAUSIUS for thermodynamics.

In HILBERT's "Foundations of Geometry",[1.4] the axiomatic standpoint has been sharpened regarding the axiomatic development of a theory: From the factual and conceptual subject matter that gives rise to the basic notions of the theory, we retain only the essence that is formulated in the axioms, and ignore[1.5] all other content. Finally, for axiomatics in the narrowest sense, the *existential*[1.6] *form* comes in as an additional factor. This marks the difference between *the axiomatic method* and the *constructive* or *genetic*[1.7] method of grounding a theory.[1] ─────────────────────────────────

───────────────────────────

[1] Compare this distinction to Anhang VI of HILBERT's "Foundations of Geometry": On the Notion of Number, 1900.[1.8]

[1.1] In German, there is no difference between the description "foundations of geometry" and the proper name "Foundations of Geometry" referring to [HILBERT, 1899]. The first alternative is more likely here because there are no quotation marks in the original text, contrary to the last paragraph on this page.

[1.2] The German original has the singular "method" instead of "methods". There are several methods however, cf. e.g. p. 36. In English, the syntactical default for an indefinite number is the plural. In German, the syntactical default for an indefinite number is the singular. Thus, in English, a syntactical singular would put an emphasis on a semantical singular — an emphasis not intended in the German original.

[1.3] The German word "methodisch" relates to method, whereas "methodologisch" relates to methodology. According to [GOVE, 1993, p.1423a, methodical, 3], it would be correct to maintain the distinction between method and methodology by translating "methodisch" as "methodical". In colloquial English, however, "methodical" is a synonym of "systematic", and "methodological" relates to both method and methodology. To avoid confusion, we never translate "methodisch" as "methodical"; we translate it as "methodological" only if the distinction is irrelevant, such as in the given case here or in "methodischer Standpunkt" ("methodological standpoint"). In the other cases, however, we translate it as "of/on the method" or as "the/our treatment of" (cf. Notes 20.5 and 44.2) or even omit it completely (cf. Note 32.1).

[1.4] Cf. the first English translation [HILBERT, 1902]. As always, the actual reference in the German original is to HILBERT's "Grundlagen der Geometrie", whose first edition is [HILBERT, 1899].

§ 1. Das Problem der Widerspruchsfreiheit in der Axiomatik als logisches Entscheidungsproblem.

Der Stand der Forschungen im Gebiete der Grundlagen der Mathematik, an den unsere Ausführungen anknüpfen, wird durch die Ergebnisse von dreierlei Untersuchungen gekennzeichnet:

1. der Ausbildung der axiomatischen Methode, insbesondere an Hand der Grundlagen der Geometrie,
2. der Begründung der Analysis nach der heutigen strengen Methode durch die Zurückführung der Größenlehre auf die Lehre von Zahlen und Zahlenmengen,
3. der Untersuchungen zur Grundlegung der Zahlen- und Mengenlehre.

An den hierdurch erreichten Standpunkt knüpft sich auf Grund einer verschärften methodischen Anforderung eine weitergehende Aufgabestellung, bei der es sich um eine neue Art der Auseinandersetzung mit dem Problem des Unendlichen handelt. Wir wollen auf diese Problemstellung von der Betrachtung der Axiomatik aus hinführen.

Der Terminus „axiomatisch" wird teils in weiterem, teils in engerem Sinne gebraucht. In der weitesten Bedeutung des Wortes nennen wir die Entwicklung einer Theorie axiomatisch, wenn die Grundbegriffe und Grundvoraussetzungen als solche an die Spitze gestellt werden und aus ihnen der weitere Inhalt der Theorie mit Hilfe von Definitionen und Beweisen logisch abgeleitet wird. In diesem Sinne ist die Geometrie von EUKLID, die Mechanik von NEWTON, die Thermodynamik von CLAUSIUS axiomatisch begründet worden.

Eine Verschärfung, welche der axiomatische Standpunkt in HILBERTS „Grundlagen der Geometrie" erhalten hat, besteht darin, daß man von dem sachlichen Vorstellungsmaterial, aus dem die Grundbegriffe einer Theorie gebildet sind, in dem axiomatischen Aufbau der Theorie nur dasjenige beibehält, was als Extrakt in den Axiomen formuliert ist, von allem sonstigen Inhalt aber abstrahiert. Bei der Axiomatik in der engsten Bedeutung kommt noch als weiteres Moment die *existentiale Form* hinzu. Durch diese unterscheidet sich *die axiomatische Methode* von der *konstruktiven* oder *genetischen* Methode der Begründung einer Theorie[1]. Während bei der konstruktiven Methode die

---

[1] Vgl. zu dieser Gegenüberstellung den Anhang VI von HILBERTS Grundlagen der Geometrie: Über den Zahlbegriff, 1900.

[1.5] We translate "abstrahieren von" semantically correct as "to ignore". None of the meanings of the straightforward translation "to abstract from" has the correct semantics, i.e. none of "to consider something separately from", "to extract something from", "to steal from", "to withdraw" would be correct here. The German intransitive verb pattern "abstrahieren von" with the given curious meaning seems to originate in German philosophy: Cf. e.g. [WOLFF, 1719, p. 30] (cited according to [SCHULZ &AL., 1995ff., Vol. I, p. 63]), [KANT, 1781, p. 26], [KANT, 1787, p. 307], [KANT, 1800, §6]. Today this verb pattern with the given curious meaning is standard usage of the verb "abstrahieren" ("to abstract") in colloquial German and the neologism "to abstract away" is sometimes used for its translation.

[1.6] The *German* word "existential" (alternative spelling: "existenzial") of the German original text is mostly used in the philosophies of human existence or Dasein ("there-being"), such as the ones of MARTIN HEIDEGGER and JEAN PAUL SARTRE, where it has to be distinguished from the German word "existenziell" (alternative spelling: "existentiell"). In logic, mathematics, and everyday German, "existenziell" is the word typically used, and use of "existential" is rare. And if the latter is used in mathematics, most of the time the former is not used at all, and the latter replaces the former synonymously. It is indeed most likely that the use of the word "existential" in the German original just replaces the word "existenziell", and that no connotation in the sense of the philosophies of human existence is intended here: For instance, on page 173 of [HILBERT, 1926], HILBERT uses the word "existential" in the term "existentiale Aussage" ("existential proposition") with the standard mathematical meaning of "there is a number of this and that property" ("es gibt eine Zahl von der und der Eigenschaft"). Moreover, BERNAYS wrote later in his article on mathematical existence and consistency:

> "Wir haben daher in der Mathematik keinen Anlass, Existenz in einem grundsätzlich anderen Sinne anzunehmen, als wir das Bestehen von gesetzlichen Beziehungen annehmen."[1.9]

> "Der Gedanke einer solchen mathematischen Tatsächlichkeit bedeutet andererseits nicht ein Zurückkommen auf die Ansicht von einer selbständigen Existenz der mathematischen Gegenstände. Es handelt sich dabei nicht um ein Dasein, sondern um beziehungsmäßige, strukturelle Bindungen und um das Hervorgehen (Induziert-werden) von ideellen Gegenständen aus anderen solchen Gegenständen."[1.10]

Furthermore, regarding a domain of individuals, the intended meaning of the German terms "existential" (English: "existential") and "existentiale Form" ("existential form") in the original text refers to "existence of the (infinite) extension of a notion", "existence of the (infinite) aggregate", or "modality of existence of the actual infinite". We will indicate each occurrence of the word "existential" with this intention in the German original with a note in the translation. On page 184 of [HILBERT, 1900a], referred to in the original Note 1 of Page 1.a, we find the less ambiguous verbalization "Existenz des Inbegriffs aller reellen Zahlen und unendlicher Mengen überhaupt" ("existence of the aggregate of the real numbers and infinite sets in general"). It seems that a disambiguation of the notion of "existence" and the closely related notion of "existence of the aggregate" (which is of a higher order) is missing in our text here. For a further understanding of this disambiguation and also for possible explanations why it is missing, see Notes 20.1 and 20.3 on Page 20, and Notes 37.2 and 37.3 on Page 37.

[1.7] Following philosophical and translational traditions (cf. e.g. [KLEENE & FEFERMAN, 1986, p. 595:1a]), we have translated the German adjective "genetisch" literally as "genetic", instead of the translational alternative "generic". Note that the adjective "generisch" ("generic") was not part of the German language at the time of writing in 1934 according to [BASLER, 1934], and that here in our text there is obviously no intention of a relation to genetics as a field of modern science.

[1.8] This reference must not be interpreted as referring to any of the first two editions [HILBERT, 1899; 1903] of HILBERT's "Foundations of Geometry" where this Appendix VI is not included. The correct reference is [HILBERT, 1900a].

[1.9] Cf. [BERNAYS, 1950, p. 23]. With old-fashioned orthography also in [BERNAYS, 1976, p. 104].
"Thus, in mathematics, we have no reasons to assume any meaning of 'existence' that would be fundamentally different from that of 'the validity of axiomatic relations'."

[1.10] Cf. [BERNAYS, 1950, p. 23]. Also in [BERNAYS, 1976, p. 105].
"The thought of such a mathematical actuality, however, does not mean a return to the opinion that mathematical objects have an independent existence" [for instance, an existence as ideal things in themselves in the sense of [KANT, 1787], independent of an observer]. "This existence is not a there-being, but consists of structural relational connections and the (inductive) generation of intellectual objects from other intellectual objects."

Gegenstände der Theorie bloß als eine *Gattung* von Dingen[1] eingeführt werden, hat man es in einer axiomatischen Theorie mit einem festen System von Dingen (bzw. mehreren solchen Systemen) zu tun, welches einen von vornherein *abgegrenzten Bereich von Subjekten* für alle Prädikate bildet, aus denen sich die Aussagen der Theorie zusammensetzen.

In der Voraussetzung einer solchen Totalität des „Individuen-Bereiches" liegt — abgesehen von den trivialen Fällen, in denen eine Theorie es ohnehin nur mit einer endlichen, festbegrenzten Gesamtheit von Dingen zu tun hat — eine idealisierende Annahme, die zu den durch die Axiome formulierten Annahmen hinzutritt.

Charakteristisch ist für die verschärfte Form der Axiomatik, wie sie sich durch die Abstraktion vom Sachgehalt und durch die existentiale Fassung ergibt — wir wollen sie kurz die „formale Axiomatik" nennen —, daß sie einen *Nachweis der Widerspruchsfreiheit* erforderlich macht, während die inhaltliche Axiomatik ihre Grundbegriffe durch den Hinweis auf bekannte Erlebnisse einführt und ihre Grundsätze entweder als evidente Tatsachen hinstellt, die man sich klarmachen kann, oder sie als Extrakt von Erfahrungskomplexen formuliert und damit dem Glauben Ausdruck gibt, daß man Gesetzen der Natur auf die Spur gekommen ist, zugleich in der Absicht, diesen Glauben durch den Erfolg der Theorie zu stützen.

Auch die formale Axiomatik bedarf sowohl zur Verfolgung der Deduktionen wie für den Nachweis der Widerspruchsfreiheit jedenfalls gewisser Evidenzen, aber mit dem wesentlichen Unterschied, daß diese Art von Evidenz nicht auf einer besonderen Erkenntnisbeziehung zu dem jeweiligen Sachgebiet beruht, vielmehr für jedwede Axiomatik ein und dieselbe ist, nämlich diejenige primitive Erkenntnisweise, welche die Vorbedingung für jede exakte theoretische Forschung überhaupt bildet. Wir werden diese Art der Evidenz noch näher zu betrachten haben.

Für die richtige Würdigung des Verhältnisses von inhaltlicher und formaler Axiomatik in ihrer Bedeutung für die Erkenntnis sind vor allem folgende Gesichtspunkte zu beachten:

Die formale Axiomatik bedarf der inhaltlichen notwendig als ihrer Ergänzung, weil durch diese überhaupt erst die Anleitung zur Auswahl der Formalismen und ferner für eine vorhandene formale Theorie auch erst die Anweisung zu ihrer Anwendung auf ein Gebiet der Tatsächlichkeit gegeben wird.

Andrerseits können wir bei der inhaltlichen Axiomatik deshalb nicht stehenbleiben, weil wir es in der Wissenschaft, wenn nicht durchweg, so doch vorwiegend mit solchen Theorien zu tun haben, die gar nicht vollkommen den wirklichen Sachverhalt wiedergeben, sondern eine

---

[1] Von BROUWER und seiner Schule wird in diesem Sinne das Wort „species" gebraucht.

While the constructive method introduces the | objects of a theory only as a *genus*[2.1] of things,[1] an axiomatic theory refers to a fixed system of things (or several such systems), which is *delimited* from the outset and constitutes a *domain of subjects* for all predicates of the propositions of the theory.

There is the assumption that the "domain of individuals" is given as a whole. Except for the trivial cases where the theory deals only with a finite and fixed set of things, this is an idealizing assumption that properly augments the assumptions formulated in the axioms.

We will call this sharpened form of axiomatics (where the subject matter is ignored and the existential[2.2] form comes in) "formal axiomatics" for short. It is characteristic of formal axiomatics that it requires a *proof of consistency*. In contrast, contentual[2.3] axiomatics introduces its basic notions by referring to common experience and presents its first principles either as evident facts (of which you can convince yourself) or formulates them as extracts from experience-complexes. Thus, contentual mathematics conveys the belief that we have actually discovered laws of nature and intend to support this belief by the success of the theory.

Just as contentual axiomatics requires certain evidences,[2.4] in formal axiomatics the pursuit[2.5] of deductions and the proof of consistency require certain evidences as well. The fundamental difference, however, is that — in the case of formal axiomatics — these evidences cannot rely on a special epistemic relation to the specific subject area; rather, they are of one and the same kind for every axiomatics; namely of the primitive kind of knowledge that is the prerequisite for every exact theoretical investigation whatsoever. We will have to consider the evidences of this kind in more detail further on.

To see the relationship between contentual and formal axiomatics in their epistemological significance, first of all, the following has to be considered:

Formal axiomatics requires contentual axiomatics as a necessary supplement. It is only the latter that provides us with some guidance for choosing the right formalisms, and with some instructions on how to apply a given formal theory to a domain of actuality.[2.6]

---

[1] BROUWER and his school use the word "species" in this sense.

[2.1] We translate the German word "Gattung" here as "genus". This is consistent with the translation of the biological classification into taxonomic ranks (such as class, order, family, genus, species), from where the word "Gattung" of the German original and the word "species" of BROUWER in Note 1 of Page 2 seem to be taken. Moreover, it goes well together with the word "genetic" on Page 1.a, cf. Note 1.7. Finally, translations such as "class", "order", "family", or "type" are all misleading in the context of mathematical logic.

[2.2] Cf. the discussion in Note 1.6.

[2.3] The word "contentual" had not been part of the English language until recently. For instance, it is not listed in the most complete Webster's [GOVE, 1993]. According to [HEIJENOORT, 1971, p. viii], this neologism was introduced by STEFAN BAUER-MENGELBERG as a translation for the word "inhaltlich" in German texts on mathematics and logic, because there was no other way to reflect the special intentions of the HILBERT school when using this word. In January 2008, "contentual" scored 6350 Google hits, 5600 of which, however, contained neither the word "HILBERT" nor the word "BERNAYS". As these hits also included a pop song, "contentual" is likely to become an English word outside science in the near future. For comparison, there were 4 million Google hits for "contentious".

[2.4] The common usage of the word "evidence" does not exactly capture the meaning of the German noun "Evidenz" in German philosophy. "Self-evidence" is sometimes used as a translation, but this does not fit here. We try to convey the deviation of BERNAYS' usage from the common English one by keeping the plural of the German original.

[2.5] We have translated the phrase "Verfolgung der Deduktionen" as "pursuit of deductions", because it may not only refer to the checking of proofs and inference rules, but also to proof search.

[2.6] From the point of view of philosophy, the "actuality" seems to be a better translation of "Tatsächlichkeit" than "objective reality", which was used in [KANT, 1787, e.g. pp. 194, 269] with a contrary meaning (namely truth within our conceptions). Moreover, "actual reality" and "factuality" are awkward alternatives, which again miss the tendency toward noumena in the negative sense [KANT, 1787, p. 307f.].

vereinfachende Idealisierung des Sachverhaltes darstellen und darin ihre Bedeutung haben. Eine derartige Theorie kann gar nicht durch Berufung auf die evidente Wahrheit ihrer Axiome oder auf Erfahrung ihre Begründung erhalten, vielmehr kann diese Begründung nur in dem Sinne geschehen, daß die in der Theorie vollzogene Idealisierung, d. h. die Extrapolation, durch welche die Begriffsbildungen und Grundsätze der Theorie die Reichweite entweder der anschaulichen Evidenz oder der Erfahrungsdaten überschreitet, als eine widerspruchsfreie eingesehen wird. Für diese Erkenntnis der Widerspruchsfreiheit nützt uns auch die Berufung auf die approximative Gültigkeit der Grundsätze nichts, denn ein Widerspruch kann ja gerade dadurch zustande kommen, daß eine Beziehung als strikte gültig angenommen wird, die nur in eingeschränktem Sinne besteht.

Wir sind also genötigt, die Widerspruchsfreiheit von theoretischen Systemen losgelöst von der Betrachtung der Tatsächlichkeiten zu untersuchen, und damit befinden wir uns bereits auf dem Standpunkt der formalen Axiomatik.

Was nun die bisherige Behandlung dieses Problems betrifft, so geschieht diese sowohl bei der Geometrie wie bei den physikalischen Disziplinen durch die *Methode der Arithmetisierung:* Man repräsentiert die Gegenstände der Theorie durch Zahlen oder Zahlensysteme und die Grundbeziehungen durch Gleichungen und Ungleichungen derart, daß auf Grund dieser Übersetzung die Axiome der Theorie entweder in arithmetische Identitäten bzw. beweisbare Sätze übergehen, wie es bei der Geometrie der Fall ist, oder aber, wie bei der Physik, in ein System von Bedingungen, deren gemeinsame Erfüllbarkeit sich auf Grund arithmetischer Existenzsätze erweisen läßt. Bei diesem Verfahren wird die Arithmetik, d. h. die Theorie der reellen Zahlen (die Analysis) als gültig vorausgesetzt, und wir kommen so zu der Frage, welcher Art diese Geltung ist.

Ehe wir uns aber mit dieser Frage beschäftigen, wollen wir zusehen, ob es nicht eine direkte Art gibt, das Problem der Widerspruchsfreiheit in Angriff zu nehmen. Wir wollen uns überhaupt einmal die Struktur dieses Problems deutlich vor Augen führen. Zugleich wollen wir uns bei dieser Gelegenheit schon mit der *logischen Symbolik* etwas vertraut machen, die sich für den vorliegenden Zweck als sehr nützlich erweist und die wir im folgenden eingehender zu betrachten haben werden.

Als Beispiel einer Axiomatik nehmen wir die *Geometrie der Ebene*, und zwar mögen der Einfachheit halber nur die Axiome für die Geometrie der Lage (welche in HILBERTS „Grundlagen der Geometrie" als „Axiome der Verknüpfung" und „Axiome der Anordnung" aufgeführt werden) nebst dem Parallelen-Axiom in Betracht gezogen werden. Dabei empfiehlt es sich für unseren Zweck, von dem HILBERTschen Axiomensystem darin abzuweichen, daß wir nicht die Punkte und die Geraden

On the other hand, we cannot rest content with contentual axiomatics, because in science we are predominantly — if not always — concerned with theories that do not reproduce the actual state of affairs completely, but whose significance consists in a | *simplifying idealization* of the actual[3.1] state of affairs. This idealization consists in the extrapolation by which the concept formations and the principles of the theory transcend the realm of experience and intuitive self-evidence. Such a theory cannot receive its grounding by referring to experience or to the evident truth of its axioms. Rather, such a grounding can only be given by realizing the consistency of the idealization. Moreover, for the realization[3.2] of consistency, it does not help to refer to the approximate validity of the principles; namely because a contradiction can arise from the very assumption of the strict validity of a relation that actually holds only in a restricted sense.

We are therefore forced to investigate the consistency of theoretical systems without considering actuality, and thus we find ourselves already at the standpoint of formal axiomatics.

Now, one usually treats this problem — both in geometry and the disciplines of physics — with the *method of arithmetization*:[3.3] The objects of a theory are represented by numbers or systems of numbers and the basic relations by equations or inequations, such that, on the basis of this translation, the axioms of the theory turn out either as arithmetical identities or provable sentences (as in geometry), or as a system of conditions whose joint satisfiability can be demonstrated via arithmetical existence sentences (as in physics). This approach[3.4] presupposes the validity of arithmetic, i.e. the theory of real numbers (analysis). And so we come to ask ourselves what kind of validity[3.5] this is.

But before we concern ourselves with this question, let us look for the possibility of tackling the problem of consistency in a direct way. Additionally, we want to obtain a better idea of the structure of this problem in general. At the same time, we want to use this opportunity to familiarize ourselves with the *logical symbolism*, which is very useful for our present purpose, and which we will have to examine in more detail later on.

[3.1] Although the word "wirklich" ("real" or "actual") is not repeated in the original text, we have added the word "actual" to the canonical translation "state of affairs" for "Sachverhalt" here, for the following two reasons: First of all, the sentence does not make sense without it, because a state of affairs that is not actual (or at least somehow instantiated) cannot help us with our overall problem of consistency. Moreover, the definite article in the German original may be understood as referring back to the first occurrence of "Sachverhalt" in this sentence.

[3.2] The German original for the word "realization" is "Erkenntnis" here, whereas the original for the word "realizing" of the previous sentence ("eingesehen") has a different root. Taking "knowledge" instead of "realization" fails to mirror the dynamic aspect of the German original in the given context, also expressed by choosing "Erkenntnis" instead of "Kenntnis" ("knowledge"). Moreover, the word "seeing" instead of "realizing" fails to mirror the absence of sense perception in the meaning of the German verb "einsehen" in the given context. We have translated the two words of different roots as two words with an identical root because they are used synonymously here: Instead of "als eine widerspruchsfreie eingesehen", the German original could have been "als widerspruchsfrei erkannt" without any change in meaning.

[3.3] The method of arithmetization will be described more formally on Page 18f.

[3.4] Although the word "Verfahren" of the German original may well be translated as "procedure" here, we translate it here and at many other places as "approach": "procedure" would be hard to understand at some occurrences (such as on Page 20) and "approach" at others (such as on Pages 26 and 29).

[3.5] Although the etymologically closely related roots of "Geltung" (in the German original of this sentence) and "gültig" (or "Gültigkeit") (in the German original of the previous sentence) are not exactly the same, they are used by BERNAYS synonymously here (and in most other places). Therefore, we non-injectively translate both roots as "validity".

als zwei Systeme von Dingen zugrunde legen, sondern *nur die Punkte als Individuen nehmen*. An die Stelle der Beziehung „die Punkte x und y bestimmen die Gerade g" tritt dann eine Beziehung zwischen *drei* Punkten: „x, y, z liegen auf einer Geraden", für die wir die Bezeichnung $Gr(x, y, z)$ anwenden. Zu dieser Beziehung kommt als zweite Grundbeziehung die des Zwischenliegens: „x liegt zwischen y und z", die wir mit $Zw(x, y, z)$ bezeichnen[1]. Ferner tritt in den Axiomen, als ein zur Logik gehöriger Begriff die Identität von x mit y auf, für die wir das übliche Gleichheitszeichen $x = y$ anwenden.

Zur symbolischen Darstellung der Axiome brauchen wir nun noch die logischen Zeichen, und zwar erstens die Zeichen für Allgemeinheit und Existenz: Ist $P(x)$ ein auf das Ding x bezügliches Prädikat, so bedeutet $(x) P(x)$: „Alle x haben die Eigenschaft $P(x)$", und $(E x) P(x)$: „Es gibt ein x von der Eigenschaft $P(x)$". (x) heißt das „Allzeichen", $(E x)$ das „Seinszeichen". Das Allzeichen und das Seinszeichen kann ebenso wie auf x auch auf irgendeine andere Variable y, z, u bezogen sein. Die zu einem solchen Zeichen gehörige Variable wird durch dieses Zeichen „gebunden", entsprechend wie die Integrationsvariable durch das Integrationszeichen, so daß die Gesamtaussage nicht von einem Werte der Variablen abhängt.

Als weitere logische Zeichen kommen hinzu die Zeichen für die Negation und für die Satzverbindungen. Die Negation einer Aussage bezeichnen wir durch Überstreichen. Dabei soll im Falle eines in der Aussage voranstehenden Allzeichens oder Seinszeichens der Negationsstrich nur über dieses Zeichen gesetzt werden, und anstatt $\overline{x = y}$ werde kürzer „$x \neq y$" geschrieben. Das Zeichen & („und") zwischen zwei Aussagen bedeutet, daß beide Aussagen zutreffen („Konjunktion"). Das Zeichen V („oder" im Sinne von „vel") zwischen zwei Aussagen bedeutet, daß mindestens eine der beiden Aussagen zutrifft („Disjunktion").

Das Zeichen $\rightarrow$ zwischen zwei Aussagen bedeutet, daß das Zutreffen der ersten das Zutreffen der zweiten nach sich zieht, oder mit anderen Worten, daß die erste Aussage nicht zutrifft, ohne daß auch die zweite zutrifft („Implikation"). Eine Implikation $\mathfrak{A} \rightarrow \mathfrak{B}$ zwischen zwei Aussagen $\mathfrak{A}, \mathfrak{B}$ ist demnach nur dann falsch, wenn $\mathfrak{A}$ wahr und $\mathfrak{B}$ falsch ist; sonst ist sie wahr.

Die Verbindung des Zeichens der Implikation mit dem Allzeichen ergibt die Darstellung der allgemeinen hypothetischen Sätze. Z. B. stellt eine Formel

$$(x)(y)(\mathfrak{A}(x, y) \rightarrow \mathfrak{B}(x, y)),$$

---

[1] Das Verfahren, die Punkte allein als Individuen zu nehmen, ist insbesondere in der Axiomatik von OSWALD VEBLEN „A system of axioms for geometry" [Trans. Amer. Math. Soc. Bd. 5 (1904) S. 343—384] zur Durchführung gebracht. Hier werden überdies alle geometrischen Beziehungen mit Hilfe der „Zwischen"-Beziehung definiert.

$^{4.1}$As an example of an axiomatics, let us take *plane geometry*. For simplicity, we will take into account only the axioms for the geometry of position$^{4.2}$ (which are listed in HILBERT's "Foundations of Geometry" as the "axioms of connection" and the "axioms of order") plus the Axiom of Parallels.$^{4.3}$ For our present purpose, it is appropriate$^{4.4}$ to deviate from HILBERT's axiom system: We do not build upon points and straight lines | as two systems of things, but *take only points as individuals*. Then, instead of the relation "the points x and y determine the straight line g", we get a relation between *three* points:

$$\text{"}x, y, z \text{ lie on a straight line",}$$

for which we will use the designation

$$Gr(x, y, z).$$

In addition to this relation, we have the second basic relation of betweenness:[1]

$$\text{"}x \text{ lies between } y \text{ and } z\text{",}$$

which will be designated by

$$Zw(x, y, z).$$

Furthermore, the axioms contain the logical notion of identity, say of x and y, for which we will use the customary equality symbol $x = y$.

For the symbolic representation of the axioms, we now also need logical symbols; first of all, the symbols for universality and existence: If $P(x)$ is a predicate of the thing x, then

$(x) P(x)$ means   "All x have the property $P(x)$", and
$(Ex) P(x)$ "   "There is an x with the property $P(x)$".

(x) is called the "universal quantifier symbol", (Ex) the "existential quantifier symbol".$^{4.5}$ The universal and the existential quantifier symbols may refer to any other variable y, z, u instead of x. The variable belonging to a universal or existential quantifier symbol is "bound" by it (just as the integration variable is bound by the integration symbol), such that the meaning of the whole proposition does not depend on any value of the variable.$^{4.6}$

In addition, there are the logical symbols for the negation and for the sentential combinations.$^{4.7}$ The negation of a proposition is designated by putting a bar over it. If the proposition begins with a universal or existential quantifier symbol, the bar is to be put over this symbol only; and instead of $\overline{x=y}$, we will write the shorter "$x \neq y$". The symbol & ("and") between two propositions means that both propositions hold ("conjunction"). The symbol ∨ ("or" in the sense of "*vel*") between two propositions means that at least one of the two propositions holds ("disjunction").

The symbol → between two propositions means that if the first proposition holds, so does the second; or, put differently, the first proposition does not hold without the second holding as well ("implication"). Accordingly, an implication $\mathfrak{A} \to \mathfrak{B}$ between two propositions $\mathfrak{A}$ and $\mathfrak{B}$ is false only if $\mathfrak{A}$ is true and $\mathfrak{B}$ is false; otherwise it is true.

---

[1]The method of taking only points as individuals was put into practice in particular in the axiomatics of OSWALD VEBLEN in [VEBLEN, 1904]. Moreover, in VEBLEN's system, all geometrical relations are defined in terms of the relation "between".$^{4.8}$

als zwei Systeme von Dingen zugrunde legen, sondern *nur die Punkte als Individuen nehmen.* An die Stelle der Beziehung „die Punkte x und y bestimmen die Gerade g" tritt dann eine Beziehung zwischen *drei* Punkten: „x, y, z liegen auf einer Geraden", für die wir die Bezeichnung $Gr(x, y, z)$ anwenden. Zu dieser Beziehung kommt als zweite Grundbeziehung die des Zwischenliegens: „x liegt zwischen y und z", die wir mit $Zw(x, y, z)$ bezeichnen[1]. Ferner tritt in den Axiomen, als ein zur Logik gehöriger Begriff die Identität von x mit y auf, für die wir das übliche Gleichheitszeichen $x = y$ anwenden.

Zur symbolischen Darstellung der Axiome brauchen wir nun noch die logischen Zeichen, und zwar erstens die Zeichen für Allgemeinheit und Existenz: Ist $P(x)$ ein auf das Ding x bezügliches Prädikat, so bedeutet $(x)P(x)$: „Alle x haben die Eigenschaft $P(x)$", und $(Ex)P(x)$: „Es gibt ein x von der Eigenschaft $P(x)$". (x) heißt das „Allzeichen", (Ex) das „Seinszeichen". Das Allzeichen und das Seinszeichen kann ebenso wie auf x auch auf irgendeine andere Variable y, z, u bezogen sein. Die zu einem solchen Zeichen gehörige Variable wird durch dieses Zeichen „gebunden", entsprechend wie die Integrationsvariable durch das Integrationszeichen, so daß die Gesamtaussage nicht von einem Werte der Variablen abhängt.

Als weitere logische Zeichen kommen hinzu die Zeichen für die Negation und für die Satzverbindungen. Die Negation einer Aussage bezeichnen wir durch Überstreichen. Dabei soll im Falle eines in der Aussage voranstehenden Allzeichens oder Seinszeichens der Negationsstrich nur über dieses Zeichen gesetzt werden, und anstatt $\overline{x = y}$ werde kürzer „$x \neq y$" geschrieben. Das Zeichen & („und") zwischen zwei Aussagen bedeutet, daß beide Aussagen zutreffen („Konjunktion"). Das Zeichen V („oder" im Sinne von „vel") zwischen zwei Aussagen bedeutet, daß mindestens eine der beiden Aussagen zutrifft („Disjunktion").

Das Zeichen → zwischen zwei Aussagen bedeutet, daß das Zutreffen der ersten das Zutreffen der zweiten nach sich zieht, oder mit anderen Worten, daß die erste Aussage nicht zutrifft, ohne daß auch die zweite zutrifft („Implikation"). Eine Implikation $\mathfrak{A} \to \mathfrak{B}$ zwischen zwei Aussagen $\mathfrak{A}, \mathfrak{B}$ ist demnach nur dann falsch, wenn $\mathfrak{A}$ wahr und $\mathfrak{B}$ falsch ist; sonst ist sie wahr.

Die Verbindung des Zeichens der Implikation mit dem Allzeichen ergibt die Darstellung der allgemeinen hypothetischen Sätze. Z. B. stellt eine Formel

$$(x)(y)(\mathfrak{A}(x, y) \to \mathfrak{B}(x, y)),$$

---

[1] Das Verfahren, die Punkte allein als Individuen zu nehmen, ist insbesondere in der Axiomatik von OSWALD VEBLEN „A system of axioms for geometry" [Trans. Amer. Math. Soc. Bd. 5 (1904) S. 343—384] zur Durchführung gebracht. Hier werden überdies alle geometrischen Beziehungen mit Hilfe der „Zwischen"-Beziehung definiert.

[4.1] §1(a)(2) (as listed in the table of contents) starts with this paragraph.

[4.2] The term "geometry of position" ("Geometrie der Lage") is used here just *ad hoc*, namely to describe that geometry which excludes considerations on parallelism, congruence, and continuity. This term is not used in HILBERT's "Foundations of Geometry". Moreover, its associated meaning here is not very closely related to historical ones, such as the meaning of *"géométrie de position"* as given by CARNOT [1803] or the meaning of "Geometrie der Lage" as given by STAUDT [1847].

[4.3] The German original refers to [HILBERT, 1899]. We translate HILBERT's names of the axioms of connection and order ("Axiome der Verknüpfung und Anordnung") and the Axiom of Parallels ("Parallelenaxiom") in accordance with HILBERT's "Foundations of Geometry" [HILBERT, 1902], the first translation of [HILBERT, 1899]. These English names are identical to the names of these axioms used in the second translation [HILBERT, 1971], with the exception of the name of the axioms of connection, which are introduced as "axioms of incidence" in [HILBERT, 1971].

[4.4] Considering the very high quality of presentation in HILBERT's "Foundations of Geometry", it may be questioned whether the ad-hoc deviation from HILBERT's axiom system is really appropriate here for the present purpose. The poor quality of this deviation is evident: See Notes 5.5 and 12.2 for the lack of faithfulness of the axioms, and Notes 6.5 and 6.7 for the sloppiness of the technical terms.

[4.5] We translate "Seinszeichen" as "existential quantifier symbol", and "Allzeichen" as "universal quantifier symbol". It does not seem to be appropriate to omit the word "symbol" here, also because the German word "Quantor" ("quantifier") is not introduced in the German original until Page 98, cf. Note 98.2. Moreover, it is too awkward to write the literal translations "all-sign" and "being-sign" all the time; and there is no difference in meaning that would justify this. Note, however, that here, just as in many other places, we witness the struggle of that time with notions and notation all too familiar to us today.

[4.6] Note that — contrary to what one would expect after the previous sentence — it is not said here that the variable *symbol* can be exchanged. Rather, it is stated that the evaluation of the proposition "$(x)\,P(x)$" does not depend on the *value* of the variable x, as opposed to the evaluation of a proposition with a free occurrence of x, such as "$x-0$".

[4.7] The German original seems to treat all of the terms "Satzverbindung" ("sentential combination") (Page 4; twice in Note 1 on Page 47), "Aussagenverbindung" ("propositional combination") (Pages 45, 48 and 52), "Aussagen-Verknüpfung" ("propositional connective") (Note 1 on Page 5; Note 1 on Page 47) and "Aussagenverknüpfung" ("propositional connective") (Pages 47, 48 and 64) roughly as synonyms. Here and at other places we are tempted to translate "Verbindung" as "connective" or "connection". The usage of "Satzverbindungen" at its second occurrence in Note 1 on Page 47 and of "Aussagenverbindung" on Page 52, however, make clear that this would not be correct: There, it becomes clear that "Verbindung" actually means the whole sentence (say "$A \vee B$") and not just the connective ("$\vee$").

[4.8] Actually, in [VEBLEN, 1904], strictly speaking, we find the notion of "order" instead of "betweenness" and, *mutatis mutandis*, instead of "x lies between y and z", we find "points x, y, z are in the order yxz".

worin $\mathfrak{A}(x, y)$, $\mathfrak{B}(x, y)$ die Darstellungen gewisser Beziehungen zwischen x und y sind, den Satz dar: „Wenn $\mathfrak{A}(x, y)$ besteht, so besteht $\mathfrak{B}(x, y)$", oder auch: „für jedes Paar von Individuen x, y, für welches $\mathfrak{A}(x, y)$ besteht, besteht auch $\mathfrak{B}(x, y)$"[1].

Zur Zusammenfassung von Formelbestandteilen wenden wir in üblicher Weise Klammern an. Dabei soll zur Ersparung von Klammern festgesetzt werden, daß für die Trennung von symbolischen Ausdrücken $\rightarrow$ den Vorrang hat vor & und $\vee$, & vor $\vee$, und daß $\rightarrow$, &, $\vee$ alle den Vorrang haben vor den Allzeichen und den Seinszeichen. Wo keine Mehrdeutigkeit in Betracht kommt, lassen wir die Klammern weg, z. B. schreiben wir an Stelle des Ausdruckes

$$(x) ((E y) R(x, y)),$$

worin $R(x, y)$ irgendeine Beziehung zwischen x und y bezeichnet, einfach $(x)(E y) R(x, y)$, da hier nur die eine Lesart in Betracht kommt: „zu jedem x gibt es ein y, für welches die Beziehung $R(x, y)$ besteht." —

Nunmehr sind wir in der Lage, das betrachtete Axiomsystem in Formeln aufzuschreiben. Zur Erleichterung soll bei den ersten Axiomen die sprachliche Fassung hinzugefügt werden.

Die Abgrenzung der Axiome entspricht nicht völlig derjenigen in HILBERTs „Grundlagen der Goemetrie". Es soll deshalb bei jeder Axiomgruppe die Beziehung der hier in Formeln aufgestellten Axiome zu den HILBERTschen Axiomen angegeben werden[2].

I. Axiome der Verknüpfung.

1) $(x) (y) Gr(x, x, y)$

„x, x, y liegen stets auf einer Geraden."

2) $(x) (y) (z) (Gr(x, y, z) \rightarrow Gr(y, x, z)\ \&\ Gr(x, z, y))$.

„Wenn x, y, z auf einer Geraden liegen, so liegen stets auch y, x, z sowie auch x, z, y auf einer Geraden."

3) $(x) (y) (z) (u) (Gr(x, y, z)\ \&\ Gr(x, y, u)\ \&\ x \neq y \rightarrow Gr(x, z, u))$.

„Wenn $x, y,$ verschiedene Punkte sind und wenn x, y, z sowie x, y, u auf einer Geraden liegen, so liegen stets auch x, z, u auf einer Geraden."

4) $(E x) (E y) (E z) \overline{Gr(x, y, z)}$.

„Es gibt Punkte x, y, z, die nicht auf einer Geraden liegen."

Von diesen Axiomen treten 1) und 2) — auf Grund des geänderten Geraden-Begriffes — an die Stelle des Axioms I 1, 3) entspricht dem Axiom I 2, und 4) dem zweiten Teil des Axioms I 3.

---

[1] Von dem Verhältnis der hier definierten Disjunktion und Implikation zu den im üblichen Sinne disjunktiven und hypothetischen Aussagen-Verknüpfungen wird im § 3 noch die Rede sein.

[2] Diese Angaben sind speziell für die Kenner von HILBERTs „Grundlagen der Geometrie" bestimmt und beziehen sich auf die 7. Auflage.

The combination of the implication symbol with the universal quantifier symbol allows the representation of general hypothetical sentences. For example, the formula

$$(x)(y)\bigl(\,\mathfrak{A}(x,y)\ \rightarrow\ \mathfrak{B}(x,y)\,\bigr),$$

(where $\mathfrak{A}(x,y)$, $\mathfrak{B}(x,y)$ are representations of certain relations between x and y) represents the sentence: "If $\mathfrak{A}(x,y)$ holds then $\mathfrak{B}(x,y)$ holds as well", alternatively: "for every pair of individuals x,y for which $\mathfrak{A}(x,y)$ holds, $\mathfrak{B}(x,y)$ holds as well".[1]

We use parentheses in the customary way to group parts of formulas. To save parentheses, we stipulate for the separation of symbolic expressions[5.1] that $\rightarrow$ has lower operator precedence than $\&$ and $\vee$, that $\&$ has lower precedence than $\vee$,[5.2] and that all of $\rightarrow$, $\&$, $\vee$ have lower precedence than the universal and the existential quantifier symbols.[5.3] Wherever there is no ambiguity, we will omit parentheses. For example, instead of

$$(x)\bigl(\,(Ey)\,R(x,y)\,\bigr),$$

where $R(x,y)$ represents some relation between x and y, we simply write $(x)(Ey)\,R(x,y)$, since only one reading comes into question here: "for each x, there is a y, for which the relation $R(x,y)$ holds". —

Now we are in the position to write down the axiom system under consideration as formulas. To make it easier, we are going to add natural language versions to the first of these axioms.

The demarcation of the axioms here does not correspond entirely to that of Hilbert's "Foundations of Geometry".[5.4] Therefore, following each group of axioms presented as formulas here, we will note their relation to Hilbert's axioms.[2]

I. Axioms of Connection.

1. $(x)(y)\,Gr(x,x,y)$.
   "x,x,y always lie on one straight line."[5.5]

2. $(x)(y)(z)\bigl(\,Gr(x,y,z)\ \rightarrow\ Gr(y,x,z)\ \&\ Gr(x,z,y)\,\bigr)$.
   "Whenever x,y,z lie on a straight line, then so do y,x,z as well as x,z,y."

3. $(x)(y)(z)(u)\bigl(\,Gr(x,y,z)\ \&\ Gr(x,y,u)\ \&\ x\neq y\ \rightarrow\ Gr(x,z,u)\,\bigr)$.
   "For any x, y, z, u, if x,y are distinct points and x,y,z as well as x,y,u lie on straight lines, then x,z,u lie on a straight line as well."

4. $(Ex)(Ey)(Ez)\,\overline{Gr(x,y,z)}$.
   "There are points x, y, z that do not lie on a straight line."

Because of the changed notion of a straight line, Axioms 1 and 2 take the place of Axiom I 1,[5.6] Axiom 3 corresponds to Axiom I 2, and Axiom 4 to the second part[5.7] of Axiom I 3.

---

[1] We will return to the relationship between the above defined disjunction and implication to the disjunctive and hypothetical propositional connectives in the customary sense in § 3.[5.8]

[2] These notes are especially for experts on Hilbert's "Grundlagen der Geometrie" and relate to the 7th edition.[5.9]

worin $\mathfrak{A}(x, y)$, $\mathfrak{B}(x, y)$ die Darstellungen gewisser Beziehungen zwischen x und y sind, den Satz dar: „Wenn $\mathfrak{A}(x, y)$ besteht, so besteht $\mathfrak{B}(x, y)$", oder auch: „für jedes Paar von Individuen x, y, für welches $\mathfrak{A}(x, y)$ besteht, besteht auch $\mathfrak{B}(x, y)$"[1].

Zur Zusammenfassung von Formelbestandteilen wenden wir in üblicher Weise Klammern an. Dabei soll zur Ersparung von Klammern festgesetzt werden, daß für die Trennung von symbolischen Ausdrücken → den Vorrang hat vor & und V, & vor V, und daß →, &, V alle den Vorrang haben vor den Allzeichen und den Seinszeichen. Wo keine Mehrdeutigkeit in Betracht kommt, lassen wir die Klammern weg, z. B. schreiben wir an Stelle des Ausdruckes

$$(x)\,((E\,y)\,R(x, y)),$$

worin $R(x, y)$ irgendeine Beziehung zwischen x und y bezeichnet, einfach $(x)(E\,y)\,R(x, y)$, da hier nur die eine Lesart in Betracht kommt: „zu jedem x gibt es ein y, für welches die Beziehung $R(x, y)$ besteht." —

Nunmehr sind wir in der Lage, das betrachtete Axiomsystem in Formeln aufzuschreiben. Zur Erleichterung soll bei den ersten Axiomen die sprachliche Fassung hinzugefügt werden.

Die Abgrenzung der Axiome entspricht nicht völlig derjenigen in HILBERTs „Grundlagen der Goemetrie". Es soll deshalb bei jeder Axiomgruppe die Beziehung der hier in Formeln aufgestellten Axiome zu den HILBERTschen Axiomen angegeben werden[2].

I. Axiome der Verknüpfung.

1) $(x)\,(y)\,Gr(x, x, y)$

„x, x, y liegen stets auf einer Geraden."

2) $(x)\,(y)\,(z)\,(Gr(x, y, z) \rightarrow Gr(y, x, z)\,\&\,Gr(x, z, y))$.

„Wenn x, y, z auf einer Geraden liegen, so liegen stets auch y, x, z sowie auch x, z, y auf einer Geraden."

3) $(x)\,(y)\,(z)\,(u)\,(Gr(x, y, z)\,\&\,Gr(x, y, u)\,\&\,x \neq y \rightarrow Gr(x, z, u))$.

„Wenn x, y, verschiedene Punkte sind und wenn x, y, z sowie x, y, u auf einer Geraden liegen, so liegen stets auch x, z, u auf einer Geraden."

4) $(E\,x)\,(E\,y)\,(E\,z)\,\overline{Gr(x, y, z)}$.

„Es gibt Punkte x, y, z, die nicht auf einer Geraden liegen."

Von diesen Axiomen treten 1) und 2) — auf Grund des geänderten Geraden-Begriffes — an die Stelle des Axioms I 1, 3) entspricht dem Axiom I 2, und 4) dem zweiten Teil des Axioms I 3.

---

[1] Von dem Verhältnis der hier definierten Disjunktion und Implikation zu den im üblichen Sinne disjunktiven und hypothetischen Aussagen-Verknüpfungen wird im § 3 noch die Rede sein.

[2] Diese Angaben sind speziell für die Kenner von HILBERTs „Grundlagen der Geometrie" bestimmt und beziehen sich auf die 7. Auflage.

$^{5.1}$We cannot translate the technical term "Ausdruck" ("expression") as "term" because it is less restrictive than "Term" ("term"), which occurs in the same contexts with restrictive intentions, such as "not a formula" (cf. [HILBERT & BERNAYS, 1968, p. 187]) or "not containing free occurrences of bound variables" (cf. the distinction between "ι-Ausdruck" and "ι-Term" [HILBERT & BERNAYS, 1968, p. 398] and between "ε-Ausdruck" and "ε-Term" [HILBERT & BERNAYS, 1970, p. 24]). The only alternative translation of "Ausdruck" seems to be "quasi term", which is less appropriate in our treatment here because we start with the notion of an "Ausdruck" and come to that of a "Term" not before Page 187.

$^{5.2}$Note that the operator precedence presented here shows only a single inversion relative to the modern first-order standard, namely that "$\vee$" has higher precedence here than "&". According to the modern standard in first-order logic as well as in programming languages such as C [KERNIGHAN & RITCHIE, 1977], the precedence of operators decreases in the following order:

$$\text{postfix operators,} \quad \begin{pmatrix} \text{prefix operators} \\ (x) \\ (Ex) \end{pmatrix}, \quad \begin{pmatrix} \text{infix operators} \\ =, \quad \&, \quad \vee, \quad \sim, \quad \rightarrow \end{pmatrix}.$$

The precedence of the German original, however, is the following (for "$\sim$" cf. p. 107):

$$(x_0, \ldots, x_\mathfrak{k}), \quad \begin{pmatrix} (x) \\ (Ex) \end{pmatrix}, \quad =, \quad \vee, \quad \&, \quad \begin{pmatrix} \sim \\ \rightarrow \end{pmatrix}.$$

The reader does not have to remember that difference, however, because we will add disambiguating parentheses to all formulas in the translation in which this difference becomes relevant.

$^{5.3}$Contrary to our translation as "to have lower precedence than", "Vorrang haben vor" literally means to "to take precedence over" or "to have higher precedence than". The latter translational alternative, however, would contradict today's standard usage of the term "operator precedence" too blatantly. The historical point of view of the German original is to be understood in terms of precedence of *separation* ("Trennung") of the parts of a formula; whereas today's usage in computer science, mathematics, and linguistics refers to precedence of operator *binding* or to precedence of handle *pruning* in parsing.

$^{5.4}$The typo "Goemetrie" for "Geometrie" appeared already in the 1st edition of 1934 and passed unnoticed into the 2nd edition of 1968.

$^{5.5}$Axiom 1 in this form is actually problematic because it is not a faithful modeling of HILBERT's "Foundations of Geometry" [HILBERT, 1971]. Cf. Note 5.7.

$^{5.6}$In [HILBERT, 1971, Chapter I, § 2] ([HILBERT, 1962]), the *axioms of connection* of HILBERT's "Foundations of Geometry" read:

I 1. "For every two points A, B there exists a line that contains each of the points A, B." ("Zu zwei Punkten A, B gibt es stets eine Gerade a, die mit jedem der beiden Punkte A, B zusammengehört.")

I 2. "For every two points A, B there exists no more than one line that contains each of the points A, B." ("Zu zwei Punkten A, B gibt es *nicht mehr als* eine Gerade, die mit jedem der beiden Punkte A, B zusammengehört.")

I 3. "There exist at least two points on a line." [I.e. on *every* line.] "There exist at least three points that do not lie on a line." ("Auf einer Geraden gibt es stets wenigstens zwei Punkte. Es gibt wenigstens drei Punkte, die nicht auf einer Geraden liegen.")

$^{5.7}$Note that the first part of Axiom I 3 of HILBERT's "Foundations of Geometry" (cf. Note 5.6) is disrespected by BERNAYS' Axiom 1 here, because it generates one-point lines in the sense of $(x) Gr(x,x,x)$. For a more faithful translation, Axiom 1 would have to be changed to the following axiom:

1'. $(x)(y)(\ x \neq y \quad \sim \quad Gr(x,x,y)\ )$.

Then, instead of the meaningless proposition $Gr(x,x,x)$, its negation would become derivable, which is clearly to be preferred for the modeling of HILBERT's "Foundations of Geometry" [HILBERT, 1971]. To achieve consistency, we then also have to weaken Axiom 3 slightly as follows:

3'. $(x)(y)(z)(u)(\ Gr(x,y,z) \ \& \ Gr(x,y,u) \ \& \ x \neq y \ \& \ z \neq u \quad \rightarrow \quad Gr(x,z,u)\ )$.

This weakening avoids the consequence of $Gr(x,z,u)$ in the now fatal case of $x = z = u$. Similarly, we have to weaken the Axiom of Parallels in Sharper Form on Page 6.a as follows:

$$(x)(y)(z)\left(\ \left\{ \begin{array}{c} \overline{Gr(x,y,z)} \\ \& \quad x \neq y \end{array} \right\} \quad \rightarrow \quad (Eu)\left\{ \begin{array}{l} Par(x,y;z,u) \\ \& \quad (v)(\ Par(x,y;z,v) \rightarrow Gr(z,u,v)\ ) \end{array} \right\}\ \right).$$

$^{5.8}$Cf. Page 47, especially Note 1.

$^{5.9}$Cf. [HILBERT, 1930a]. For an English translation we have to refer to the second translation [HILBERT, 1971]. The first translation [HILBERT, 1902] is not helpful here.

II. Axiome der Anordnung.

1) $(x)(y)(z)(Zw(x,y,z) \to Gr(x,y,z))$
2) $(x)(y)\overline{Zw(x,y,y)}$.
3) $(x)(y)(z)(Zw(x,y,z) \to Zw(x,z,y) \& \overline{Zw(y,x,z)})$.
4) $(x)(y)(x \neq y \to (Ez)Zw(x,y,z))$.

„Wenn x und y verschiedene Punkte sind, so gibt es stets einen Punkt z derart, daß x zwischen y und z liegt."

5) $(x)(y)(z)(u)(v)(\overline{Gr(x,y,z)} \& Zw(u,x,y) \& \overline{Gr(v,x,y)}$
   $\& \overline{Gr(z,u,v)}) \to (Ew)\{Gr(u,v,w) \& Zw(w,x,z) \lor Zw(w,y,z)\})$.

1) und 2) bilden zusammen den ersten Teil des HILBERTschen Axioms II 1; 3) vereinigt den letzten Teil des HILBERTschen Axioms II 1 mit II 3; 4) ist das Axiom II 2, und 5) das Axiom der ebenen Anordnung II 4.

III. Parallelen-Axiom.

Da wir die Kongruenz-Axiome beiseite lassen, so müssen wir hier dem Parallelen-Axiom die erweiterte Fassung geben: „Zu jeder Geraden gibt es durch einen ausserhalb gelegenen Punkt stets eine und nur eine sie nicht schneidende Gerade[1]."

Zur Erleichterung der symbolischen Formulierung werde das Symbol

$$Par(x,y;u,v)$$

als Abkürzung angewendet für den Ausdruck:

$$\overline{(Ew)}\,(Gr(x,y,w) \& Gr(u,v,w))$$

„Es gibt keinen Punkt w, der sowohl mit x und y wie mit u und v auf einer Geraden liegt".

Das Axiom lautet dann:

$$(x)(y)(z)(\overline{Gr(x,y,z)} \to (Eu)\{Par(x,y;z,u) \& (v)(Par(x,y;z,v) \to Gr(z,u,v))\}).$$

Denken wir uns die aufgezählten Axiome der Reihe nach durch & verbunden, so erhalten wir eine einzige logische Formel, welche eine Aussage über die Prädikate Gr, Zw darstellt und die wir mit

$$\mathfrak{A}(Gr, Zw)$$

bezeichnen wollen.

In entsprechender Weise können wir einen Lehrsatz der ebenen Geometrie, welcher nur die Lagen- und Anordnungsbeziehungen betrifft, durch eine Formel

$$\mathfrak{S}(Gr, Zw)$$

darstellen.

Diese Darstellung entspricht aber noch der inhaltlichen Axiomatik, bei welcher die Grundbeziehungen als etwas in der Erfahrung oder in

---

[1] Vgl. in HILBERTS Grundlagen der Geometrie S. 83.

II. Axioms of Order.

1. $(x)(y)(z)\bigl(\ Zw(x,y,z)\ \to\ Gr(x,y,z)\ \bigr).$
2. $(x)(y)\ \overline{Zw(x,y,y)}.$
3. $(x)(y)(z)\bigl(\ Zw(x,y,z)\ \to\ Zw(x,z,y)\ \&\ \overline{Zw(y,x,z)}\ \bigr).$
4. $(x)(y)\bigl(\ x\neq y\ \to\ (Ez)\,Zw(x,y,z)\ \bigr).$

   "If x and y are distinct points, then in any case there is a point z such that x lies between y and z."

5. $(x)(y)(z)(u)(v)\left(\left\{\begin{array}{l}\overline{Gr(x,y,z)}\\ \&\ Zw(u,x,y)\\ \&\ \overline{Gr(v,x,y)}\\ \&\ Gr(z,u,v)\end{array}\right\}\ \to\ (Ew)\left\{\begin{array}{l}Gr(u,v,w)\\ \&\ \begin{pmatrix}Zw(w,x,z)\\ \vee\ Zw(w,y,z)\end{pmatrix}\end{array}\right\}\right).$ [6.1]

Axioms 1 and 2 together constitute the first part of HILBERT's Axiom II 1;[6.2] Axiom 3 combines the last part of HILBERT's Axiom II 1 with Axiom II 3; Axiom 4 is HILBERT's Axiom II 2; Axiom 5 is HILBERT's Axiom II 4 (Plane Axiom of Order).[6.3]

III. Axiom of Parallels.

Since we omit the congruence axioms, we have to take the Axiom of Parallels in its sharper form:[6.4] "For every straight line, through every point outside of it, there is one and only one non-intersecting straight line."[1]

To simplify the symbolic formulation, the symbol
$$Par(x,y;u,v)$$
will be used as an abbreviation for:
$$\overline{(Ew)}\bigl(\ Gr(x,y,w)\ \&\ Gr(u,v,w)\ \bigr)$$

"There is no point w that is on a straight line with x and y as well as with u and v."

The axiom then reads:
$$(x)(y)(z)\left(\ \overline{Gr(x,y,z)}\ \to\ (Eu)\left\{\begin{array}{l}Par(x,y;z,u)\\ \&\ (v)\bigl(\ Par(x,y;z,v)\ \to\ Gr(z,u,v)\ \bigr)\end{array}\right\}\right)$$

If we consider all the axioms listed here to be consecutively conjoined with &, we obtain a single formula representing a proposition about the predicates Gr, Zw. This formula will be designated by
$$\mathfrak{A}(Gr,Zw).$$

Similarly, we can represent any theorem of plane geometry that is only about the relations of position and order[6.5] as a formula
$$\mathfrak{S}(Gr,Zw).$$

Yet, this representation still conforms to contentual axiomatics. And in contentual axiomatics, the basic relations are viewed as something to be found in our experience or | intuitive conception, and thus as something contentually[6.6] determined, about which the sentences of the theory contain assertions.

---

[1] Cf. HILBERT's "Grundlagen der Geometrie", p. 83.[6.7]

II. Axiome der Anordnung.

1) $(x)\,(y)\,(z)\,(Zw(x,y,z) \to Gr(x,y,z))$
2) $(x)\,(y)\,\overline{Zw(x,y,y)}$.
3) $(x)\,(y)\,(z)\,(Zw(x,y,z) \to Zw(x,z,y)\,\&\,\overline{Zw(y,x,z)})$.
4) $(x)\,(y)\,(x \neq y \to (Ez)\,Zw(x,y,z))$.

„Wenn x und y verschiedene Punkte sind, so gibt es stets einen Punkt z derart, daß x zwischen y und z liegt."

5) $(x)\,(y)\,(z)\,(u)\,(v)\,\bigl(\overline{Gr(x,y,z)}\,\&\,Zw(u,x,y)\,\&\,\overline{Gr(v,x,y)}$
$\&\,\overline{Gr(z,u,v)} \to (Ew)\,\{Gr(u,v,w)\,\&\,Zw(w,x,z)\,\vee\,Zw(w,y,z)\}\bigr)$.

1) und 2) bilden zusammen den ersten Teil des HILBERTschen Axioms II 1; 3) vereinigt den letzten Teil des HILBERTschen Axioms II 1 mit II 3; 4) ist das Axiom II 2, und 5) das Axiom der ebenen Anordnung II 4.

III. Parallelen-Axiom.

Da wir die Kongruenz-Axiome beiseite lassen, so müssen wir hier dem Parallelen-Axiom die erweiterte Fassung geben: „Zu jeder Geraden gibt es durch einen ausserhalb gelegenen Punkt stets eine und nur eine sie nicht schneidende Gerade[1]."

Zur Erleichterung der symbolischen Formulierung werde das Symbol

$$Par(x,y;u,v)$$

als Abkürzung angewendet für den Ausdruck:

$$\overline{(Ew)}\,(Gr(x,y,w)\,\&\,Gr(u,v,w))$$

„Es gibt keinen Punkt w, der sowohl mit x und y wie mit u und v auf einer Geraden liegt".

Das Axiom lautet dann:

$$(x)(y)(z)\bigl(\overline{Gr(x,y,z)} \to (Eu)\{Par(x,y;z,u)\,\&\,(v)(Par(x,y;z,v) \to Gr(z,u,v))\}\bigr).$$

Denken wir uns die aufgezählten Axiome der Reihe nach durch & verbunden, so erhalten wir eine einzige logische Formel, welche eine Aussage über die Prädikate Gr, Zw darstellt und die wir mit

$$\mathfrak{A}\,(Gr, Zw)$$

bezeichnen wollen.

In entsprechender Weise können wir einen Lehrsatz der ebenen Geometrie, welcher nur die Lagen- und Anordnungsbeziehungen betrifft, durch eine Formel

$$\mathfrak{S}\,(Gr, Zw)$$

darstellen.

Diese Darstellung entspricht aber noch der inhaltlichen Axiomatik, bei welcher die Grundbeziehungen als etwas in der Erfahrung oder in

---

[1] Vgl. in HILBERTS Grundlagen der Geometrie S. 83.

6.1 The alternatives of $\begin{pmatrix} Zw(w,x,z) \\ \vee \ Zw(w,y,z) \end{pmatrix}$ occurring in Axiom 5 can be illustrated with the following diagram where we have $Zw(w_1,x,z)$ and $Zw(w_2,y,z)$:

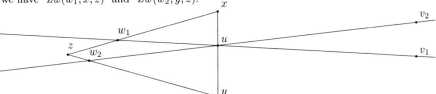

Note that xyz is the triangle that determines the plane to which HILBERT's Plane Axiom of Order refers; cf. Notes 6.2 and 6.3.

6.2 In [HILBERT, 1971, Chapter I, § 3], the *axioms of order* of HILBERT's "Foundations of Geometry" read:

II 1. "If a point B lies between a point A and a point C then the points A, B, C are three distinct points of a line, and B then also lies between C and A."

II 2. "For two points A and C, there always exists at least one point B on the line AC such that C lies between A and B."

II 3. "Of any three points on a line there exists no more than one that lies between the other two."

II 4. *(HILBERT's "Plane Axiom of Order")*
"Let A, B, C be three points that do not lie on a line and let a be a line in the plane ABC which does not meet any of the points A, B, C. If the line a passes through a point of the segment AB, it also passes through a point of the segment AC, or through a point of the segment BC."
See Note 6.1 for an illustration, where A, B, C are named x, y, z, and where the line a is either the line uv_1 or the line uv_2.

6.3 We translate the name of the Plane Axiom of Order ("Axiom der ebenen Anordnung") in accordance with Axiom II 5 of [HILBERT, 1902] as well as Axiom II 4 of [HILBERT, 1971], both occurring in Chapter I, § 3, of the respective publication.

6.4 We translate the name of the Axiom of Parallels in Sharper Form in accordance with [HILBERT, 1971]. The Axiom of Parallels in Sharper Form is contained neither in the *first* English translation [HILBERT, 1902] of [HILBERT, 1899], nor in the later editions of it, still in print in 2008. It is contained, however, in the *second* English translation [HILBERT, 1971, Chapter V, § 22, p. 72]:

IV*. *(HILBERT's "Axiom of Parallels in Sharper Form")*
"Let a be any line and A a point off a. Then there exists in the plane determined by a and A *one and only one* line that passes through A and does not intersect a."

Cf. also Note 6.7.

6.5 The usage of the German noun "Lehrsatz" ("theorem") (instead of simply "Satz" ("sentence")) points to a mathematical content here (beside mere syntax). In mathematical terms, "theorem of plane geometry that is only about the relations of position and order" ("Lehrsatz der ebenen Geometrie, welcher nur die Lagen- und Anordnungsbeziehungen betrifft,") is an *ad hoc* term (just as explained in Note 4.2), which has to mean something like "theorem of plane geometry that depends neither on congruence nor on continuity". This, however, is not consistent with the actual wording, which — in the terminology of HILBERT's "Foundations of Geometry" — fails to indicate the inclusion of any axiom of parallels. Therefore, it is much more likely that BERNAYS just wanted to express here that all sentences involving only the relation of being positioned on a line and being ordered by betweenness can be *expressed* in the first-order language of Gr and Zw.

6.6 Note that native German speakers may easily overlook the common word "inhaltlich" ("contentually") here and may understand something like "definite in content" instead of "contentually determined". In the English translation, however, the rare word "contentually" stands out more prominently.

6.7 For the correct English reference see Note 6.4. The German original refers to HILBERT's "Grundlagen der Geometrie" [HILBERT, 1899], but the reference is correct only with regard to the later editions. The actually intended reference of the German original when literally referring to HILBERT's "Grundlagen der Geometrie" [HILBERT, 1899] is always to the 7th edition [HILBERT, 1930a], where the axiom occurs for the first time in the long sequence of editions of [HILBERT, 1899]. There, and in all following editions, the Axiom of Parallels in Sharper Form appears on page 83 in Chapter V, § 22, with the label IV*, but it is actually called "Parallelenaxiom in schärferer Fassung" ("axiom of parallels in sharper form") instead of the "erweiterte Fassung" ("extended form") of our German original here.

anschaulicher Vorstellung Aufweisbares und somit inhaltlich Bestimmtes angesehen werden, worüber die Sätze der Theorie Behauptungen enthalten.

In der formalen Axiomatik dagegen werden die Grundbeziehungen nicht als von vornherein inhaltlich bestimmt angenommen, vielmehr erhalten sie erst implizite ihre Bestimmung durch die Axiome; und es wird auch in allen Überlegungen einer axiomatischen Theorie nur dasjenige von den Grundbeziehungen benutzt, was in den Axiomen ausdrücklich formuliert ist.

Wenn somit in der axiomatischen Geometrie die der anschaulichen Geometrie entsprechenden Beziehungsnamen wie „liegen auf", „zwischen" für die Grundbeziehungen gebraucht werden, so geschieht das nur als eine Konzession an das Gewohnte und um die Anknüpfung der Theorie an die anschaulichen Tatsachen zu erleichtern. In Wahrheit aber haben für die formale Axiomatik die Grundbeziehungen die Rolle von *variablen* Prädikaten.

Dabei verstehen wir „Prädikat" hier sowie im folgenden stets in dem weiteren Sinne, daß auch Prädikate mit zwei oder mehreren Subjekten inbegriffen sind. Je nach der Anzahl der Subjekte sprechen wir von „einstelligen", „zweistelligen" ... Prädikaten.

In dem von uns betrachteten Teil der axiomatischen Geometrie handelt es sich um zwei variable dreistellige Prädikate

$$R(x, y, z), \quad S(x, y, z).$$

Das Axiomensystem besteht in einer Anforderung an zwei solche Prädikate, welche sich ausdrückt durch die logische Formel $\mathfrak{A}(R, S)$, die wir aus $\mathfrak{A}(Gr, Zw)$ erhalten, indem wir $Gr(x, y, z)$ durch $R(x, y, z)$, $Zw(x, y, z)$ durch $S(x, y, z)$ ersetzen. In dieser Formel tritt neben den variablen Prädikaten noch die inhaltlich zu deutende Identitätsbeziehung $x = y$ auf. Daß wir diese in inhaltlicher Bestimmtheit zulassen, ist nicht etwa ein Verstoß gegen unseren methodischen Standpunkt. Denn die Inhaltsbestimmung der Identität — die im eigentlichen Sinne überhaupt gar keine Beziehung ist — wird ja nicht dem besonderen Vorstellungskreis des axiomatisch zu untersuchenden Sachgebietes entnommen, sondern betrifft lediglich die Sonderung der Individuen, die mit der Zugrundelegung eines Individuenbereiches jedenfalls als gegeben angenommen werden muß.

Einem Satz von der Form $\mathfrak{S}(Gr, Zw)$ entspricht im Sinne dieser Auffassung die Feststellung logischen Inhalts, daß für *irgendwelche* Prädikate $R(x, y, z)$, $S(x, y, z)$, die der Anforderung $\mathfrak{A}(R, S)$ genügen, auch die Beziehung $\mathfrak{S}(R, S)$ besteht, daß also für irgend zwei Prädikate $R(x, y, z)$, $S(x, y, z)$ die Formel

$$\mathfrak{A}(R, S) \rightarrow \mathfrak{S}(R, S)$$

eine wahre Aussage darstellt. Ein geometrischer Satz wird auf diese Weise transformiert in einen Satz der reinen Prädikaten-Logik.

$^{7.1}$In formal axiomatics, however, the basic relations are not conceived to be contentually determined from the outset; rather, they receive their determination only *implicitly*$^{7.2}$ through the axioms. And any considerations within an axiomatic theory may make use only of those aspects of the basic relations that are explicitly formulated in the axioms.

Thus, in axiomatic geometry, whenever we use names that correspond to intuitive geometry — such as "lie on" or "lie between" — this is just a concession to custom, and to ease the connection of the theory with the intuitive facts. Actually, however, in formal axiomatics, the basic relations play the rôle of *variable* predicates.

Here and in what follows, we understand "predicate" in the broader sense that includes predicates with two or more subjects as well. According to the number of subjects we refer to them as "singulary",$^{7.3}$ "binary", "ternary", ... or simply as "n-ary" predicates.$^{7.4}$

The part of axiomatic geometry we considered so far contains two variable ternary predicates
$$R(x,y,z), \quad S(x,y,z).$$

The axiom system consists of a requirement on two such predicates, which is expressed in the logical formula $\mathfrak{A}(R,S)$, obtained from $\mathfrak{A}(Gr,Zw)$ by replacing $Gr(x,y,z)$ with $R(x,y,z)$, and $Zw(x,y,z)$ with $S(x,y,z)$. In addition to the variable predicates, this formula contains the identity relation $x=y$, which is to be interpreted contentually.$^{7.5}$ The fact that we accept this predicate to be contentually$^{7.5}$ determined is no violation of our methodological standpoint; namely because the contentual determination of identity — which is no relation in the proper sense anyway$^{7.6}$ — does not depend on any particular conception in the field of our axiomatic investigation. Rather, it concerns only the separation of individuals, which has to be taken for granted in any case if a domain of individuals is taken as a basis.$^{7.7}$

According to this view, a sentence of the form $\mathfrak{S}(Gr,Zw)$ corresponds to a statement of the following logical content: For *arbitrary* predicates $R(x,y,z)$, $S(x,y,z)$ satisfying the condition $\mathfrak{A}(R,S)$, the relation $\mathfrak{S}(R,S)$ holds as well, i.e., for any two predicates $R(x,y,z)$, $S(x,y,z)$, the formula
$$\mathfrak{A}(R,S) \quad \rightarrow \quad \mathfrak{S}(R,S)$$
is a true proposition. In this way, a geometrical sentence is transformed into a sentence of pure predicate logic.

$^{7.1}$§ 1(a)(3) (as listed in the table of contents) starts on this page, probably with this paragraph.

$^{7.2}$There are two modes of emphasizing text in the German original, italics and stretching. We see no pattern in their use, so in translation we use italics for both.

$^{7.3}$Following [QUINE, 1981, p.13, Note 1], we prefer the correct word "singulary" for "1-ary" to the following alternatives occasionally found in the literature: "unary", "uninary", "unitary", "one-place", "monadic".

$^{7.4}$The following texts were added in the translation: ", 'ternary', ", "or simply as 'n-ary'".

$^{7.5}$Actually, the word "semantically" may be a better translation of the German "inhaltlich" here than the word "contentually". As it would be problematic in many respects to translate a German word that is most crucial in the given context with different English words at different places, we use the word "contentually" here as well.

$^{7.6}$Why equality is rather a logical notion than a semantical relation is explained in [TARSKI, 1986].

$^{7.7}$Note that the German original text does not implicitly assume that we have to make a choice among the possible domains of individuals, nor even that we need any domain of individuals as a basis at all, but just that for any domain of individuals that might be considered as a basis, a notion of equality is part and parcel of the conception of such a domain.

§ 1. Das Problem der Widerspruchsfreiheit in der Axiomatik.

Ganz entsprechend stellt sich von diesem Standpunkt auch die Frage der Widerspruchsfreiheit als ein Problem der reinen Prädikaten-Logik dar. Nämlich es handelt sich darum, ob zwei dreistellige Prädikate $R(x, y, z)$, $S(x, y, z)$ den in der Formel $\mathfrak{A}(R, S)$ zusammengefaßten Bedingungen genügen können[1] oder ob im Gegenteil die Annahme, daß die Formel $\mathfrak{A}(R, S)$ für ein gewisses Prädikatenpaar erfüllt ist, zu einem Widerspruch führt, so daß also allgemein für jedes Prädikatenpaar R, S die Formel $\overline{\mathfrak{A}(R, S)}$ eine richtige Aussage darstellt.

Eine solche Frage wie die hier vorliegende fällt unter das „*Entscheidungsproblem*". Hierunter versteht man in der neueren Logik das Problem der Auffindung allgemeiner Methoden zur Entscheidung über die „Allgemeingültigkeit" bzw. über die „Erfüllbarkeit" logischer Formeln[2].

Dabei sind die zu untersuchenden Formeln solche, die aus Prädikaten-Variablen und Gleichungen — nebst den an den Subjektstellen stehenden Variablen, welche wir als „Individuen-Variablen" bezeichnen — mit Hilfe der logischen Zeichen zusammengesetzt sind, wobei jede der Individuen-Variablen durch ein Allzeichen oder ein Seinszeichen gebunden ist.

Eine Formel dieser Art heißt allgemeingültig, wenn sie für *jede* Bestimmung der variablen Prädikate eine wahre Aussage darstellt, sie heißt erfüllbar, wenn sie bei *geeigneter* Bestimmung der variablen Prädikate eine wahre Aussage darstellt.

Einfache Beispiele von allgemeingültigen Formeln sind folgende:

$$(x) F(x) \,\&\, (x) G(x) \rightarrow (x)(F(x) \,\&\, G(x))$$
$$(x) P(x, x) \rightarrow (x)(E y) P(x, y)$$
$$(x)(y)(z)(P(x, y) \,\&\, y = z \rightarrow P(x, z)).$$

Beispiele erfüllbarer Formeln sind:

$$(E x) F(x) \,\&\, (E x) \overline{F(x)}$$
$$(x)(y)(P(x, y) \,\&\, P(y, x) \rightarrow x = y)$$
$$(x)(E y) P(x, y) \,\&\, (E y)(x) \overline{P(x, y)}.$$

Diese ergeben z. B. für den Individuenbereich der Zahlen 1, 2 wahre Aussagen, wenn in der ersten Formel für $F(x)$ das Prädikat „x ist geradzahlig", in der zweiten Formel für $P(x, y)$ das Prädikat $x \leq y$, in der dritten Formel für $P(x, y)$ das Prädikat $x \leq y \,\&\, y \neq 1$ gesetzt wird.

Zu beachten ist, daß man zugleich mit der Bestimmung der Prädikate auch den *Individuenbereich* festzulegen hat, auf den sich die Varia-

---

[1] Diese hier noch unscharfe Form der Fragestellung wird später verschärft werden.

[2] Diese Erklärung trifft allerdings nur für das Entscheidungsproblem in engerer Bedeutung zu. Auf die weitere Fassung des Entscheidungsproblems brauchen wir in diesem Zusammenhang nicht einzugehen.

| Accordingly, from this standpoint, the issue of consistency presents itself as a problem of pure predicate logic. Namely, it is the following:[1] Can two ternary predicates $R(x,y,z)$, $S(x,y,z)$ satisfy the conditions expressed in the formula $\mathfrak{A}(R,S)$? Or does, on the contrary, the assumption that a certain pair of predicates satisfies $\mathfrak{A}(R,S)$ lead to a contradiction (which means that, in general, for every pair of predicates R, S, the formula $\overline{\mathfrak{A}(R,S)}$ represents a correct proposition) ?

[8.1] A question such as the one presented here is subsumed by the *"Entscheidungsproblem"*. This problem is understood in the more recent logic as the problem of finding general methods for deciding "validity"[8.2] or "satisfiability" of logical formulas.[2]

The formulas to be investigated in that context are composed with the help of the logical symbols out of predicate variables and equations[8.3] — together with variables in subject positions, which we call "individual variables"[8.4] — and it is assumed that every individual variable is bound by a universal or existential quantifier symbol.

A formula of this kind is called valid if it represents a true proposition for *every* determination[8.5] of the variable predicates; it is called satisfiable if it represents a true proposition for *some suitable* determination[8.5] of the variable predicates.

Simple examples of valid formulas are the following:

$$(x)\, F(x) \;\&\; (x)\, G(x) \quad \to \quad (x)\big(\, F(x) \;\&\; G(x)\, \big)$$

$$(x)\, P(x,x) \quad \to \quad (x)(Ey)\, P(x,y)$$

$$(x)(y)(z)\big(\, P(x,y) \;\&\; y=z \quad \to \quad P(x,z)\, \big)$$

Examples of satisfiable formulas are:

$$(Ex)\, F(x) \;\&\; (Ex)\, \overline{F(x)}$$

$$(x)(y)\big(\, P(x,y) \;\&\; P(y,x) \quad \to \quad x=y\, \big)$$

$$(x)(Ey)\, P(x,y) \;\&\; (Ey)(x)\, \overline{P(x,y)}$$

These formulas result in true propositions for the domain of individuals consisting of the numbers 1 and 2, e.g., if

- in the first formula we take "x is even" for $F(x)$,
- in the second formula we take $x \leq y$ for $P(x,y)$, and
- in the third formula we take $x \leq y \;\&\; y \neq 1$ for $P(x,y)$.

---

[1] This loose formulation of the question will be sharpened in what follows.

[2] This explanation, however, only applies to the Entscheidungsproblem in the narrower sense.[8.6] We need not address the Entscheidungsproblem in the broader sense in this context.

[8.1] § 1(b) (as listed in the table of contents) starts on this page, probably with this paragraph. In [HILBERT & BERNAYS, 2001], however, the beginning of § 1(b) is indicated one paragraph later.

[8.2] We translate both "Allgemeingültigkeit" (which has its first occurrence in the German original here, beside the table of contents) and "Gültigkeit" as "validity". Cf. Note 9.1.

[8.3] Actually, instead of "Gleichungen" ("equations"), the German original should have "dem Gleichheitszeichen" ("the equality symbol") (cf. p. 4.a) here.

[8.4] We translate "Individuen-Variable" briefly with the standard term "individual variable", instead of the clearer alternative "variable for an individual". Be aware that an individual variable is a variable ranging over the domain of individuals, but not necessarily a variable to which the adjective "individual" applies.

[8.5] The German word "Bestimmung" belongs to traditional mathematical language; so we translate it as "determination" instead of "assignment" or "substitution", which belong to the terminology of modern logic.

[8.6] In 1934, the year of publication of the 1st ed. of this Vol. I, this Entscheidungsproblem in the narrower sense ("Entscheidungsproblem in engerer Bedeutung") (i.e. the decision problem for first-order logic), defined in [HILBERT & ACKERMANN, 1928], had not yet received its negative answer of non-co-semi-decidability of validity given by CHURCH [1936] and TURING [1936/7].

blen $x, y, \ldots$ beziehen sollen. Dieser geht gewissermaßen als *versteckte Variable* in die logische Formel ein. Allerdings verhält sich die logische Formel in bezug auf ihre Erfüllbarkeit invariant gegenüber einer umkehrbar eindeutigen Abbildung des Individuenbereiches auf einen anderen, da ja die Individuen nur als variable Subjekte in den Formeln auftreten, und somit ist die einzige wesentliche Bestimmung des Individuenbereiches die *Anzahl der Individuen*.

Wir haben demnach betreffs der Allgemeingültigkeit und Erfüllbarkeit folgende Fragen zu unterscheiden:

1. Die Frage nach der Allgemeingültigkeit für *jeden* Individuenbereich bzw. der Erfüllbarkeit für *irgendeinen* Individuenbereich.

2. Die Frage nach der Allgemeingültigkeit bzw. Erfüllbarkeit bei gegebener Anzahl der Individuen.

3. Die Frage, für welche Anzahlen von Individuen Allgemeingültigkeit bzw. Erfüllbarkeit besteht.

Bemerkt sei, daß man gut tut, die Individuenzahl 0 grundsätzlich auszuschließen, da die 0-zahligen Individuenbereiche formal eine Sonderstellung einnehmen und da andererseits ihre Betrachtung trivial und für die Anwendungen wertlos ist[1].

Des weiteren hat man zu beachten, daß es bei der Bestimmung eines Prädikates nur auf seinen „Wertverlauf" ankommt, das heißt darauf, für welche Werte der (an den Subjektstellen auftretenden) Variablen das Prädikat zutrifft bzw. nicht zutrifft („wahr" bzw. „falsch" ist).

Dieser Umstand hat zur Folge, daß für eine *gegebene endliche* Individuenzahl die Allgemeingültigkeit bzw. die Erfüllbarkeit einer vorgelegten logischen Formel einen rein *kombinatorischen Sachverhalt* darstellt, den man durch elementares Durchprobieren aller Fälle feststellen kann.

Ist nämlich n die Anzahl der Individuen und k die Anzahl der Subjekte („Stellen") eines Prädikates, so ist n^k die Anzahl der verschiedenen Wertsysteme der Variablen; und da für jedes dieser Wertsysteme das Prädikat entweder wahr oder falsch ist, so gibt es

$$2^{(n^k)}$$

verschiedene mögliche Wertverläufe für ein k-stelliges Prädikat.

---

[1] Die Festsetzung, daß jeder Individuenbereich mindestens ein Ding enthalten soll, so daß also ein wahres allgemeines Urteil für mindestens ein Ding zutreffen muß, darf nicht verwechselt werden mit der in der ARISTOTELischen Logik herrschenden Konvention, wonach ein Urteil von der Form „alle S sind P" nur als wahr gilt, wenn überhaupt Dinge von der Eigenschaft S vorhanden sind. Diese Konvention wird in der neueren Logik fallen gelassen. Ein Urteil von jener Art stellt sich symbolisch dar in der Form $(x)(S(x) \rightarrow P(x))$ und gilt als wahr, wenn ein Ding x, sofern es die Eigenschaft $S(x)$ besitzt, auch stets die Eigenschaft $P(x)$ besitzt — unabhängig davon, ob es überhaupt Dinge von der Eigenschaft $S(x)$ gibt. Wir werden hierauf beim deduktiven Aufbau der Prädikatenlogik nochmals zu sprechen kommen. (Siehe § 4 S. 105—106.)

Note that simultaneously with the determination of the predicates, the *domain of individuals* (over which the | variables $x, y, \ldots$ range) has to be fixed as well. This domain of individuals enters the logical formula as a kind of *hidden variable*. The satisfiability of a logical formula is invariant, however, with respect to a one-to-one mapping of the domain of individuals onto another, because the individuals enter into the formulas only as variable subjects. Thus, the only essential determination for a domain of individuals is the *number of the individuals*.

Therefore, we have to distinguish the following questions concerning validity and satisfiability:

1. The question of validity for *every* domain of individuals (or of satisfiability for *some arbitrary* domain of individuals).
2. The question of validity[9.1] (or satisfiability) for a given number of the individuals.
3. The question: For which numbers of the individuals is a formula valid (or satisfiable)?

[9.2]Note that we had better exclude domains of 0 individuals on principle, because of their special formal status and, moreover, because their consideration is trivial and worthless for applications.[1]

Furthermore, note that the determination of a predicate depends only on its "graph",[9.3] i.e., it is just relevant for which values of the variables appearing in the subject positions the predicate holds or does not hold (i.e., is "true" or "false").

This circumstance has the consequence that, for a *given finite* domain of individuals, validity (or satisfiability) of a certain logical formula turns out to be a purely *combinatorial state of affairs*, which can be verified by an elementary checking of all cases.

That is to say that if n is the number of individuals and k the number of subjects of a predicate ("arity"), then n^k is the number of the different systems of values of the variables; and there are

$$2^{(n^k)}$$

possible graphs for a k-ary predicate, because the predicate is either true or false for each of these systems. |

---

[1]The requirement that every domain of individuals should contain at least one thing (which means that a true general judgment must hold for at least one thing) must not be confused with the convention prevalent in ARISTOTELIAN logic, where a judgment of the form "all S are P" is considered true only if things with the property S are given at all. This convention is dropped in the more recent logic. A judgment of this kind is represented symbolically in the form $(x)(S(x) \to P(x))$, and it is taken to be true if the following holds: If a thing x has the property $S(x)$, then, in any case, it has the property $P(x)$ as well — no matter whether there are things with the property $S(x)$ at all. We will return to this in the context of the deductive construction of predicate logic. (Cf. § 4, pp. 105–106.)

[9.1]This occurrence of "Allgemeingültigkeit" in the German original should be translated as "validity", but neither as "general validity" nor as "logical validity". As a consequence we have decided to translate "Allgemeingültigkeit" always as "validity". Cf. Note 8.2.

[9.2]§ 1(b)(2) (as listed in the table of contents) starts on this page, probably with this paragraph. In [HILBERT & BERNAYS, 2001], however, the beginning of § 1(b)(2) is indicated two paragraphs later.

[9.3]The term "Wertverlauf" originates in GOTTLOB FREGE (1848–1925); cf. Note 87.2 on Page 87. Nowadays, "graph" is the standard translation for the German word "Wertverlauf" with the given meaning here. The alternative, more literal translation "course of values", may be misleading here; see the discussion on Page 32 of the introduction to the French translation [HILBERT & BERNAYS, 2001], which emphasizes literal correctness of translation, but where nevertheless the French word *"graphe"* is chosen instead of the more literal translation *"parcours des valeurs"*. In some English translations of the writings of FREGE, however, we also find "course-of-values"; cf. e.g. [FREGE, 1964a].

§ 1. Das Problem der Widerspruchsfreiheit in der Axiomatik.

Sind also
$$R_1, \ldots, R_t$$
die in einer vorgelegten Formel vorkommenden verschiedenen Prädikaten-Variablen,
$$k_1, \ldots, k_t$$
ihre Stellenzahlen, so ist
$$2^{(n^{k_1} + n^{k_2} + \cdots + n^{k_t})}$$
die Anzahl der in Betracht kommenden Systeme von Wertverläufen, oder, wie wir kurz sagen wollen, die Anzahl der verschiedenen möglichen Prädikatensysteme.

Hiernach bedeutet die Allgemeingültigkeit der Formel, daß für alle diese
$$2^{(n^{k_1} + \cdots + n^{k_t})}$$
explizite aufzählbaren Prädikatensysteme die Formel eine wahre Aussage darstellt, und ihre Erfüllbarkeit bedeutet, daß für eines unter diesen Prädikatensystemen die Formel eine wahre Aussage darstellt; dabei ist für ein festes Prädikatensystem die Wahrheit oder Falschheit der durch die Formel dargestellten Aussage wiederum durch ein endliches Ausprobieren entscheidbar, da ja für die mit Allzeichen oder Seinszeichen verbundenen Variablen nur n Werte in Betracht kommen, so daß das „alle" gleichbedeutend ist mit einer n-gliedrigen Konjunktion, das „es gibt" gleichbedeutend mit einer n-gliedrigen Disjunktion.

Nehmen wir als Beispiel die beiden vorher genannten Formeln
$$(x)\,P(x, x) \rightarrow (x)(E\,y)\,P(x, y)$$
$$(x)(y)(P(x, y) \,\&\, P(y, x) \rightarrow x = y),$$
von denen die erste als allgemeingültige, die zweite als erfüllbare Formel angeführt wurde, und beziehen wir diese Formeln auf einen zweizahligen Individuenbereich.

Die beiden Individuen können wir durch die Ziffern 1, 2 bezeichnen. Wir haben hier $t = 1$, $n = 2$, $k_1 = 2$, also ist die Anzahl der verschiedenen Prädikatensysteme
$$2^{(2^2)} = 2^4 = 16.$$
An Stelle von $(x)\,P(x, x)$ können wir setzen
$$P(1, 1) \,\&\, P(2, 2),$$
an Stelle von $(x)(E\,y)\,P(x, y)$
$$P(1, 1) \vee P(1, 2) \,\&\, P(2, 1) \vee P(2, 2),$$
so daß die erste der beiden Formeln übergeht in
$$P(1, 1) \,\&\, P(2, 2) \;\rightarrow\; P(1, 1) \vee P(1, 2) \,\&\, P(2, 1) \vee P(2, 2).$$

| Thus, if
$$R_1, \ldots R_t$$
are the distinct predicate variables in a given formula, with arities
$$k_1, \ldots k_t,$$
respectively, then
$$2^{(n^{k_1}+n^{k_2}+\cdots+n^{k_t})}$$
is the number of the systems of graphs to be considered, or — as we will say for short — it is the number of the distinct possible *predicate systems*.

Accordingly, validity of a formula means that, for all
$$2^{(n^{k_1}+\cdots+n^{k_t})}$$
explicitly enumerable predicate systems, the formula represents a true proposition. And satisfiability means that for at least one of these predicate systems the formula represents a true proposition. And in this context, the truth or falsity of a proposition represented by a formula for a given fixed predicate system is again decidable by a finite checking of cases. The reason is that only n values need to be considered for a variable bound by a universal or existential quantifier symbol, such that the "for all" is equivalent to an n-element conjunction and the "there is" is equivalent to an n-element disjunction.

For example, let us take the formulas
$$(x)\,P(x,x) \quad \to \quad (x)(Ey)\,P(x,y)$$
$$(x)(y)(\ P(x,y)\ \&\ P(y,x) \quad \to \quad x=y\ )$$
of which the first was mentioned before as a valid and the second as a satisfiable formula. Let us assume a domain of two individuals for these formulas.

The two individuals can be designated by the numerals 1 and 2. We have here $t=1$, $n=2$, $k_1=2$. Thus, the number of distinct predicate systems is
$$2^{(2^2)} = 2^4 = 16.$$

In place of $(x)\,P(x,x)$ we can put
$$P(1,1)\ \&\ P(2,2),$$
and in place of $(x)(Ey)\,P(x,y)$ we can put
$$\left(\begin{array}{c}P(1,1)\\ \vee\ P(1,2)\end{array}\right)\ \&\ \left(\begin{array}{c}P(2,1)\\ \vee\ P(2,2)\end{array}\right),$$
whereby the first of the two formulas is turned into[10.1]
$$P(1,1)\ \&\ P(2,2) \quad \to \quad \left(\begin{array}{c}P(1,1)\\ \vee\ P(1,2)\end{array}\right)\ \&\ \left(\begin{array}{c}P(2,1)\\ \vee\ P(2,2)\end{array}\right).$$ |

[10.1] This procedure is called HERBRAND *expansion* today, after JACQUES HERBRAND (1908–1931), cf. e.g. [WIRTH &AL., 2009]. It goes back, however, already to the first elaborate description of first-order logic in [PEIRCE, 1885], following the invention of quantifiers in [FREGE, 1879]. (Note that quantification does not require quantifiers, neither in ancient logics, nor in the formal logic of [WIRTH, 2011].)

Diese Implikation ist nun für diejenigen Prädikate P wahr, für welche
$$P(1,1) \& P(2,2)$$
falsch, sowie auch für diejenigen, bei welchen
$$P(1,1) \lor P(1,2) \& P(2,1) \lor P(2,2)$$
wahr ist. Man kann nun verifizieren, daß bei jedem der 16 Wertverläufe, die man erhält, indem man jedem der 4 Wertepaare
$$(1,1), (1,2), (2,1), (2,2)$$
je einen der Wahrheitswerte „wahr", „falsch" zuordnet, eine von jenen beiden Bedingungen erfüllt ist, so daß jedesmal die ganze Aussage den Wert „wahr" erhält. [Die Verifikation vereinfacht sich bei diesem Beispiel dadurch, daß zur Feststellung der Richtigkeit der Aussage bereits die Bestimmung der Werte von $P(1,1)$ und $P(2,2)$ genügt.] Auf diese Weise läßt sich die Allgemeingültigkeit unserer ersten Formel für zweizahlige Individuenbereiche durch direktes Ausprobieren feststellen.

Die zweite der genannten Formeln ist für zweizahlige Individuenbereiche gleichbedeutend mit der Konjunktion

$$(P(1,1) \& P(1,1) \to 1=1) \& (P(2,2) \& P(2,2) \to 2=2)$$
$$\& (P(1,2) \& P(2,1) \to 1=2) \& (P(2,1) \& P(1,2) \to 2=1).$$

Da $1=1$ und $2=2$ wahr ist, so sind die beiden ersten Konjunktionsglieder stets wahre Aussagen; die beiden letzten Glieder sind dann und nur dann wahr, wenn
$$P(1,2) \& P(2,1)$$
falsch ist.

Zur Erfüllung der betrachteten Formel hat man also nur diejenigen Wertbestimmungen von P auszuschließen, bei welchen die Paare $(1,2)$ und $(2,1)$ beide mit dem Werte „wahr" versehen sind. Jede andere Wertbestimmung liefert eine wahre Aussage. Die Formel ist also für einen zweizahligen Individuenbereich erfüllbar.

Diese Beispiele sollen uns den rein kombinatorischen Charakter verdeutlichen, den das Entscheidungsproblem im Falle einer gegebenen endlichen Anzahl von Individuen besitzt. Aus diesem kombinatorischen Charakter ergibt sich insbesondere, daß für eine vorgeschriebene endliche Anzahl von Individuen die Allgemeingültigkeit einer Formel $\mathfrak{F}$ gleichbedeutend ist mit der Unerfüllbarkeit der Formel $\overline{\mathfrak{F}}$ und die Erfüllbarkeit der Formel $\mathfrak{F}$ gleichbedeutend damit, daß die Formel $\overline{\mathfrak{F}}$ nicht allgemeingültig ist. In der Tat stellt ja $\mathfrak{F}$ für diejenigen Prädikatensysteme eine richtige Aussage dar, für welche $\overline{\mathfrak{F}}$ eine falsche Aussage darstellt und umgekehrt.

Wenden wir uns nun zu unserer Frage der Widerspruchsfreiheit eines Axiomensystems zurück[1]. Denken wir uns das Axiomensystem,

---
[1] Vgl. S. 2f.

This implication is now true for those predicates P for which

$$P(1,1) \ \& \ P(2,2)$$

is false, as well as for those for which

$$\left(\begin{array}{c} P(1,1) \\ \vee \ P(1,2) \end{array} \right) \ \& \ \left(\begin{array}{c} P(2,1) \\ \vee \ P(2,2) \end{array} \right)$$

is true. It can be verified now that one of these two conditions is satisfied for each of the 16 graphs obtained by assigning to each of the 4 pairs of values

$$(1,1), \quad (1,2), \quad (2,1), \quad (2,2)$$

one of the truth values "true" or "false". Thus, the whole proposition receives the value "true" every time. [The verification can be simplified in this example, because the determination of truth values for $P(1,1)$ and $P(2,2)$ already suffices to verify the correctness of the proposition.] In this way we can verify the validity of our first formula for domains with two elements by directly trying it out.

For domains of two individuals the second of the given formulas has the same meaning as the conjunction

$$\left(\begin{array}{cc} P(1,1) \ \& \ P(1,1) & \to \ 1=1 \\ P(1,2) \ \& \ P(2,1) & \to \ 1=2 \end{array} \right) \ \& \ \left(\begin{array}{cc} P(2,2) \ \& \ P(2,2) & \to \ 2=2 \\ P(2,1) \ \& \ P(1,2) & \to \ 2=1 \end{array} \right).$$

Since $1=1$ and $2=2$ are true, the first two elements of the conjunction are true in any case; the last two elements of the conjunction are true if and only if

$$P(1,2) \ \& \ P(2,1)$$

is false.

Thus, to satisfy the considered formula, we only have to eliminate those determinations of values of P in which both of the pairs $(1,2)$ and $(2,1)$ get the value "true". Any other determination of values yields a true proposition. Thus, the formula is satisfiable in a domain of two individuals.

These examples should serve to clarify the purely combinatorial character of the Entscheidungsproblem in case of a given finite number of individuals. One result of this combinatorial character is that, for a given finite number of individuals, the validity of a formula $\mathfrak{F}$ is equivalent to the unsatisfiability of the formula $\overline{\mathfrak{F}}$. And likewise the satisfiability of a formula $\mathfrak{F}$ is equivalent to the formula $\overline{\mathfrak{F}}$ not being valid. Indeed, $\mathfrak{F}$ represents a true proposition for those predicate systems for which $\overline{\mathfrak{F}}$ represents a false proposition and vice versa.[11.1]

[11.2]Let us now return to our question of the consistency of an axiom system.[1] And let us assume the axiom system | to be written down symbolically and combined into one single formula as in the example we have considered.

---

[1]Cf. p. 2f.[11.3]

[11.1]The last two negation bars over "$\mathfrak{F}$" are present in the 1st edition of 1934, but lacking in the German original of the 2nd edition of 1968, probably because the reprinting technique was having problems with very thin lines. Our copy, however, was corrected anonymously and previously to our copying process. Moreover, these typos are in the official list of typos (delivered with [HILBERT & BERNAYS, 1970]).

[11.2]§1(b)(3) (as listed in the table of contents) starts with this paragraph.

[11.3]Footnote added in the 2nd edition of 1968.

§ 1. Das Problem der Widerspruchsfreiheit in der Axiomatik.

wie in dem betrachteten Beispiel, symbolisch aufgeschrieben und in eine Formel zusammengefaßt.

Die Frage der Erfüllbarkeit dieser Formel läßt sich dann für eine vorgeschriebene endliche Anzahl von Individuen, wenigstens grundsätzlich, durch Ausprobieren zur Entscheidung bringen. Angenommen nun, es sei für eine bestimmte endliche Anzahl von Individuen die Erfüllbarkeit der Formel festgestellt; dann erhalten wir dadurch einen Nachweis für die Widerspruchsfreiheit des Axiomensystems, und zwar nach der *Methode der Aufweisung*, indem der endliche Individuenbereich zusammen mit den (zur Erfüllung der Formel) gewählten Wertverläufen der Prädikate ein Modell bildet, an dem wir das Erfülltsein der Axiome konkret aufzeigen können.

Es sei ein Beispiel einer solchen Aufweisung aus der geometrischen Axiomatik vorgebracht. Wir gehen aus von dem anfangs aufgestellten Axiomensystem. Hierin ersetzen wir das Axiom I 4), welches die Existenz dreier nicht auf einer Geraden liegenden Punkte fordert, durch das schwächere Axiom

I 4') $(Ex)(Ey)(x \neq y)$.

„Es gibt zwei verschiedene Punkte."

Ferner lassen wir das Axiom der ebenen Anordnung II 5) weg, nehmen aber dafür zwei Sätze, die mit Hilfe von II 5) beweisbar sind, unter die Axiome auf[1] indem wir erstens II 4) erweitern zu

II 4') $(x)(y)\{x \neq y \rightarrow (Ez)Zw(z, x, y) \& (Ez)Zw(x, y, z)\}$

und zweitens hinzufügen

II 5') $(x)(y)(z)\{x \neq y \& x \neq z \& y \neq z$
$\rightarrow Zw(x, y, z) \vee Zw(y, z, x) \vee Zw(z, x, y)\}$.

Das Parallelen-Axiom behalten wir bei. Das so entstehende Axiomensystem, dem an Stelle der früheren Formel $\mathfrak{A}(R, S)$ jetzt eine Formel $\mathfrak{A}'(R, S)$ entspricht, läßt sich, wie O. VEBLEN bemerkt hat[2], mit einem Individuenbereich von 5 Dingen erfüllen. Die Wahl der Wertverläufe für die Prädikate R, S — wir können hier ohne Gefahr eines Mißverständnisses wieder die Bezeichnungen „Gr", „Zw" gebrauchen — geschieht so, daß zunächst das Prädikat Gr so bestimmt wird, daß es für jedes Werte-Tripel x, y, z wahr ist. Dann sind, wie man sofort sieht, alle Axiome I, ferner II 1) und III erfüllt. Damit die Axiome II 2), 3), 5'), 4') erfüllt werden, ist es notwendig und auch hinreichend, an das Prädikat Zw folgende drei Forderungen zu stellen:

---

[1] Diese beiden Sätze wurden in den früheren Auflagen von HILBERTS „Grundlagen der Geometrie" als Axiome aufgeführt. Es erwies sich, daß sie mittels des Axioms der ebenen Anordnung beweisbar sind. Vgl. hierzu 7. Auflage, S. 5—6.

[2] In der bereits erwähnten Untersuchung „A system of axioms for geometry", Trans. Amer. Math. Soc. Bd. 5, S. 350.

Then, for a given finite number of individuals, the question of the satisfiability of this formula can be decided by trying it out, at least in principle. Suppose now that the satisfiability of a formula has been verified for a given finite number of individuals. Then, we thereby obtain a proof of the consistency of the axiom system, namely by the *method of exhibition*,[12.1] according to which the finite domain of individuals together with the graphs chosen for the predicates (to satisfy the formula) constitutes a model in which we can concretely point out that the axioms are satisfied.

Let us give an example of such an exhibition from geometric axiomatics. We start from the axiom system presented at the beginning.[12.2] We replace Axiom I 4,[12.3] which postulates the existence of three points not lying on a straight line, with the weaker axiom

I 4'. $(Ex)(Ey)\,(x \neq y)$.
"There are two distinct points."

Furthermore, we drop the Plane Axiom of Order (i.e. Axiom II 5), but add two sentences[1] instead, which are provable from Axiom II 5: First, we extend Axiom II 4 to

II 4'. $(x)(y)\bigl(\ x \neq y \quad \rightarrow \quad (Ez)\,Zw(z,x,y) \ \&\ (Ez)\,Zw(x,y,z)\ \bigr)$.

Second, we add

II 5'. $(x)(y)(z)\left(\left(\begin{array}{c} x \neq y \\ \&\ \ x \neq z \\ \&\ \ y \neq z \end{array}\right) \rightarrow \left(\begin{array}{c} Zw(x,y,z) \\ \vee\ Zw(y,z,x) \\ \vee\ Zw(z,x,y) \end{array}\right)\right)$. [12.4]

We keep the Axiom of Parallels unchanged. The resulting axiom system (in which the formula $\mathfrak{A}'(R,S)$ now takes the place of the previous formula $\mathfrak{A}(R,S)$) is satisfiable in a domain of 5 individuals, as was noticed by OSWALD VEBLEN.[2] The graphs for the predicates R, S — here we can use the notation "Gr", "Zw" without any danger of misunderstanding — are chosen in a such a way that, first of all, the predicate Gr is determined to be true for every value triple x, y, z. It is then obvious that all Axioms I, as well as Axiom II 1 and Axiom III are satisfied. To satisfy Axioms II 2, II 3, II 5', and II 4', it is necessary and also sufficient to impose the following three conditions on the predicate Zw:

---

[1] Both these sentences were listed in the earlier editions of HILBERT's "Grundlagen der Geometrie" as axioms.[12.5] It turned out that they are provable using the Plane Axiom of Order. Cf. the 7th edition,[12.6] pp. 5–6.

[2] On page 350 of the previously mentioned[12.7] investigation [VEBLEN, 1904].

wie in dem betrachteten Beispiel, symbolisch aufgeschrieben und in eine Formel zusammengefaßt.

Die Frage der Erfüllbarkeit dieser Formel läßt sich dann für eine vorgeschriebene endliche Anzahl von Individuen, wenigstens grundsätzlich, durch Ausprobieren zur Entscheidung bringen. Angenommen nun, es sei für eine bestimmte endliche Anzahl von Individuen die Erfüllbarkeit der Formel festgestellt; dann erhalten wir dadurch einen Nachweis für die Widerspruchsfreiheit des Axiomensystems, und zwar nach der *Methode der Aufweisung*, indem der endliche Individuenbereich zusammen mit den (zur Erfüllung der Formel) gewählten Wertverläufen der Prädikate ein Modell bildet, an dem wir das Erfülltsein der Axiome konkret aufzeigen können.

Es sei ein Beispiel einer solchen Aufweisung aus der geometrischen Axiomatik vorgebracht. Wir gehen aus von dem anfangs aufgestellten Axiomensystem. Hierin ersetzen wir das Axiom I 4), welches die Existenz dreier nicht auf einer Geraden liegenden Punkte fordert, durch das schwächere Axiom

I 4') $(E x) (E y) (x \neq y)$.

„Es gibt zwei verschiedene Punkte."

Ferner lassen wir das Axiom der ebenen Anordnung II 5) weg, nehmen aber dafür zwei Sätze, die mit Hilfe von II 5) beweisbar sind, unter die Axiome auf[1] indem wir erstens II 4) erweitern zu

II 4') $(x)(y)\{x \neq y \rightarrow (Ez) Zw(z, x, y) \& (Ez) Zw(x, y, z)\}$

und zweitens hinzufügen

II 5') $(x)(y)(z)\{x \neq y \& x \neq z \& y \neq z$
$\rightarrow Zw(x, y, z) \lor Zw(y, z, x) \lor Zw(z, x, y)\}$.

Das Parallelen-Axiom behalten wir bei. Das so entstehende Axiomensystem, dem an Stelle der früheren Formel $\mathfrak{A}(R, S)$ jetzt eine Formel $\mathfrak{A}'(R, S)$ entspricht, läßt sich, wie O. VEBLEN bemerkt hat[2], mit einem Individuenbereich von 5 Dingen erfüllen. Die Wahl der Wertverläufe für die Prädikate R, S — wir können hier ohne Gefahr eines Mißverständnisses wieder die Bezeichnungen „Gr", „Zw" gebrauchen — geschieht so, daß zunächst das Prädikat Gr so bestimmt wird, daß es für jedes Werte-Tripel x, y, z wahr ist. Dann sind, wie man sofort sieht, alle Axiome I, ferner II 1) und III erfüllt. Damit die Axiome II 2), 3), 5'), 4') erfüllt werden, ist es notwendig und auch hinreichend, an das Prädikat Zw folgende drei Forderungen zu stellen:

---

[1] Diese beiden Sätze wurden in den früheren Auflagen von HILBERTS „Grundlagen der Geometrie" als Axiome aufgeführt. Es erwies sich, daß sie mittels des Axioms der ebenen Anordnung beweisbar sind. Vgl. hierzu 7. Auflage, S. 5—6.

[2] In der bereits erwähnten Untersuchung „A system of axioms for geometry", Trans. Amer. Math. Soc. Bd. 5, S. 350.

$^{12.1}$We translate the term "Methode der Aufweisung" as "method of exhibition". The alternative translation "method of demonstration" would fail to mirror the emphasis on actuality of the rarely used German word "Aufweisung". Cf. also Note 15.1.

$^{12.2}$This axiom system (found on Pages 5.a and 6.a), however, does not have a finite model. The necessity of infinity is mainly caused by Axiom I 4 on Page 5.a, the Plane Axiom of Order (i.e. Axiom II 5 on p. 6.a), and the Axiom of Parallels in Sharper Form (i.e. Axiom III on p. 6.a). Therefore, the idea is to satisfy the latter two axioms trivially, namely by restricting all points to be positioned on one single straight line. Then the premises of Axioms II 5 and III cannot be satisfied. Thus, these axioms become vacuously true. For positioning all points on a single straight line, however, Axiom I 4 has to be dropped. Moreover, we have to think this single straight line to be somehow non-linear, because otherwise Axiom II 4 on Page 6.a would not admit a finite number of points. Finally, this non-linear structure needs more than three points and some non-primitive kind of betweenness, because otherwise Axiom II 3 on Page 6.a would not be satisfied. In the further discussion on Page 12.a, it remains unclear why the Plane Axiom of Order (i.e. Axiom II 5 on p. 6.a) should be dropped as well. At least, this change is not required for the existence of a finite model, as shown by the model discussed on Page 13 and in Note 13.1.

Anyway, the whole discussion of geometry is not an elaborated part of this volume. In fact, HILBERT and BERNAYS, the authors and editors of the seminal series of editions of HILBERT's "Foundations of Geometry" ([HILBERT, 1899; 1903; 1909; 1913; 1922a; 1923b], [HILBERT, 1930a; 1956; 1962; 1968], [HILBERT, 1972]), were just looking for some simple examples from geometry here.

$^{12.3}$Note that the numbering of these axioms does not refer to HILBERT's "Foundations of Geometry", but to the axiom system presented on Pages 5.a and 6.a.

$^{12.4}$Contrary to what is stated in Note 1 on Page 12.a, this Axiom II 5′ is actually a flawed formalization of a part of Axiom II 3 of [HILBERT, 1899],$^{12.8}$ which is omitted in [HILBERT, 1930a; 1968] (English version: [HILBERT, 1971]). A flawless formalization of that part of Axiom II 3 is

$$\text{II 5}''. \quad (x)(y)(z)\left(\left(\begin{array}{l} x \neq y \\ \&\ x \neq z \\ \&\ y \neq z \\ \&\ Gr(x,y,z) \end{array}\right) \rightarrow \left(\begin{array}{l} Zw(x,y,z) \\ \vee\ Zw(y,z,x) \\ \vee\ Zw(z,x,y) \end{array}\right)\right).$$

The reason why this flaw was not noticed in the preparation of the 2nd edition of 1968 may be that $Gr(x,y,z)$ is constantly true in the model that is constructed on Pages 12 and 13.

$^{12.5}$This refers to the 1st edition [HILBERT, 1899]$^{12.8}$ of HILBERT's "Grundlagen der Geometrie". More precisely, Axiom II 4′ is a formalization of Axiom II 2 of [HILBERT, 1899],$^{12.8}$ and Axiom II 5′ (after the necessary correction indicated in Note 12.4) is a partial formalization of Axiom II 3 of [HILBERT, 1899].$^{12.8}$

$^{12.6}$Cf. [HILBERT, 1930a].

$^{12.7}$Cf. Note 1 of Page 4.a.

$^{12.8}$English translation of [HILBERT, 1899] is [HILBERT, 1902].

1. Für ein Tripel x, y, z mit zwei übereinstimmenden Elementen ist Zw stets falsch.

2. Haben wir eine Kombination von drei verschiedenen unserer 5 Individuen, so sind unter den 6 möglichen Anordnungen der Elemente 2 Anordnungen mit gemeinsamem ersten Element, für welche Zw wahr ist, während für die übrigen 4 Anordnungen Zw falsch ist.

3. Jedes Paar von verschiedenen Elementen kommt sowohl als vorderes wie auch als hinteres Paar in je einem derjenigen Tripel vor, für welche Zw wahr ist.

Die erste Forderung läßt sich direkt durch Festsetzung erfüllen. Die gemeinsame Erfüllung der beiden anderen Forderungen geschieht in folgender Weise: Wir bezeichnen die 5 Elemente durch die Ziffern 1, 2, 3, 4, 5. Die Zahl der Werte-Tripel aus drei verschiedenen Elementen, für die wir Zw noch zu definieren haben, ist gleich $5 \cdot 4 \cdot 3 = 60$. Je sechs davon gehören zu einer Kombination; für zwei von diesen soll Zw wahr, für die übrigen falsch sein. Wir müssen also die 20 Tripel unter den 60 angeben, für die Zw als wahr definiert wird. Das sind diejenigen, die man aus den vier Tripeln

$$(1\ 2\ 5),\ (1\ 5\ 2),\ (1\ 3\ 4),\ (1\ 4\ 3)$$

durch Anwendung der zyklischen Permutation $(1\ 2\ 3\ 4\ 5)$ erhält.

Man verifiziert leicht, daß hierdurch allen Forderungen genügt wird. Auf diese Weise wird das Axiomensystem durch die Methode der Aufweisung als widerspruchsfrei erkannt[1].

Die an diesem Beispiel vorgeführte Methode der Aufweisung findet in neueren axiomatischen Untersuchungen sehr mannigfache Anwendungen. Sie dient hier vor allem zur Ausführung von *Unabhängigkeitsbeweisen*. Die Behauptung der Unabhängigkeit eines Satzes $\mathfrak{S}$ von einem Axiomensystem $\mathfrak{A}$ ist gleichbedeutend mit der Behauptung der Widerspruchsfreiheit des Axiomensystems

$$\mathfrak{A}\ \&\ \overline{\mathfrak{S}},$$

welches wir erhalten, indem wir die Negation des Satzes $\mathfrak{S}$ als Axiom zu $\mathfrak{A}$ hinzunehmen. Ist dieses Axiomensystem für einen endlichen Individuenbereich erfüllbar, so kann seine Widerspruchsfreiheit nach der Methode der Aufweisung festgestellt werden[2]. Somit liefert diese Me-

---

[1] Aus der Tatsache, daß das modifizierte Axiomensystem $\mathfrak{A}'$ mit einem 5-zahligen Individuenbereich erfüllbar ist, folgt auch sofort, daß die in diesem Axiomensystem enthaltenen Axiome die lineare Anordnung nicht vollständig bestimmen.

[2] Eine Fülle von Beispielen für dieses Verfahren findet man in den Abhandlungen über lineare und zyklische Ordnung von E. V. Huntington und seinen Mitarbeitern. Siehe insbesondere die Abhandlung „A new set of postulates for betweenness with proof of complete independence", Trans. Amer. Math. Soc. Bd. 26 (1924) S. 257—282. Dort sind auch die vorausgehenden Abhandlungen angegeben.

1. For a triple x, y, z in which two elements coincide, Zw is always false.

2. If we have a combination of three distinct individuals of the 5, then among the 6 possible orders of the elements there are 2 orders with the same first element for which Zw is true, whereas Zw is false for the other 4.

3. Each pair of distinct elements occurs as an initial as well as a final pair in one of the triples for which Zw is true.

The first demand can be satisfied directly by stipulation. The joint satisfaction of the other two conditions is achieved as follows: The 5 elements will be designated by the numerals $1, 2, 3, 4, 5$. The number of value triples of three distinct elements for which we still have to define Zw is equal to $5 \cdot 4 \cdot 3 = 60$. Every six of those belong to one combination. For two of these, Zw should be true, and false for the others. Thus, we have to indicate those 20 triples among the 60 for which Zw is defined to be true. They are those which one obtains from the four triples

$$(1\ 2\ 5),\ (1\ 5\ 2),\ (1\ 3\ 4),\ (1\ 4\ 3)$$

by application of the cyclic permutation $(1\ 2\ 3\ 4\ 5)$.

It is easy to verify that all demands are satisfied by this.[13.1] In this way, the axiom system is recognized as consistent by the method of exhibition.[1]

The method of exhibition as demonstrated in this example has very diverse applications in more recent axiomatic investigations, where it is especially used for *independence proofs*. The assertion that a sentence $\mathfrak{S}$ is independent of an axiom system $\mathfrak{A}$ is equivalent with the assertion that the axiom system

$$\mathfrak{A}\ \&\ \overline{\mathfrak{S}}$$

(obtained by adding the negation of the sentence $\mathfrak{S}$ to $\mathfrak{A}$ as an axiom) is consistent. If this axiom system is satisfiable in a finite domain of individuals, then its consistency can be established by the method of exhibition.[2] Thus, this method | provides a complement to the method of progressive inference. This complement is sufficient for many fundamental investigations in the sense that inferences show the provability, and exhibitions show the unprovability of sentences from certain axioms.

---

[1] It follows immediately from the fact that the modified axiom system $\mathfrak{A}'$ is satisfiable in a domain with 5 individuals that the axioms of this system do not completely determine linear order.

[2] Examples for this method can be found in abundance in the publications on linear and cyclic order by EDWARD V. HUNTINGTON and his collaborators. Note especially the article [HUNTINGTON, 1924], where also the references to the previous publications are to be found.

[13.1] The solution with 5 elements presented here is best visualized with the help of a pentagon circumscribing a pentagram:

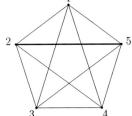

Now $Gr(x, y, z)$ is true for all $x, y, z \in \{1, 2, 3, 4, 5\}$. Moreover, $Zw(x, y, z)$ is true iff yxz form an isosceles triangle where the sides yx and xz have equal length in the above diagram. For instance, we have $Zw(1, 2, 5)$ and $Zw(1, 3, 4)$, but neither $Zw(2, 3, 4)$ nor $Zw(2, 3, 5)$. This obviously satisfies all axioms of the Pages 5.a and 6.a, except Axiom I 4. Moreover, this finite model also satisfies all axioms on Page 12.a.

thode für viele grundsätzliche Untersuchungen die ausreichende Ergänzung zur Methode des progressiven Schließens, indem durch das Schließen die Beweisbarkeit, durch die Aufweisungen die Unbeweisbarkeit von Sätzen aus gewissen Axiomen dargetan wird.

Ist nun die Methode der Aufweisung in ihrer Anwendung auf die endlichen Individuenbereiche beschränkt? Aus unserer bisherigen Überlegung können wir das noch nicht folgern. Wir sehen allerdings sogleich, daß bei einem unendlichen Individuenbereich die möglichen Prädikatensysteme nicht mehr eine überblickbare Mannigfaltigkeit bilden und daß von einem Durchprobieren aller Wertverläufe keine Rede sein kann. Gleichwohl könnten wir doch bei bestimmten vorgelegten Axiomen in der Lage sein, ihre Erfüllung durch bestimmte Prädikate aufzuweisen. Und das ist auch tatsächlich der Fall. Nehmen wir z. B. das System der drei Axiome

$$(x)\overline{R(x, x)},$$
$$(x)(y)(z)(R(x, y) \& R(y, z) \rightarrow R(x, z)),$$
$$(x)(Ey)R(x, y).$$

Machen wir uns klar, was diese besagen: Wir gehen aus von einem Ding a des Individuenbereiches. Gemäß dem dritten Axiom muß es ein Ding b geben, für das $R(a, b)$ wahr ist, und dieses ist auf Grund des ersten Axioms von a verschieden. Zu b muß es wieder ein Ding c geben, für das $R(b, c)$ wahr ist, und auf Grund des zweiten Axioms ist auch $R(a, c)$ wahr; gemäß dem ersten Axiom ist c von a und von b verschieden. Zu c muß es wieder ein Ding d geben, für welches $R(c, d)$ wahr ist. Für dieses ist auch $R(a, d)$ und $R(b, d)$ wahr, und d ist von a, b, c verschieden. Das Verfahren dieser Überlegung kommt nicht zum Abschluß, und wir ersehen daraus, daß wir mit einem endlichen Individuenbereich die Axiome nicht erfüllen können. Andererseits aber können wir leicht eine Erfüllung durch einen unendlichen Individuenbereich aufweisen: Wir nehmen als Individuen die ganzen Zahlen und setzen für $R(x, y)$ die Beziehung „x ist kleiner als y"; dann ergibt sich sofort, daß alle drei Axiome erfüllt werden.

Der gleiche Fall liegt vor bei den Axiomen

$$(Ex)(y)\overline{S(y, x)},$$
$$(x)(y)(u)(v)(S(x, u) \& S(y, u) \& S(v, x) \rightarrow S(v, y)),$$
$$(x)(Ey)S(x, y).$$

Für diese weist man auch leicht nach, daß sie mit einem endlichen Individuenbereich nicht erfüllt werden können. Andererseits aber sind sie im Bereich der positiven ganzen Zahlen erfüllt, wenn wir für $S(x, y)$ die Beziehung setzen: „y folgt unmittelbar auf x".

An diesen Beispielen bemerken wir aber, daß durch die gegebene Aufweisung die Frage der Widerspruchsfreiheit für die betrachteten

^{14.1}Now, is the application of the method of exhibition restricted to finite domains of individuals? We cannot conclude this from our considerations so far. We do see immediately, however, that in the case of infinite domains of individuals the possible predicate systems no longer constitute a comprehensible manifold,[14.2] and thus testing all graphs is out of the question. Nevertheless, we might be able to exhibit the satisfiability of a certain set of axioms by given predicates. And this is indeed the case. For example, let us take the system of the three axioms

$$(x)\,\overline{R(x,x)},$$

$$(x)(y)(z)\bigl(\,R(x,y)\;\&\;R(y,z)\;\rightarrow\;R(x,z)\,\bigr),$$

$$(x)(Ey)\,R(x,y).$$

Let us clarify what these axioms mean: We start with a thing a in the domain of individuals. According to the third axiom there must be a thing b for which $R(a,b)$ is true, and it must be distinct from a because of the first axiom. For b, there again must be a thing c for which $R(b,c)$ is true, and $R(a,c)$ must be true as well because of the second axiom. According to the first axiom, c is distinct from a as well as from b. For c, there again must be a thing d for which $R(c,d)$ is true. Again $R(a,d)$ and $R(b,d)$ must be true for d, and d is distinct from a,b,c. This procedure does not terminate; thus, it is evident that the axioms are not satisfiable in a finite domain of individuals. On the other hand, however, we can easily exhibit satisfiability in an infinite domain of individuals: We take the integers as the individuals, and set $R(x,y)$ to the relation "x is smaller than y". Then all three axioms are immediately satisfied.

The same is the case for the axioms

$$(Ex)(y)\,\overline{S(y,x)},$$

$$(x)(y)(u)(v)\bigl(\,S(x,u)\;\&\;S(y,u)\;\&\;S(v,x)\;\rightarrow\;S(v,y)\,\bigr),$$

$$(x)(Ey)\,S(x,y).$$

For these axioms, it is again easy to show that they are not satisfiable in a finite domain of individuals. On the other hand, however, they are satisfied in the domain of positive integers, if we set $S(x,y)$ to the relation "y directly succeeds x".

---

^{14.1}§§ 1(c) and 1(c)(1) (as listed in the table of contents) start with this paragraph.

^{14.2}We translate the German word "Mannigfaltigkeit" — which was introduced into German philosophy by KANT [1781] and which was intensively used by KANT and later German philosophers — as "manifold", because this is the translation that is etymologically closest. "Manifold" is also the translation in [KANT, 2008] as well as in several other translations of HILBERT and BERNAYS, cf. e.g. [SIEG, 2009], [MUELLER, 2006]. Alternatives could be "variety", "multiplicity", "plurality", but not "aggregate", because we use this already for the translation of "Inbegriff".

Axiome gar nicht endgültig erledigt, sondern vielmehr nur auf die nach der *Widerspruchsfreiheit der Zahlentheorie zurückgeführt* wird. Wir hatten auch bei dem früheren Beispiel einer endlichen Aufweisung ganze Zahlen als Individuen genommen. Das geschah aber dort nur zum Zweck der einfachen Bezeichnung der Individuen. Wir hätten statt der Zahlen auch andere Dinge, etwa Buchstaben nehmen können. Auch war dasjenige, was von den Zahlen gebraucht wurde, derart, daß es durch konkreten Nachweis festgestellt werden konnte.

Im vorliegenden Falle kommen wir aber mit einer konkreten Zahlenvorstellung nicht aus; denn wir brauchen wesentlich die Voraussetzung, daß die *ganzen Zahlen einen Individuenbereich,* also eine fertige Gesamtheit bilden.

Diese Voraussetzung ist uns allerdings sehr geläufig, da wir in der neueren Mathematik dauernd mit ihr operieren, und man ist geneigt, sie für selbstverständlich zu halten. Es war zuerst FREGE, der mit allem Nachdruck und mit scharfer, witziger Kritik die Forderung zur Geltung brachte, daß die Vorstellung der Zahlenreihe als einer fertigen Gesamtheit durch einen Nachweis ihrer Widerspruchsfreiheit gesichert werden müsse[1]. Ein solcher Nachweis war nach FREGES Meinung nur im Sinne einer Aufweisung, als Existenzbeweis, zu führen, und er glaubte, die Objekte für eine solche Aufweisung im Bereich der Logik zu finden. Sein Verfahren der Aufweisung kommt darauf hinaus, daß er die Gesamtheit der Zahlen definiert mit Hilfe der als existierend vorausgesetzten Gesamtheit aller überhaupt denkbaren einstelligen Prädikate. Aber die hierbei zugrunde gelegte Voraussetzung, welche ohnehin schon einer unbefangenen Betrachtung als sehr verdächtig erscheint, hat sich durch die berühmten, von RUSSELL und ZERMELO entdeckten logischen und mengentheoretischen Paradoxien als unhaltbar erwiesen. Und das Mißlingen des FREGEschen Unternehmens hat mehr noch als FREGES Dialektik das Problematische an der Annahme der Totalität der Zahlenreihe zum Bewußtsein gebracht.

Wir können nun angesichts dieser Problematik versuchen, an Stelle der Zahlenreihe einen anderen unendlichen Individuenbereich für die Zwecke der Widerspruchsfreiheitsbeweise zu verwenden, der nicht wie die Zahlenreihe ein reines Gedankengebilde, sondern aus dem Gebiet der sinnlichen Wahrnehmung oder aber der realen Wirklichkeit entnommen ist. Sehen wir aber näher zu, so werden wir gewahr, daß überall, wo wir im Gebiet der Sinnesqualitäten oder in der physikalischen Wirklichkeit unendliche Mannigfaltigkeiten anzutreffen glauben, von einem eigentlichen Vorfinden einer solchen Mannigfaltigkeit keine Rede ist, daß vielmehr die Überzeugung von dem Vorhandensein einer solchen Mannigfaltigkeit auf einer gedanklichen Extrapolation beruht, deren

---

[1] GOTTLOB FREGE „Grundlagen der Arithmetik", Breslau 1884, sowie „Grundgesetze der Arithmetik", Jena 1893.

In these examples we notice, however, that the given exhibition does not conclusively settle the issue of consistency of the considered | axioms. Rather, this question is only *reduced to the issue of consistency of number theory.* In the previous example of a finite exhibition, we took integers as individuals as well. There, however, this was only for the purpose of designating individuals in a simple way. Instead of numbers, we could also have taken something else, letters for example. Moreover, in the previous example of a finite exhibition, the required properties of the numbers were concretely verifiable.

In the present case, however, a concrete conception of numbers is not sufficient because we essentially need the presupposition that *the integers constitute a domain of individuals,* i.e. a completed totality.

We are actually quite familiar with this presupposition as we are constantly working with it in more recent mathematics, and we are inclined to take it for granted. FREGE was the first who — with all insistence and with a sharp and witty critique — brought to bear the demand that the conception of the number series as a completed totality must be ensured with a proof of its consistency.[1] According to FREGE, such a proof could only be carried out in the sense of an exhibition, as an *existence proof*; and he believed he could find the objects for such an exhibition in the field of logic. His procedure of exhibition[15.1] amounts to defining the totality of numbers on the basis of the presupposed existence of the totality of all generally conceivable singular predicates. This underlying presupposition, however, which seems in any case highly suspect to unbiased examination already, was shown to be untenable[15.2] by the famous logical and set-theoretical paradoxes discovered by RUSSELL and ZERMELO. And even more than FREGE's dialectic, the failure of the FREGEan endeavor has made us aware of the problematic nature of the assumption of the totality of the number series.

---

[1]Cf. [FREGE, 1884] and [FREGE, 1893/1903, Vol. I].

[15.1]We translate "Verfahren der Aufweisung" of the original text as "procedure of exhibition", which seems to be more appropriate here than "approach of exhibition". Note, however, that at all places where we write "method of exhibition", the German original has indeed "Methode der Aufweisung". Cf. also Note 12.1.

[15.2]Note that this is not to say that FREGE's definition of the natural numbers in [FREGE, 1893/1903, Vol. I] would be "untenable" ("unhaltbar"); cf. e.g. [QUINE, 1981] for the potential consistency of this definition. FREGE reduces the existence of the set of natural numbers to the even more problematic existence of a universal class of singular predicates, which, however, does not have to contain the predicate of RUSSELL's Paradox for this reduction — and must not contain this predicate for consistency reasons. The following two aspects of FREGE's definition of the natural numbers are untenable, however: the belief that the reduction establishes the existence of the completed totality of all natural numbers, and the presupposition that every specifiable singular predicate belongs to the completed totality of "all generally conceivable singular predicates".

Berechtigung jedenfalls ebensosehr der Prüfung bedarf wie die Vorstellung von der Totalität der Zahlenreihe.

Ein typisches Beispiel hierfür bildet diejenige Unendlichkeit, welche zu der bekannten Paradoxie des ZENO Anlaß gegeben hat. Wird eine Strecke in endlicher Zeit durchlaufen, so sind in dieser Durchlaufung nacheinander unendlich viele Teilvorgänge enthalten: die Durchlaufung der ersten Hälfte, dann die des folgenden Viertels, des folgenden Achtels usw. Haben wir es mit einer wirklichen Bewegung zu tun, so müssen diese Teildurchlaufungen lauter reale Prozesse sein, welche nach einander erfolgen.

Man pflegt diese Paradoxie mit dem Argument abzuweisen, daß die Summe von unendlich vielen Zeitintervallen doch konvergieren, also eine endliche Zeitdauer ergeben kann. Dadurch wird aber ein wesentlicher Punkt der Paradoxie nicht getroffen, nämlich das Paradoxe, was darin liegt, daß eine unendliche Aufeinanderfolge, deren Vollendung wir in der Vorstellung nicht nur faktisch, sondern auch grundsätzlich nicht vollziehen können, in der Wirklichkeit abgeschlossen vorliegen soll.

Tatsächlich gibt es auch eine viel radikalere Lösung der Paradoxie. Diese besteht in der Erwägung, daß wir keineswegs genötigt sind, zu glauben, daß die mathematische raum-zeitliche Darstellung der Bewegung für beliebig kleine Raum- und Zeitgrößen noch physikalisch sinnvoll ist, vielmehr allen Grund haben zu der Annahme, daß jenes mathematische Modell die Tatsachen eines gewissen Erfahrungsbereiches, eben die Bewegungen innerhalb der unserer Beobachtung bisher zugänglichen Größenordnungen, im Sinne einer einfachen Begriffsbildung extrapoliert, ähnlich wie die Mechanik der Kontinua eine Extrapolation vollzieht, indem sie die Vorstellung einer kontinuierlichen Erfüllung des Raumes mit Materie zugrunde legt: so wenig wie eine Wassermenge bei unbegrenzter räumlicher Teilung immer wieder Wassermengen ergibt, ebensowenig wird es bei einer Bewegung der Fall sein, daß durch ihre Teilung ins Unbegrenzte immer wieder etwas entsteht, das sich als Bewegung charakterisieren läßt. Geben wir dieses zu, so schwindet die Paradoxie.

Das mathematische Modell der Bewegung hat ungeachtet dessen als *idealisierende Begriffsbildung* zum Zweck der vereinfachten Darstellung seinen bleibenden Wert. Für diesen Zweck muß es außer der approximativen Übereinstimmung mit der Wirklichkeit noch die Bedingung erfüllen, daß die in ihm vollzogene Extrapolation auch in sich widerspruchsfrei ist. Unter diesem Gesichtspunkt wird unsere mathematische Vorstellung von der Bewegung durch die ZENOsche Paradoxie nicht im geringsten erschüttert; das genannte mathematische Gegenargument hat hierfür seine volle Geltung. Eine andere Frage aber ist, ob wir einen wirklichen Nachweis für die Widerspruchsfreiheit der mathematischen Theorie der Bewegung besitzen. Diese Theorie beruht wesentlich auf

[16.1]In view of this difficulty in proving consistency, we could now try to use some other infinite domain of individuals which is not a mere product of thought (such as the number series), but is taken from the realm of sense perception or physical reality. If we take a closer look, however, we will realize that wherever we believe that we encounter infinite manifolds in the realm of qualia or in physical reality, there can be no actual detection of such a manifold:[16.2] The conviction of the existence of such a manifold actually rests on a mental extrapolation, | which requires an examination of its justification at least as necessarily as the conception of the totality of the number series.

A typical example of this is that kind of infinity which gave rise to the well-known paradox of ZENO OF ELEA: If some distance is traversed in finite time, then this traversal includes infinitely many successive sub-processes: traversing the first half, then the next quarter, then the next eighth, and so on. If we are dealing with a real motion, then these sub-traversals must all be real processes which succeed one another.

Usually, one dismisses this paradox with the argument that the sum of infinitely many time intervals may yet converge, and thus result in a finite duration. This argument, however, misses an essential point of the paradox: Namely the paradoxical aspect that lies in the fact that an infinite succession whose completion we cannot accomplish in our conception — neither factually, nor in principle — should be existent in reality as a completed process.

Actually, there is a more radical solution to the paradox. It consists in the deliberation that we are not at all forced to believe that the mathematical space–time representation of motion remains physically meaningful for arbitrarily small spatio–temporal segments. Rather, we have every reason to assume that our mathematical model — for the purpose of a simple conceptual structure — extrapolates the facts of a certain domain of experience; namely the facts of motion within those ranges of magnitudes that have been accessible to our observation up until now. This extrapolation is similar to the one carried out in continuum mechanics by referring to the conception that space is continuously filled with matter. Just as unbounded spatial division of a quantity of water does not always produce quantities of water, it will also not be the case that unbounded division of a motion always produces something characterizable as motion.[16.3] If we admit this, the paradox wanes.

Despite this, our mathematical model of motion has its lasting value as an *idealizing concept formation* for the purpose of a simplified representation. For this purpose, beside the approximate agreement of the model with reality, the extrapolation carried out in the model also has to be self-consistent. From this viewpoint, the mathematical conception of motion is not in the least shaken by ZENO's Paradox, for which the mentioned mathematical counterargument is perfectly valid. It is, however, another question whether we have an actual proof of the consistency of the mathematical theory of motion. This theory is based essentially on | the mathematical theory of the continuum, and this in turn relies essentially on the conception of the set of all integers as a completed totality. Therefore, we return, via detours, to our original problem, which we tried to bypass by referring to the facts of motion.

[16.1]§ 1(c)(2) (as listed in the table of contents) starts with this paragraph. In [HILBERT & BERNAYS, 2001], however, the beginning of § 1(c)(2) is indicated one paragraph before.

[16.2]The non-existence of the infinite in physical reality is also discussed in detail in [HILBERT, 1926].

[16.3]Cf. e.g. [LANDSMANN, 2006] and [JAMMER, 1974] for the question of the traceability of motion, and for the historical development of quantum mechanics contemporary to the writing of the German original.

der mathematischen Theorie des Kontinuums, diese wiederum stützt sich wesentlich auf die Vorstellung von der Menge aller ganzen Zahlen als einer fertigen Gesamtheit. Wir kommen also auf Umwegen zu dem Problem zurück, das wir durch den Hinweis auf die Tatsachen der Bewegung zu umgehen versuchten.

Ähnlich verhält es sich in all den Fällen, wo man glaubt, direkt eine Unendlichkeit als durch Erfahrung oder durch Anschauung gegeben aufweisen zu können, wie etwa die Unendlichkeit der von Oktave zu Oktave ins Unendliche fortschreitenden Tonreihe oder die stetige unendliche Mannigfaltigkeit beim Übergang von einer Farbenqualität zu einer anderen. Die nähere Betrachtung zeigt jeweils, daß eine Unendlichkeit uns hier tatsächlich gar nicht gegeben ist, sondern erst durch einen gedanklichen Prozeß interpoliert oder extrapoliert wird.

Auf Grund dieser Überlegungen kommen wir zu der Einsicht, daß die Frage nach der Existenz einer unendlichen Mannigfaltigkeit durch eine Berufung auf außermathematische Objekte nicht entschieden werden kann, sondern innerhalb der Mathematik selbst gelöst werden muß. Wie soll aber eine solche Lösung angesetzt werden? Auf den ersten Blick scheint es, daß hiermit etwas Unmögliches verlangt wird: Unendlich viele Individuen vorzuführen, ist grundsätzlich unmöglich; daher kann ein unendlicher Individuenbereich als solcher nur durch seine Struktur gekennzeichnet werden, d. h. durch Beziehungen, welche zwischen seinen Elementen bestehen. Mit anderen Worten: es muß der Nachweis erbracht werden, daß für ihn gewisse formale Relationen sich erfüllen lassen. Die Existenz eines unendlichen Individuenbereiches läßt sich also *gar nicht anders darstellen als durch die Erfüllbarkeit gewisser logischer Formeln;* das sind dann aber gerade Formeln von der Art wie diejenigen, durch deren Untersuchung wir auf die Frage nach der Existenz eines unendlichen Individuenbereiches geführt worden sind und deren Erfüllbarkeit eben durch die Aufweisung eines unendlichen Individuenbereiches dargetan werden sollte. Der Versuch, die Methode der Aufweisung auf die betrachteten Formeln anzuwenden, führt uns also auf einen Circulus vitiosus.

Nun sollte uns aber die Aufweisung nur als Mittel dienen zum Nachweis der Widerspruchsfreiheit von Axiomensystemen. Auf dieses Verfahren brachte uns die Betrachtung von Individuenbereichen mit einer gegebenen endlichen Anzahl von Individuen, indem wir erkannten, daß für einen solchen Bereich die Widerspruchsfreiheit einer Formel gleichbedeutend ist mit ihrer Erfüllbarkeit.

Im Falle unendlicher Individuenbereiche ist der Sachverhalt komplizierter. Es gilt hier zwar auch noch, daß ein durch eine Formel $\mathfrak{A}$ dargestelltes Axiomensystem dann und nur dann widerspruchsvoll ist, wenn die Formel $\overline{\mathfrak{A}}$ allgemeingültig ist. Aber wir können jetzt, da wir es nicht mehr mit einem überblickbaren Vorrat von Wertverläufen für die

In every case where one thinks that an infinity can be directly exhibited as given in experience or intuition (for example: the infinity of the tone series extending from octave to octave up to infinity, or the continuous infinite manifold involved in the passage from one color quality to another), the situation is much the same as given in our more radical solution to ZENO's Paradox.[17.1] Closer inspection always reveals that such an infinity is in fact not given here at all; it rather results from a mental process of interpolation or extrapolation.

[17.2]On the basis of these considerations, we come to the insight that the question of the existence of an infinite manifold cannot be decided by appealing to extra-mathematical objects. Rather, the question must be solved within mathematics itself. But how should such a solution be approached? At first glance, it seems that something impossible is asked for: To present infinitely many individuals is impossible in principle; thus, an infinite domain of individuals can only be characterized by its structure, i.e. via the relations that hold between its elements. In other words: A proof must be given that certain formal relations can be satisfied for this domain. The existence of an infinite domain of individuals *cannot be represented in any other way but by the satisfiability of certain logical formulas.* But then these formulas are of exactly the same kind as the ones whose investigation has led us to the question of the existence of an infinite domain of individuals, and whose satisfiability was to be shown by the exhibition of an infinite domain of individuals. Thus, the attempt to apply the method of exhibition to the formulas under consideration leads us into a *circulus vitiosus*.[17.3]

The exhibition, however, was only to serve as a means to prove the consistency of axiom systems. We came to this approach after recognizing that — for domains with a given finite number of individuals — the consistency of a formula is equivalent to its satisfiability.

In the case of infinite domains of individuals the situation is more complicated: It still holds that an axiom system represented by a formula $\mathfrak{A}$ is inconsistent if and only if the formula $\overline{\mathfrak{A}}$ is valid. However, we no longer have a comprehensible supply of combinations of graphs | for the variable predicates. Thus, from the fact that $\overline{\mathfrak{A}}$ is not valid, we can no longer conclude that we also have some model of $\mathfrak{A}$ at our disposal.

[17.1]In the translation we have extended the simple phrase "Ähnlich" of the German original ("much the same") with the clarifying phrase "as given in our more radical solution to ZENO's Paradox", because it does not refer to the previous paragraph (on the potential (set-theoretical) impossibility of mathematical model construction), but to the one before the previous paragraph (which nicely illustrates the transcendental impossibility of identifying models of physics with the things in themselves in the sense of KANT [1787]).

[17.2]§ 1(c)(3) (as listed in the table of contents) starts on this page, probably with this paragraph. In [HILBERT & BERNAYS, 2001], however, the beginning of § 1(c)(3) is indicated one paragraph before.

[17.3]We do not translate "Circulus vitiosus" as "vicious circle" here because the German original does not use the standard German word "Teufelskreis", either.

variablen Prädikate zu tun haben, nicht mehr folgen, daß uns, falls $\overline{\mathfrak{A}}$ nicht allgemeingültig ist, auch schon ein Modell zur Erfüllung des Axiomensystems $\mathfrak{A}$ zu Gebote steht.

Die Erfüllbarkeit eines Axiomensystems ist demnach, wenn ein unendlicher Individuenbereich in Frage kommt, zwar eine hinreichende Bedingung für seine Widerspruchsfreiheit, aber als notwendige Bedingung ist sie nicht erwiesen. Wir können daher nicht erwarten, daß sich allgemein der Nachweis der Widerspruchsfreiheit durch einen Nachweis der Erfüllbarkeit erbringen lasse. Andrerseits aber sind wir auch gar nicht genötigt, die Widerspruchsfreiheit durch die Feststellung der Erfüllbarkeit zu erweisen, vielmehr können wir bei der ursprünglichen negativen Bedeutung der Widerspruchsfreiheit stehenbleiben. Das heißt — wenn wir uns das Axiomensystem wieder durch eine Formel $\mathfrak{A}$ dargestellt denken —, wir brauchen nicht die Erfüllbarkeit der Formel $\mathfrak{A}$ zu zeigen, sondern vielmehr nur nachzuweisen, daß die Annahme der Erfüllung von $\mathfrak{A}$ durch gewisse Prädikate nicht auf einen logischen Widerspruch führen kann.

Um das Problem von dieser Seite in Angriff zu nehmen, werden wir darauf ausgehen, uns einen Überblick über die möglichen logischen Schlüsse zu verschaffen, welche aus einem Axiomensystem gezogen werden können. Als das geeignete Mittel hierfür bietet sich die Methode der *Formalisierung des logischen Schließens*, wie sie von FREGE, SCHRÖDER, PEANO und RUSSELL ausgebildet worden ist.

Wir gelangen somit zu folgender Aufgabestellung: 1. die Prinzipien des logischen Schließens streng zu formalisieren und dadurch zu einem völlig überblickbaren System von Regeln zu machen; 2. für ein vorgelegtes Axiomensystem $\mathfrak{A}$ (das als widerspruchsfrei erwiesen werden soll) den Nachweis zu führen, daß beim Ausgehen von diesem *System* $\mathfrak{A}$ *mittels logischer Deduktionen kein Widerspruch zustande kommen kann*, d. h. daß nicht zwei Formeln beweisbar werden, von denen die eine die Negation der andern ist.

Wir brauchen nun aber diesen Nachweis nicht für jedes Axiomensystem einzeln auszuführen, sondern können uns die bereits am Anfang unserer Betrachtung erwähnte Methode der *Arithmetisierung*[1] zunutze machen. Diese läßt sich von unserem jetzigen Standpunkt so charakterisieren: Wir suchen uns ein Axiomensystem $\mathfrak{A}$, welches einerseits eine so übersichtliche Struktur hat, daß wir den Nachweis seiner Widerspruchsfreiheit (im Sinne der gestellten Aufgabe 2) erbringen können, das aber andererseits auch so reichhaltig ist, daß wir aus einer *als vorhanden vorausgesetzten* Erfüllung dieses Axiomensystems durch ein System $\mathfrak{S}$ von Dingen und Beziehungen Erfüllungen für die Axiomensysteme der geometrischen und physikalischen Disziplinen ableiten können in der Weise, daß wir die Gegenstände eines solchen Axiomensystems $\mathfrak{B}$ durch Individuen aus $\mathfrak{S}$ oder Komplexe solcher Individuen

---
[1] Vgl. S. 3.

Thus, although the satisfiability of an axiom system is a sufficient condition for its consistency, it has not been shown to be a necessary condition if an infinite domain is in question. Therefore, in general, we cannot expect that a proof of consistency can be accomplished by means of a proof of satisfiability. On the other hand, we are not at all forced to prove consistency by establishing satisfiability; rather, we may retain the original negative meaning of consistency. That is to say — if we again suppose the axiom system to be represented by a formula $\mathfrak{A}$ — we do not need to show the satisfiability of the formula $\mathfrak{A}$, but we only need to prove that the assumption that $\mathfrak{A}$ is satisfied by certain predicates cannot lead to a logical contradiction.

To attack the problem from this side, we are going to gain an overview of the possible logical inferences which can be drawn from an axiom system. The *formalization of logical inference*, as developed by FREGE, SCHRÖDER, PEANO, and RUSSELL, presents itself as an appropriate means to this end.

We thus arrive at the following tasks:[18.1]

1. to formalize the principles of logical inference rigorously, and thereby to turn them into a completely comprehensible system of rules;
2. to prove, for a given axiom system $\mathfrak{A}$ (whose consistency is to be proved), that *no contradiction can be derived from this system $\mathfrak{A}$ by logical deductions*; i.e. that no two formulas become provable, one of which is the negation of the other.

But now, we do not have to carry out this proof for every axiom system individually; rather, we can make use of the method of *arithmetization*, which we already mentioned at the beginning of our treatment.[1] From our current standpoint, this method can be characterized as follows: We are looking for an axiom system $\mathfrak{A}$ with the following properties:[18.2]

- The structure of $\mathfrak{A}$ has to be comprehensible to such an extent that we can prove its consistency (in the sense of the above Task 2).
- $\mathfrak{A}$ has to be sufficiently comprehensive in the following sense. For an axiom system $\mathfrak{B}$ of the disciplines of geometry or physics, and from the *assumed existence*[18.3] of a model[18.4] $\mathfrak{S}$ of $\mathfrak{A}$, we have to be able to derive a model[18.5] of $\mathfrak{B}$ whose individuals are represented by individuals or complexes of individuals from $\mathfrak{S}$, | and whose basic relations are predicates formed from the basic relations of $\mathfrak{S}$ by logical operations.

---

[1] Cf. p. 3.[18.6]

[18.1] The itemization (vertical list) was added in the translation.

[18.2] The itemization and the bullets were added in the translation.

[18.3] The German original does not actually speak of "the assumed existence of a model", but of "einer als vorhanden vorausgesetzten Erfüllung", which also may be translated as "a satisfaction, presupposed to be given", or as "a satisfying system, presupposed to be existent".

[18.4] The German original does not actually speak of "models" but of "Erfüllungen" ("satisfactions"), not to be confused with "Erfüllbarkeiten" ("satisfiabilities"). Moreover, the German original does not speak of "a model $\mathfrak{S}$ of $\mathfrak{A}$", but of "Erfüllung dieses Axiomensystems durch ein System $\mathfrak{S}$ von Dingen und Beziehungen" ("satisfaction of this axiom system by a system $\mathfrak{S}$ of things and relations").

[18.5] Neither the phrase "model of $\mathfrak{B}$" nor any of the references to it (represented by the two occurrences of the pronoun "whose" in this sentence) occur explicitly in the German original, where the second reference is completely omitted and the first actually refers to the axiom system $\mathfrak{B}$. It is not clear whether this is a problem of use and mention (denotation and syntax), or just a sloppy contraction, or a lapse that resulted from calling both syntactical and model-semantical representations just "systems" ("axiom systems" and "systems of things and relations").

[18.6] Footnote added in the 2nd edition of 1968.

repräsentieren und für die Grundbeziehungen solche Prädikate setzen, die sich durch logische Operationen aus den Grundbeziehungen von $\mathfrak{S}$ bilden lassen.

Damit ist dann das betreffende Axiomensystem $\mathfrak{B}$ in der Tat als widerspruchsfrei erwiesen; denn ein Widerspruch, der sich als Folgerung aus diesem Axiomensystem ergäbe, würde sich ja als ein aus dem Axiomensystem $\mathfrak{A}$ ableitbarer Widerspruch darstellen, während doch das Axiomensystem $\mathfrak{A}$ als widerspruchsfrei erkannt ist.

Als ein solches Axiomensystem $\mathfrak{A}$ bietet sich die (axiomatisch aufgebaute) Arithmetik dar.

Diese „Methode der Zurückführung" axiomatischer Theorien auf die Arithmetik erfordert nicht, daß die Arithmetik einen anschaulich vorweisbaren Tatbestand bilde, vielmehr braucht hierzu die Arithmetik nichts anderes zu sein als eine Ideenbildung, die wir als widerspruchsfrei nachweisen können und welche einen systematischen Rahmen liefert, in den die Axiomensysteme der theoretischen Wissenschaften sich einordnen lassen, so daß die in ihnen vollzogenen Idealisierungen des tatsächlich Gegebenen durch diese Einordnung ebenfalls als widerspruchsfrei erwiesen werden. —

Es seien die Ergebnisse der letzten Betrachtung kurz zusammengefaßt: Das Problem der Erfüllbarkeit eines Axiomensystems (bzw. einer logischen Formel), welches im Falle endlicher Individuenbereiche positiv durch Aufweisung gelöst werden kann, ist im Falle, wo zur Erfüllung der Axiome nur ein unendlicher Individuenbereich in Betracht kommt, nicht nach dieser Methode lösbar, weil die Existenz unendlicher Individuenbereiche nicht als ausgemacht gelten kann, vielmehr die Einführung solcher unendlichen Bereiche erst durch einen Nachweis der Widerspruchsfreiheit für ein das Unendliche charakterisierendes Axiomensystem gerechtfertigt wird.

Angesichts des Versagens der positiven Entscheidungsmethode bleibt uns nur der Weg, den Nachweis der Widerspruchsfreiheit im negativen Sinne, das heißt als *Unmöglichkeitsbeweis* zu führen, wozu eine Formalisierung des logischen Schließens erforderlich wird. —

Wenn wir nun an die Aufgabe eines solchen Unmöglichkeitsbeweises herantreten, so müssen wir uns darüber klar sein, daß dieser nicht wieder mit der Methode des axiomatisch-existentialen Schließens geführt werden kann. Wir dürfen vielmehr nur solche Schlußweisen anwenden, die von idealisierenden Existenzannahmen frei sind.

Auf Grund dieser Erwägung stellt sich aber sogleich folgender Gedanke ein: Wenn wir ohne axiomatisch-existentiale Annahmen den Unmöglichkeitsbeweis führen können, sollte es dann nicht auch möglich sein, auf solche Weise direkt die ganze Arithmetik zu begründen und damit jenen Unmöglichkeitsbeweis ganz überflüssig zu machen? Dieser Frage wollen wir uns im folgenden Paragraphen zuwenden.

This then proves that the axiom system $\mathfrak{B}$ under consideration is in fact consistent: Any contradiction that arose as a conclusion from the axiom system $\mathfrak{B}$ would, of course, represent a contradiction derivable from the axiom system $\mathfrak{A}$ — even though the axiom system $\mathfrak{A}$ is known to be consistent.

Arithmetic (in axiomatic formulation) presents itself as such an axiom system $\mathfrak{A}$.

This "method of reduction" of axiomatic theories to arithmetic does not demand that arithmetic consists of intuitively producible facts. Rather, arithmetic does not have to be anything else but a formation of ideas that we can prove to be consistent and that provides a systematic framework into which the axiom systems of the theoretical sciences can be integrated in such a way that their idealizations of the actually given are shown to be consistent as well. —

Let us summarize the results of the preceding discussion briefly: In case of a finite domain of individuals, the problem of satisfiability of an axiom system (or a logical formula) may be given a positive answer by the method of exhibition. If the axioms can only be satisfied in an infinite domain, this problem is not solvable with this method because the existence[19.1] of infinite domains of individuals cannot be taken for granted. Rather, the introduction of such infinite domains is justified only by a proof of the consistency of an axiom system that characterizes the infinite.

In view of the failure of the positive decision method, the only way left is to prove consistency in the negative sense, i.e. via an *impossibility proof*.[19.2] This requires a formalization of logical inference. —

When we now approach the task of such an impossibility proof, we have to be aware of the fact that we cannot again execute this proof with the method of axiomatic-existential[19.1] inference. Rather, we may only apply modes of inference that are free from idealizing existence[19.1] assumptions.

Yet, as a result of this deliberation, the following idea suggests itself right away: If we can conduct the impossibility proof without making any axiomatic-existential[19.1] assumptions, should it then not be possible to provide a grounding for the whole of arithmetic directly in this way, whereby that impossibility proof would become entirely superfluous? We will turn to this question in the following section.

[19.1] With respect to this notion of "existence", see the discussion on Pages 1.a and 2, and especially in Note 1.6 on Page 1.b.

[19.2] Note that an impossibility proof is not necessarily a proof by contradiction.

§ 2. Die elementare Zahlentheorie. — Das finite Schließen und seine Grenzen.

Die zum Schluß des vorigen Paragraphen aufgeworfene Frage, ob wir nicht durch eine von der Axiomatik unabhängige Methode direkt die Arithmetik begründen und dadurch einen besonderen Nachweis der Widerspruchsfreiheit entbehrlich machen können, gibt uns Anlaß, uns darauf zu besinnen, daß ja die Methode der verschärften Axiomatik, insbesondere das existentiale Schließen, unter Zugrundelegung eines festumgrenzten Individuenbereiches, gar nicht das ursprüngliche Verfahren der Mathematik ist.

Die Geometrie wurde zwar von vornherein axiomatisch aufgebaut. Aber die Axiomatik EUKLIDs ist inhaltlich und anschaulich gemeint. Es wird hier nicht von der anschaulichen Bedeutung der Figuren abstrahiert. Ferner haben auch die Axiome nicht die existentiale Form. EUKLID setzt nicht voraus, daß die Punkte sowie die Geraden je einen festen Individuenbereich bilden. Er stellt deshalb auch nicht Existenz-Axiome auf, sondern Konstruktions-Postulate.

Ein solches Postulat ist z. B., daß man zwei Punkte durch eine Gerade verbinden kann; ferner, daß man um einen gegebenen Punkt einen Kreis mit vorgeschriebenem Radius ziehen kann.

Dieser methodische Standpunkt ist jedoch nur dann durchführbar, wenn die Postulate als der Ausdruck einer bekannten Tatsächlichkeit oder einer unmittelbaren Evidenz angesehen werden. Die hiermit sich erhebende Frage nach dem Geltungsbereich der geometrischen Axiome ist bekanntermaßen sehr heikel und strittig, und es besteht gerade ein wesentlicher Vorzug der formalen Axiomatik darin, daß sie die Begründung der Geometrie von der Entscheidung dieser Frage unabhängig macht.

Von dieser Problematik, welche mit dem besonderen Charakter der geometrischen Erkenntnis zusammenhängt, sind wir frei im Gebiet der Arithmetik, und in der Tat ist auch hier, in den Disziplinen der elementaren Zahlenlehre und der Algebra, der Standpunkt der direkten inhaltlichen, ohne axiomatische Annahmen sich vollziehenden Überlegung am reinsten ausgebildet.

Das Kennzeichnende für diesen methodischen Standpunkt ist, daß die Überlegungen in der Form von *Gedankenexperimenten* an Gegenständen angestellt werden, die als *konkret vorliegend* angenommen werden. In der Zahlentheorie handelt es sich um Zahlen, die als vorliegend gedacht werden, in der Algebra um vorgelegte Buchstabenausdrücke mit gegebenen Zahl-Koeffizienten.

Wir wollen das Verfahren genauer betrachten, und die Anfangsgründe methodisch etwas verschärfen. In der Zahlentheorie haben wir ein Ausgangsobjekt und einen Prozeß des Fortschreitens. Beides müssen

§ 2. Elementary Number Theory. — Finitistic Inference and its Limits

At the end of the previous section, we raised the question of whether we can give a grounding of arithmetic directly (by a method independent of axiomatics) and thereby dispense with a special consistency proof. This gives us occasion to recall that the method of sharpened axiomatics (in particular existential[20.1] inference on the basis of a well-determined and fixed domain of individuals) is not at all the original approach to mathematics.

Geometry was, in fact, built up axiomatically from the very beginning; but EUCLID's axiomatics was intended to be contentual and intuitive, and the intuitive meaning of the figures is not ignored in [EUCLID, ca. 300 B.C.].[20.2] Furthermore, its axioms are not in existential[20.1] form either: EUCLID does not presuppose that points and lines constitute fixed domains of individuals. Therefore, he does not state any existence axioms either, but construction postulates.[20.3]

Such a postulate is for instance that we can join two points by a straight line, or that it is possible to draw a circle around a given point with a prescribed radius.

This methodological standpoint, however, is feasible only if the postulates are seen as an expression of a known actuality or of an immediate self-evidence. It is well-known that the resultant question on the scope of the validity of the geometrical axioms is very delicate and controversial; and a substantial advantage of formal axiomatics is that it makes the grounding of geometry independent of deciding this question.

In the field of arithmetic, we are not concerned with problematic issues connected with the question of the specific character of geometrical knowledge; and it is indeed also here, in the disciplines of elementary number theory and algebra, that we find the purest manifestation of the standpoint of direct contentual thought that has evolved without axiomatic assumptions.

[20.4]This methodological standpoint is characterized by *thought experiments* with things that are assumed to be *concretely present*, such as numbers in number theory, or expressions of letters with given numerical coefficients in algebra.

We will analyze this approach more closely, and we will sharpen the treatment of its foundational elements to a certain degree.[20.5]

[20.1] For this notion of "existence", cf. the discussion on Pages 1.a and 2, especially in Note 1.6 on Page 1.b.

[20.2] The explicit reference "[EUCLID, ca. 300 B.C.]" was added in the translation.

[20.3] If this means that existence sentences are stronger than construction postulates, then we can conclude that construction is seen here as a state-changing process. To the contrary, from a static point of view, an existence sentence always follows from the axiomatization of the related construction postulate; whereas the replacement of existence sentences with construction postulates (SKOLEMization) may actually provide problems in certain logics; cf. e.g. [WIRTH, 2004, § 1.2.5]. In formal axiomatization, the difference between an (non-constructive) *existence axiom* and a *construction postulate* is that the latter introduces a constructor for the object, whereas the existence axiom just states existence without presenting any witnesses. For instance, the 4th axiom of connection on Page 5.a is such a non-constructive existence axiom.

It is important to note that, although the meaning of "existence" here is different from that other form of existence or existential form of Note 20.1, that other form of existence is actually the answer to the question of what makes an existence sentence stronger than a construction postulate. Thus, the line of thinking developed here may be the source for calling that other form simply "existence", although "existence of the (infinite) extension of a notion" or "modality of existence of the actual infinite" may be more appropriate designations here. See also the discussions in Note 1.6 on Page 1.b, and in Note 37.2 on Page 37.

[20.4] This paragraph break was introduced in the 2nd edition of 1968, together with the correction of the typo "Überlegung" of the 1st edition of 1934 (instead of "Überlegungen") in the sentence that follows.

[20.5] This is one of the rare cases where we translate "methodisch" as "the treatment of". Cf. also Notes 1.3 and 44.2.

wir in bestimmter Weise anschaulich festlegen. Die besondere Art der Festlegung ist dabei unwesentlich, nur muß die einmal getroffene Wahl für die ganze Theorie beibehalten werden. Wir wählen als Ausgangsding die Ziffer 1 und als Prozeß des Fortschreitens das Anhängen von 1.

Die Dinge, die wir, ausgehend von der Ziffer 1, durch Anwendung des Fortschreitungsprozesses erhalten, wie z. B.

$$1, \; 11, \; 1111$$

sind Figuren von folgender Art: sie beginnen mit 1, sie enden mit 1; auf jede 1, die nicht schon das Ende der Figur bildet, folgt eine angehängte 1. Sie werden durch Anwendung des Fortschreitungsprozesses, also durch einen konkret zum Abschluß kommenden *Aufbau* erhalten, und dieser Aufbau läßt sich daher auch durch einen schrittweisen *Abbau* rückgängig machen.

Wir wollen diese Figuren, mit einer leichten Abweichung vom gewohnten Sprachgebrauch, als „*Ziffern*" bezeichnen.

Was die genaue figürliche Beschaffenheit der Ziffern betrifft, so denken wir uns, wie üblich, für diese einen gewissen Spielraum gelassen, d. h. kleine Unterschiede in der Ausführung, sowohl was die Form der 1 wie ihre Größe, wie auch den Abstand beim Ansetzen einer 1 betrifft, sollen nicht in Betracht gezogen werden. Was wir als wesentlich brauchen, ist nur, daß wir sowohl in der 1 wie in der Anfügung ein anschauliches Objekt haben, das sich in eindeutiger Weise wiedererkennen läßt, und daß wir an einer Ziffer stets die diskreten Teile, aus denen sie aufgebaut ist, überblicken können.

Neben den Ziffern führen wir noch anderweitige Zeichen ein, Zeichen „zur Mitteilung", die von den Ziffern, welche die *Objekte* der Zahlentheorie bilden, grundsätzlich zu unterscheiden sind.

Ein Zeichen zur Mitteilung ist für sich genommen auch eine Figur, von der wir auch voraussetzen, daß sie sich eindeutig wiedererkennen läßt und bei der es auf geringe Unterschiede in der Ausführung nicht ankommt. Innerhalb der Theorie wird es aber nicht zum Gegenstand der Betrachtung gemacht, sondern bildet hier ein Hilfsmittel zur kurzen und deutlichen Formulierung von Tatsachen, Behauptungen und Annahmen.

Wir gebrauchen in der Zahlentheorie folgende Arten von Zeichen zur Mitteilung:

1. Kleine deutsche Buchstaben zur Bezeichnung für irgendeine nicht festgelegte Ziffer;

2. die üblichen Nummern zur Abkürzung für bestimmte Ziffern, z. B. 2 für 11, 3 für 111;

3. Zeichen für gewisse Bildungsprozesse und Rechenoperationen, durch die wir aus gegebenen Ziffern andere gewinnen. Diese können sowohl auf bestimmte wie auch auf unbestimmt gelassene Ziffern angewandt werden, wie z. B. in $\mathfrak{a}+11$;

In number theory, we have an initial thing and a process of progression. | We must fix both intuitively in some definite way. It is inessential how we do this;[21.1] but the choice once made, must be maintained throughout the theory. We choose the initial thing to be the numeral 1, and the process of progression to be suffixing 1.

The things we obtain, starting from the numeral 1, by applying the process of progression, such as

$$1, \quad 11, \quad 1111,$$

are figures of the following kind: they begin with 1, they end with 1; every 1 which is not yet the end of the figure is followed by a suffixed 1. They are obtained by the process of progression, i.e. by a concretely terminating *construction*, and this construction can thus be reversed by a stepwise *deconstruction*.

Slightly deviating from the common usage of language, we will call these figures *"numerals"*.[21.2]

As customary, we assume that a certain tolerance is admitted concerning the exact shape of the numerals. This means that small differences in the realization will not be taken into account, such as the shape of the 1, its size, or the distance at which a 1 is suffixed. What we essentially need is only that the numeral 1 and the suffix 1 are intuitive objects which can be recognized unambiguously, and that we can clearly see in any case out of which discrete parts a numeral is constructed.

Beside the numerals, we introduce further symbols "for communication", which, as a matter of principle, must be distinguished from the numerals, which constitute the *objects* of number theory.

By itself, a symbol for communication is also a figure; and we require that it be unambiguously recognizable, and that minor differences in its realization are irrelevant. Within the theory, however, it is not made into an object of consideration, but it merely serves as a means for the concise and clear formulation of facts, assertions, and assumptions.

In number theory, we use the following kinds of symbols for communication:

1. small German[21.3] letters for designating an arbitrary indeterminate numeral;[21.4]

2. the customary number symbols as abbreviations for definite numerals, such as 2 for 11, and 3 for 111;

3. symbols for certain formation processes and operations of calculation with which we get new numerals from given numerals, and which may be applied to definite and indefinite numerals as well, such as in $\mathfrak{a} + 11$;[21.5] |

[21.1] In the 1st edition of 1934, the page break from Page 20 to Page 21 comes roughly here, exactly one line later than in the 2nd edition of 1968.

[21.2] Note that this sentence is significantly different from the 1st edition of 1934, where it still reads as follows: "Diese Figuren bilden eine Art von Ziffern; wir wollen hier das Wort „Ziffer" schlechtweg zur Bezeichnung *dieser* Figuren gebrauchen." ("These figures constitute a kind of numeral; and we will simply use the word *"numeral"* to denote just *these* figures.")

[21.3] Here and in the following, "German letters" will mean letters in Frankfurter Fraktur (Frankfurt Gothic typeface), such as 𝔄, 𝔅, ℭ, 𝔇, 𝔈, 𝔉, 𝔊, 𝔥, ℑ, 𝔍, 𝔎, 𝔏, 𝔐, 𝔑, 𝔒, 𝔓, 𝔔, 𝔑, 𝔖, 𝔗, 𝔘, 𝔙, 𝔚, 𝔛, 𝔜, 𝔷, 𝔞, 𝔟, 𝔠, 𝔡, 𝔢, 𝔣, 𝔤, 𝔥, 𝔦, 𝔧, 𝔨, 𝔩, 𝔪, 𝔫, 𝔬, 𝔭, 𝔮, 𝔯, 𝔰 (𝔰, 𝔥), 𝔱, 𝔲, 𝔳, 𝔴, 𝔵, 𝔶, 𝔷.

[21.4] This somewhat strange phrase can be understood as "small German letters as variables for numerals", but we prefer the more literal translation given here because there is a small chance that it might refer to something like the arbitrary objects of [FINE, 1985].

[21.5] It is interesting to note that the 1st edition of 1934 stresses the aspect of calculation as an activity in its Item 3 here: "3. Zeichen für gewisse Handlungen, die wir mit den Ziffern vornehmen, und gewisse Bildungsprozesse, durch welche wir aus gegebenen Ziffern andere gewinnen." ("3. symbols for certain operations we carry out with the numerals and certain formation processes with which we get new numerals from given ones;") (complete quotation of the whole item; no example was given). Today, "Handlung" would not be used for a mathematical operation in this context anymore, but only for an action.

4. das Zeichen $=$ zur Mitteilung der figürlichen Übereinstimmung, das Zeichen $\neq$ zur Mitteilung der Verschiedenheit zweier Ziffern: die Zeichen $<, >$ zur Bezeichnung der noch zu erklärenden Größenbeziehung zwischen Ziffern.

5. Klammern als Zeichen für die Art der Aufeinanderfolge von Prozessen, wo diese nicht ohne weiteres deutlich ist.

In welcher Weise mit den eingeführten Zeichen operiert wird, und wie die inhaltlichen Überlegungen anzustellen sind, wird am besten deutlich, wenn wir die Zahlentheorie ein Stück weit in den Hauptzügen entwickeln.

Das erste, was wir an den Ziffern feststellen, ist die Größenbeziehung. Es sei eine Ziffer $\mathfrak{a}$ von einer Ziffer $\mathfrak{b}$ verschieden. Überlegen wir, wie das sein kann. Beide beginnen mit 1, und der Aufbau schreitet für $\mathfrak{a}$ wie für $\mathfrak{b}$ in ganz derselben Weise fort, sofern nicht eine der Ziffern endigt, während der Aufbau der anderen noch weiter geht. Dieser Fall muß also einmal eintreten, und somit stimmt die eine Ziffer mit einem *Teilstück* der anderen überein, oder genauer ausgedrückt: der Aufbau der einen Ziffer stimmt mit einem Anfangsstück von dem Aufbau der anderen überein.

Wir sagen in dem Falle, wo eine Ziffer $\mathfrak{a}$ mit einem Teilstück von $\mathfrak{b}$ übereinstimmt, daß $\mathfrak{a}$ kleiner ist als $\mathfrak{b}$ oder auch daß $\mathfrak{b}$ größer ist als $\mathfrak{a}$, und wenden dafür die Bezeichnung

$$\mathfrak{a} < \mathfrak{b}, \quad \mathfrak{b} > \mathfrak{a}$$

an. Aus unserer Überlegung geht hervor, daß für eine Ziffer $\mathfrak{a}$ und eine Ziffer $\mathfrak{b}$ stets eine der Beziehungen

$$\mathfrak{a} = \mathfrak{b}, \quad \mathfrak{a} < \mathfrak{b}, \quad \mathfrak{b} < \mathfrak{a}$$

stattfinden muß, und andererseits ist aus der anschaulichen Bedeutung ersichtlich, daß diese Beziehungen einander ausschließen. Desgleichen ergibt sich unmittelbar, daß, falls $\mathfrak{a} < \mathfrak{b}$ und $\mathfrak{b} < \mathfrak{c}$, auch stets $\mathfrak{a} < \mathfrak{c}$ ist.

In engem Zusammenhang mit der Größenbeziehung der Ziffern steht die *Addition*. Wenn eine Ziffer $\mathfrak{b}$ mit einem Teilstück von $\mathfrak{a}$ übereinstimmt, so ist das Reststück wiederum eine Ziffer $\mathfrak{c}$; man erhält also die Ziffer $\mathfrak{a}$, indem man $\mathfrak{c}$ an $\mathfrak{b}$ ansetzt, in der Weise, daß die 1, mit welcher $\mathfrak{c}$ beginnt, an die 1, mit welcher $\mathfrak{b}$ endigt, nach der Art des Fortschreitungsprozesses angehängt wird. Diese Art der Zusammensetzung von Ziffern bezeichnen wir als *Addition* und wenden dafür das Zeichen $+$ an.

Aus dieser Definition der Addition entnehmen wir direkt: Wenn $\mathfrak{b} < \mathfrak{a}$, so gewinnt man durch Vergleichung von $\mathfrak{b}$ mit $\mathfrak{a}$ eine Darstellung von $\mathfrak{a}$ in der Form $\mathfrak{b} + \mathfrak{c}$, wobei $\mathfrak{c}$ wieder eine Ziffer ist. Geht man andererseits von irgendwelchen Ziffern $\mathfrak{b}, \mathfrak{c}$ aus, so liefert die Addition wiederum eine Ziffer $\mathfrak{a}$, so daß

$$\mathfrak{a} = \mathfrak{b} + \mathfrak{c},$$

4. the symbol = to communicate figural coincidence and the symbol $\neq$ to communicate difference of numerals;[22.1] the[22.2] symbols[22.3] < and > to designate the relation of magnitude between numerals,[22.4] which still is to be explained;[22.5]

5. parentheses as symbols for the order of processes, where this order is not clear without further explanation.

The way in which one operates with the introduced symbols and how the contentual considerations are to be carried out is best made clear when we now develop number theory and its main features a bit further.

The first thing we establish for the numerals is the relation of magnitude. Let a numeral $\mathfrak{a}$ be different from a numeral $\mathfrak{b}$. Let us reflect on how this is possible. Both begin with a 1, and the construction of $\mathfrak{a}$ and $\mathfrak{b}$ continues in exactly the same way, unless one of the numerals ends whereas the construction of the other still continues. This must be the case at some time, and thus one numeral coincides with a *segment* of the other, or more precisely: the construction of one numeral coincides with a proper initial segment of the construction of the other.

If a numeral $\mathfrak{a}$ coincides with a segment of $\mathfrak{b}$, we say either that $\mathfrak{a}$ is smaller than $\mathfrak{b}$, or that $\mathfrak{b}$ is greater than $\mathfrak{a}$, and for this we use the notation

$$\mathfrak{a} < \mathfrak{b}, \quad \mathfrak{b} > \mathfrak{a}.$$

It follows from our considerations that for a numeral $\mathfrak{a}$ and a numeral $\mathfrak{b}$ one of the following relations must always hold:

$$\mathfrak{a} = \mathfrak{b}, \quad \mathfrak{a} < \mathfrak{b}, \quad \mathfrak{b} < \mathfrak{a}.$$

Moreover, it is apparent from the intuitive meaning that these relations are mutually exclusive. Furthermore, it immediately follows that, whenever $\mathfrak{a} < \mathfrak{b}$ and $\mathfrak{b} < \mathfrak{c}$, then $\mathfrak{a} < \mathfrak{c}$ holds as well.

Closely related to the relation of magnitude between numerals is *addition*. If a numeral $\mathfrak{b}$ coincides with a segment of a numeral $\mathfrak{a}$, then the remaining segment is again a numeral $\mathfrak{c}$; and thus we obtain the numeral $\mathfrak{a}$ by attaching[22.6] $\mathfrak{c}$ to $\mathfrak{b}$ by suffixing the 1 with which $\mathfrak{c}$ begins according to our process of progression to the 1 with which $\mathfrak{b}$ ends. This kind of concatenation of numerals we call *addition*, and we use the symbol + for it.

Directly from this definition of addition, we get the following: If $\mathfrak{b} < \mathfrak{a}$, by comparing $\mathfrak{b}$ to $\mathfrak{a}$, one obtains a representation of $\mathfrak{a}$ of the form $\mathfrak{b}+\mathfrak{c}$, where $\mathfrak{c}$ is again a numeral. In turn, if one is given two arbitrary numerals $\mathfrak{b}$ and $\mathfrak{c}$, then addition produces another numeral $\mathfrak{a}$, such that

$$\mathfrak{a} = \mathfrak{b}+\mathfrak{c},$$

[22.1] The colon ":" in the German original of 1968 is a typo, probably caused by the fact that the semicolon ";" in the 1st edition of 1934 looks like a colon to the naked eye.

[22.2] In the 1st edition of 1934, the page break from Page 21 to Page 22 is here (i.e. between "das" ("the") and "Zeichen" ("symbol")), the page break from Page 20 to Page 21 is exactly 1 line after the current one, and the page breaks from Page 22 to Page 23 from Page 23 to Page 24 are exactly 2 lines before the current ones.

[22.3] In the 1st edition of 1934 this reads "das Zeichen <" ("the symbol <") instead of "die Zeichen <, >" ("the symbols < and >").

[22.4] In the 1st edition of 1934, this reads "Zahlzeichen" ("number symbols") instead of "Ziffern" ("numerals").

[22.5] The full-stop "." in the German original is a typo of the 2nd edition (1968) only.

[22.6] We have generally translated "anhängen" as "to suffix". Here we translate "ansetzen" as "to attach". There is no semantical difference between these German words here. Moreover, "ansetzen" with the given transitive verb pattern is used only in tailoring.

und es ist dann
$$\mathfrak{b} < \mathfrak{a}.$$
Allgemein gilt also:
$$\mathfrak{b} < \mathfrak{b} + \mathfrak{c}.$$
Auf Grund der eingeführten Definitionen ergibt sich die Bedeutung der numerischen Gleichungen und Ungleichungen, wie
$$2 < 3, \ 2 + 3 = 5.$$
$2 < 3$ besagt, daß die Ziffer 11 mit einem Teilstück von 111 übereinstimmt;

$2 + 3 = 5$ besagt, daß durch Ansetzen von 111 an 11 die Ziffer 11111 entsteht.

Wir haben hier beide Male die Darstellung einer richtigen Aussage, während z. B.
$$2 + 3 = 4$$
die Darstellung einer falschen Aussage ist.

Für die anschaulich definierte Addition haben wir nunmehr die Gültigkeit der Rechengesetze festzustellen.

Diese werden hier als Sätze über beliebig vorgelegte Ziffern aufgefaßt und als solche durch anschauliche Überlegung eingesehen.

Unmittelbar aus der Definition der Addition entnimmt man das assoziative Gesetz, wonach, wenn $\mathfrak{a}, \mathfrak{b}, \mathfrak{c}$ irgendwelche Ziffern sind, stets
$$\mathfrak{a} + (\mathfrak{b} + \mathfrak{c}) = (\mathfrak{a} + \mathfrak{b}) + \mathfrak{c}.$$
Nicht so direkt ergibt sich das kommutative Gesetz, welches besagt, daß stets
$$\mathfrak{a} + \mathfrak{b} = \mathfrak{b} + \mathfrak{a}$$
ist. Wir gebrauchen hier die Beweismethode der *vollständigen Induktion*. Machen wir uns zunächst klar, wie diese Schlußweise von unserem elementaren Standpunkt aufzufassen ist: Es werde irgendeine Aussage betrachtet, die sich auf eine Ziffer bezieht und die einen elementar anschaulichen Inhalt besitzt. Die Aussage treffe für 1 zu, und man wisse auch, daß die Aussage, falls sie für eine Ziffer $\mathfrak{n}$ zutrifft, dann auch jedenfalls für die Ziffer $\mathfrak{n} + 1$ zutrifft. Hieraus folgert man, daß die Aussage für jede vorgelegte Ziffer $\mathfrak{a}$ zutrifft.

In der Tat ist ja die Ziffer $\mathfrak{a}$ aufgebaut, indem man, von 1 beginnend, den Prozeß des Anhängens von 1 anwendet. Konstatiert man nun zunächst das Zutreffen der betrachteten Aussage für 1, und bei jedem Anhängen einer 1, auf Grund der gemachten Voraussetzung, das Zutreffen der Aussage für die neu erhaltene Ziffer, so gelangt man beim fertigen Aufbau von $\mathfrak{a}$ zu der Feststellung, daß die Aussage für $\mathfrak{a}$ zutrifft.

Wir haben es also hier nicht mit einem selbständigen Prinzip zu tun, sondern mit einer Folgerung, die wir aus dem konkreten Aufbau der Ziffern entnehmen.

Mit Hilfe dieser Schlußweise können wir nun nach der üblichen Art zeigen, daß für jede Ziffer
$$1 + \mathfrak{a} = \mathfrak{a} + 1,$$

| and thus
$$\mathfrak{b} < \mathfrak{a}.$$
Therefore, in general we have[23.1]
$$\mathfrak{b} < \mathfrak{b}+\mathfrak{c}.$$

On the basis of the definitions introduced so far, the meaning of numerical equations and inequations becomes clear, such as
$$2 < 3, \qquad 2+3 = 5.$$
$2 < 3$ means that the numeral 11 coincides with a segment of 111;[23.2] $2+3 = 5$ means that attaching 111 to 11 results in the numeral 11111. In both cases we have the representation of a correct proposition, whereas
$$2+3 = 4$$
is the representation of a false proposition.

[23.3]We now have to establish the validity of the laws of calculation for this intuitively defined addition.

These laws are regarded here as sentences about arbitrarily given numerals, and they are as such seen to hold by intuitive consideration.

Immediately from our definition of addition, we obtain the associative law, according to which, if $\mathfrak{a}, \mathfrak{b}, \mathfrak{c}$ are arbitrary numerals, we always have
$$\mathfrak{a} + (\mathfrak{b}+\mathfrak{c}) = (\mathfrak{a}+\mathfrak{b}) + \mathfrak{c}.$$
The law of commutativity, according to which we always have
$$\mathfrak{a}+\mathfrak{b} = \mathfrak{b}+\mathfrak{a},$$
does not arise quite so directly. Here we use the proof method of *mathematical induction*.[23.4] Let us first clarify how this kind of inference is to be understood from our elementary standpoint: Suppose we are given any proposition that refers to a numeral and that has an elementarily intuitive content. Further, suppose that the proposition holds for 1, and that it is also known that, if the proposition holds for a numeral $\mathfrak{n}$, then it must also hold for the numeral $\mathfrak{n}+1$. From all this, one infers that the proposition holds for every given numeral $\mathfrak{a}$.

Indeed, the numeral $\mathfrak{a}$ is constructed by applying the process of suffixing 1, starting from 1. If one observes now that the proposition considered holds for 1, and that it holds as well for the new numeral resulting from suffixing 1 under the given hypothesis, then one realizes by the completion of the construction of $\mathfrak{a}$ that the proposition holds for $\mathfrak{a}$.

Thus, we are not dealing with an independent principle here,[23.5] but with a consequence that we take from the concrete construction of the numerals.

[23.1]This sentence was added in the 2nd edition of 1968. Because of a slight difference in page breaking, Page 23 starts in the 1st edition of 1934 right with the following senctence.

[23.2]The following two paragraph indentations of the German original (which would turn the remainder of the sentence into a separate paragraph) seem to be a misprints of both the 1st and the 2nd edition.

[23.3]§ 2(a)(2) (as listed in the table of contents) starts with this paragraph.

[23.4]We translate the German term "vollständige Induktion" (literally translated: "complete induction") as "mathematical induction". "Complete induction" is a term hardly used in English, where it typically means *course-of-values induction* (cf. e.g. http://en.wikipedia.org/wiki/Mathematical_induction#Complete_induction), which is different from the *structural induction* to which the German original actually refers.

[23.5]Seen as a statement in favor of contentual mathematics as a closed system, this remark may be misleading: No matter in which way we try to approach it, induction is definitely a principle that is separate from deduction. What is indeed the case, however, and probably meant here, is that the principle of induction is already implicitly required for the definition of the numerals.

und auf Grund hiervon weiter, daß stets

ist. $$\mathfrak{a} + \mathfrak{b} = \mathfrak{b} + \mathfrak{a}$$

Es werde nun noch kurz die Einführung der Multiplikation, der Division und der anschließenden Begriffsbildungen angegeben.

Die *Multiplikation* kann folgendermaßen definiert werden: $\mathfrak{a} \cdot \mathfrak{b}$ bedeutet die Ziffer, die man aus der Ziffer $\mathfrak{b}$ erhält, indem man beim Aufbau immer die 1 durch die Ziffer $\mathfrak{a}$ ersetzt, so daß man also zunächst $\mathfrak{a}$ bildet und anstatt jedes in der Bildung von $\mathfrak{b}$ vorkommenden Anfügens von 1 das Ansetzen von $\mathfrak{a}$ ausführt.

Aus dieser Definition ergibt sich unmittelbar das assoziative Gesetz der Multiplikation, ferner das distributive Gesetz, wonach stets

$$\mathfrak{a} \cdot (\mathfrak{b} + \mathfrak{c}) = (\mathfrak{a} \cdot \mathfrak{b}) + (\mathfrak{a} \cdot \mathfrak{c}).$$

Das andere distributive Gesetz, wonach stets

$$(\mathfrak{b} + \mathfrak{c}) \cdot \mathfrak{a} = (\mathfrak{b} \cdot \mathfrak{a}) + (\mathfrak{c} \cdot \mathfrak{a}),$$

wird auf Grund der Gesetze der Addition mit Hilfe der vorhin beschriebenen vollständigen Induktion eingesehen. Durch diese Beweismethode erhält man dann auch das kommutative Gesetz der Multiplikation.

Um zur Division zu gelangen, stellen wir zunächst eine Vorbetrachtung an. Der Aufbau einer Ziffer ist so beschaffen, daß beim Anhängen von 1 jedesmal eine neue Ziffer gewonnen wird. Die Bildung einer Ziffer $\mathfrak{a}$ geschieht also auf dem Wege der Bildung einer konkreten Reihe von Ziffern, die mit 1 beginnt, mit $\mathfrak{a}$ endigt und wo jede Ziffer aus der vorhergehenden durch Anhängen von 1 entsteht. Man sieht auch sogleich, daß diese Reihe außer $\mathfrak{a}$ selbst nur solche Ziffern enthält, die $< \mathfrak{a}$ sind, und daß eine Ziffer, welche $< \mathfrak{a}$ ist, auch in dieser Reihe vorkommen muß. Wir nennen diese Aufeinanderfolge von Ziffern kurz „die Reihe der Ziffern von 1 bis $\mathfrak{a}$".

Sei nun $\mathfrak{b}$ eine von 1 verschiedene Ziffer, die $< \mathfrak{a}$ ist. $\mathfrak{b}$ hat dann die Form $1 + \mathfrak{c}$, und daher ist

$$\mathfrak{b} \cdot \mathfrak{a} = (1 \cdot \mathfrak{a}) + (\mathfrak{c} \cdot \mathfrak{a}) = \mathfrak{a} + (\mathfrak{c} \cdot \mathfrak{a}),$$

also

$$\mathfrak{a} < \mathfrak{b} \cdot \mathfrak{a}.$$

Multiplizieren wir nun $\mathfrak{b}$ nacheinander mit den Ziffern aus der Reihe von 1 bis $\mathfrak{a}$, so ist in der entstehenden Reihe von Ziffern

$$\mathfrak{b} \cdot 1, \quad \mathfrak{b} \cdot 11, \quad \ldots, \mathfrak{b} \cdot \mathfrak{a}$$

die erste $< \mathfrak{a}$ und die letzte $> \mathfrak{a}$. Gehen wir nun in der Reihe dieser Ziffern so weit, bis wir zuerst auf eine solche treffen, die $> \mathfrak{a}$ ist; die vorherige, welche $\mathfrak{b} \cdot \mathfrak{q}$ sei, ist dann entweder $= \mathfrak{a}$ oder $< \mathfrak{a}$, während

$$\mathfrak{b} \cdot (\mathfrak{q} + 1) = (\mathfrak{b} \cdot \mathfrak{q}) + \mathfrak{b} > \mathfrak{a}$$

ist. Somit ist entweder $\mathfrak{a} = \mathfrak{b} \cdot \mathfrak{q}$,

oder wir haben eine Darstellung

$$\mathfrak{a} = (\mathfrak{b} \cdot \mathfrak{q}) + \mathfrak{r},$$

With the help of this method of inference, we can now show in the customary way that for each numeral we have
$$1+\mathfrak{a} = \mathfrak{a}+1,$$
and from of this we can further show that
$$\mathfrak{a}+\mathfrak{b} = \mathfrak{b}+\mathfrak{a}$$
always holds.[24.1]

We will now briefly sketch the introduction of multiplication, division, and the subsequent concept formations.

Multiplication can be defined as follows: $\mathfrak{a}\cdot\mathfrak{b}$ denotes the numeral one obtains from the numeral $\mathfrak{b}$ by replacing each 1 in its construction with the numeral $\mathfrak{a}$. Thus, one starts with the formation of $\mathfrak{a}$ and suffixes $\mathfrak{a}$ instead of any 1 in the construction of $\mathfrak{b}$.

From this definition, the associative law of multiplication follows immediately, as well as the distributive law according to which
$$\mathfrak{a}\cdot(\mathfrak{b}+\mathfrak{c}) = (\mathfrak{a}\cdot\mathfrak{b}) + (\mathfrak{a}\cdot\mathfrak{c}).$$
The other distributive law, according to which
$$(\mathfrak{b}+\mathfrak{c})\cdot\mathfrak{a} = (\mathfrak{b}\cdot\mathfrak{a}) + (\mathfrak{c}\cdot\mathfrak{a}),$$
is seen to be true by the laws of addition and with the help of mathematical induction as described above. By this proof method, we can also get the commutative law of multiplication.

Regarding division, we start with a preliminary consideration. The construction of a numeral is such that with each suffixing of 1 a new numeral is obtained. The formation of a numeral $\mathfrak{a}$ then involves the formation of a concrete series of numerals, starting with 1 and ending with $\mathfrak{a}$, in which each numeral is obtained from its predecessor by suffixing a 1. One also immediately sees that, except for $\mathfrak{a}$ itself, this series contains only numerals $<\mathfrak{a}$, and that any numeral $<\mathfrak{a}$ must occur in the series. We call this sequence of numerals "the series of numerals from 1 to $\mathfrak{a}$" for short.

Now let $\mathfrak{b}$ be a numeral different from 1 and $<\mathfrak{a}$. Then $\mathfrak{b}$ is of the form $1+\mathfrak{c}$, and therefore we have
$$(\mathfrak{b}\cdot\mathfrak{a}) = (1\cdot\mathfrak{a}) + (\mathfrak{c}\cdot\mathfrak{a}) = \mathfrak{a} + (\mathfrak{c}\cdot\mathfrak{a}),$$
thus
$$\mathfrak{a} < \mathfrak{b}\cdot\mathfrak{a}.$$
If we now multiply $\mathfrak{b}$ successively with the numerals in the series from 1 to $\mathfrak{a}$, then in the resulting series of numerals
$$\mathfrak{b}\cdot 1, \quad \mathfrak{b}\cdot 11, \quad \ldots, \quad \mathfrak{b}\cdot \mathfrak{a}.$$
the first is $<\mathfrak{a}$ and the last is $>\mathfrak{a}$. Now we run through this series up to the point where we first find a numeral $>\mathfrak{a}$; its predecessor, say $\mathfrak{b}\cdot\mathfrak{q}$, must then be $=\mathfrak{a}$ or $<\mathfrak{a}$, whereas
$$\mathfrak{b}\cdot(\mathfrak{q}+1) = (\mathfrak{b}\cdot\mathfrak{q}) + \mathfrak{b} > \mathfrak{a}$$
holds. Thus, either
$$\mathfrak{a} = \mathfrak{b}\cdot\mathfrak{q},$$
or we have a representation
$$\mathfrak{a} = (\mathfrak{b}\cdot\mathfrak{q}) + \mathfrak{r},$$

[24.1] Because of a slight difference in page breaking, Page 24 starts in the 1st edition of 1934 right with the following senctence, i.e. three lines later than in the 2nd edition of 1968.

und dabei ist
$$(\mathfrak{b} \cdot \mathfrak{q}) + \mathfrak{r} < (\mathfrak{b} \cdot \mathfrak{q}) + \mathfrak{b},$$
also
$$\mathfrak{r} < \mathfrak{b}.$$

Im ersten Falle ist $\mathfrak{a}$ „durch $\mathfrak{b}$ teilbar" („$\mathfrak{b}$ geht in $\mathfrak{a}$ auf"), im zweiten Fall haben wir die Division mit Rest.

Wir nennen allgemein $\mathfrak{a}$ durch $\mathfrak{b}$ teilbar, wenn in der Reihe

$$\mathfrak{b} \cdot 1, \quad \mathfrak{b} \cdot 11, \quad \ldots, \quad \mathfrak{b} \cdot \mathfrak{a}$$

die Ziffer $\mathfrak{a}$ vorkommt. Dies trifft zu für $\mathfrak{b} = 1$, für $\mathfrak{b} = \mathfrak{a}$ und sonst in dem eben erhaltenen ersten Fall.

Aus der Definition der Teilbarkeit folgt unmittelbar, daß, falls $\mathfrak{a}$ durch $\mathfrak{b}$ teilbar ist, mit der Feststellung der Teilbarkeit auch eine Darstellung

$$\mathfrak{a} = \mathfrak{b} \cdot \mathfrak{q}$$

gegeben ist. Aber es gilt auch die Umkehrung, daß aus einer Gleichung

$$\mathfrak{a} = \mathfrak{b} \cdot \mathfrak{q}$$

stets die Teilbarkeit von $\mathfrak{a}$ durch $\mathfrak{b}$ (in dem definierten Sinne) folgt, da die Ziffer $\mathfrak{q}$ der Reihe der Ziffern von 1 bis $\mathfrak{a}$ angehören muß.

Ist $\mathfrak{a} \neq 1$ und kommt in der Reihe der Ziffern von 1 bis $\mathfrak{a}$ außer 1 und $\mathfrak{a}$ kein Teiler von $\mathfrak{a}$ vor, so daß jedes der Produkte $\mathfrak{m} \cdot \mathfrak{n}$, worin $\mathfrak{m}$ und $\mathfrak{n}$ der Reihe der Ziffern von 2 bis $\mathfrak{a}$ angehören, von $\mathfrak{a}$ verschieden ist, so nennen wir $\mathfrak{a}$ eine *Primzahl*.

Ist $\mathfrak{n}$ eine von 1 verschiedene Ziffer, so gibt es in der Reihe der Ziffern 1 bis $\mathfrak{n}$ jedenfalls eine erste, welche die Eigenschaft hat, von 1 verschieden und Teiler von $\mathfrak{n}$ zu sein. Von diesem „kleinsten von 1 verschiedenen Teiler von $\mathfrak{n}$" zeigt man leicht, daß er eine Primzahl ist.

Nun können wir auch nach dem Verfahren von EUKLID den Satz beweisen, daß zu jeder Ziffer $\mathfrak{a}$ eine Primzahl $> \mathfrak{a}$ bestimmt werden kann: Man multipliziere die Zahlen der Reihe von 1 bis $\mathfrak{a}$ miteinander, addiere 1 und nehme von der so erhaltenen Ziffer den kleinsten von 1 verschiedenen Teiler $\mathfrak{t}$. Dieser ist dann eine Primzahl, und man erkennt leicht, daß $\mathfrak{t}$ nicht in der Reihe der Zahlen von 1 bis $\mathfrak{a}$ vorkommen kann, mithin $> \mathfrak{a}$ ist.

Von hier aus ist der weitere Aufbau der elementaren Zahlentheorie ersichtlich; nur noch ein Punkt bedarf hier der grundsätzlichen Erörterung, das Verfahren der *rekursiven Definition*. Vergegenwärtigen wir uns, worin dieses Verfahren besteht: Ein neues Funktionszeichen, etwa φ wird eingeführt, und die Definition der Funktion geschieht durch zwei Gleichungen, welche im einfachsten Falle die Form haben:

$$\varphi(1) = \mathfrak{a},$$

$$\varphi(\mathfrak{n} + 1) = \psi(\varphi(\mathfrak{n}), \mathfrak{n}).$$

| where
$$(\mathfrak{b}\cdot\mathfrak{q}) + \mathfrak{r} \;<\; (\mathfrak{b}\cdot\mathfrak{q}) + \mathfrak{b},$$
thus
$$\mathfrak{r} \;<\; \mathfrak{b}.$$
In the first case, $\mathfrak{a}$ "is divisible by $\mathfrak{b}$" ("$\mathfrak{b}$ divides $\mathfrak{a}$"). In the second case, we have a division with remainder.

Generally, we call $\mathfrak{a}$ divisible by $\mathfrak{b}$ if the numeral $\mathfrak{a}$ occurs in the series
$$\mathfrak{b}\cdot 1, \quad \mathfrak{b}\cdot 11, \quad \ldots, \quad \mathfrak{b}\cdot\mathfrak{a}.$$
This holds for $\mathfrak{b}=1$, for $\mathfrak{b}=\mathfrak{a}$, and also in the first of the two cases just obtained.

If $\mathfrak{a}$ is divisible by $\mathfrak{b}$, then it follows immediately from the definition of divisibility that we also have a representation as
$$\mathfrak{a} \;=\; \mathfrak{b}\cdot\mathfrak{q}.$$
But the converse holds as well: From an equation
$$\mathfrak{a} \;=\; \mathfrak{b}\cdot\mathfrak{q},$$
the divisibility of $\mathfrak{a}$ by $\mathfrak{b}$ (in the sense defined) follows since the numeral $\mathfrak{q}$ must belong to the series of numerals from 1 to $\mathfrak{a}$.

If $\mathfrak{a}\neq 1$ and no divisor of $\mathfrak{a}$ other than 1 and $\mathfrak{a}$ occurs in the series of numerals from 1 to $\mathfrak{a}$, which means that every product $\mathfrak{m}\cdot\mathfrak{n}$ in which $\mathfrak{m}$ and $\mathfrak{n}$ belong to the series from 2 to $\mathfrak{a}$ is distinct from $\mathfrak{a}$, then we call $\mathfrak{a}$ a *prime number*.

If $\mathfrak{n}$ is a numeral distinct from 1, then, in the series of numerals from 1 to $\mathfrak{n}$, there is a first numeral that is distinct from 1 and a divisor of $\mathfrak{n}$. It is easy to show that this "smallest divisor of $\mathfrak{n}$ distinct from 1" is a prime number.

By the method of EUCLID, we can now prove the sentence that, for any numeral $\mathfrak{a}$, we can determine a prime number $>\mathfrak{a}$: Multiply the numbers in the series from 1 to $\mathfrak{a}$, add 1, and take the smallest divisor $\mathfrak{t}$ of that numeral distinct from 1. This $\mathfrak{t}$ is then a prime number and one easily recognizes that $\mathfrak{t}$ cannot occur in the series of numbers from 1 to $\mathfrak{a}$ and thus it is $>\mathfrak{a}$.

Hierbei ist $\mathfrak{a}$ eine Ziffer und ψ eine Funktion, die aus bereits bekannten Funktionen durch Zusammensetzung gebildet ist, so daß $\psi(\mathfrak{b}, \mathfrak{c})$ für gegebene Ziffern $\mathfrak{b}, \mathfrak{c}$ berechnet werden kann und als Wert wieder eine Ziffer liefert.

Z. B. kann die Funktion
$$\varrho(\mathfrak{n}) = 1 \cdot 2 \ldots \mathfrak{n}$$
definiert werden durch die Gleichungen:
$$\varrho(1) = 1,$$
$$\varrho(\mathfrak{n} + 1) = \varrho(\mathfrak{n}) \cdot (\mathfrak{n} + 1).$$
Es ist nicht ohne weiteres klar, welcher Sinn diesem Definitionsverfahren zukommt. Zur Erklärung ist zunächst der Funktionsbegriff zu präzisieren. Unter einer *Funktion* verstehen wir hier eine anschauliche Anweisung, auf Grund deren einer vorgelegten Ziffer, bzw. einem Paar, einem Tripel, ... von Ziffern, wieder eine Ziffer zugeordnet wird. Ein Gleichungspaar der obigen Art — wir nennen ein solches eine „Rekursion" — haben wir anzusehen als eine *abgekürzte Mitteilung* folgender Anweisung:

Es sei $\mathfrak{m}$ irgendeine Ziffer. Wenn $\mathfrak{m} = 1$ ist, so werde $\mathfrak{m}$ die Ziffer $\mathfrak{a}$ zugeordnet. Andernfalls hat $\mathfrak{m}$ die Form $\mathfrak{b} + 1$. Man schreibe dann zunächst schematisch auf:
$$\psi(\varphi(\mathfrak{b}), \mathfrak{b}).$$
Ist nun $\mathfrak{b} = 1$, so ersetze man hierin $\varphi(\mathfrak{b})$ durch $\mathfrak{a}$; andernfalls hat wieder $\mathfrak{b}$ die Form $\mathfrak{c} + 1$, und man ersetze dann $\varphi(\mathfrak{b})$ durch
$$\psi(\varphi(\mathfrak{c}), \mathfrak{c}).$$
Nun ist wieder entweder $\mathfrak{c} = 1$ oder $\mathfrak{c}$ von der Form $\mathfrak{d} + 1$. Im ersten Fall ersetze man $\varphi(\mathfrak{c})$ durch $\mathfrak{a}$, im zweiten Fall durch
$$\psi(\varphi(\mathfrak{d}), \mathfrak{d}).$$

Die Fortsetzung dieses Verfahrens führt jedenfalls zu einem Abschluß. Denn die Ziffern
$$\mathfrak{b}, \mathfrak{c}, \mathfrak{d}, \ldots,$$
welche wir der Reihe nach erhalten, entstehen durch den *Abbau der Ziffer* $\mathfrak{m}$, und dieser muß ebenso wie der Aufbau von $\mathfrak{m}$ zum Abschluß gelangen. Wenn wir beim Abbau bis zu 1 gekommen sind, dann wird $\varphi(1)$ durch $\mathfrak{a}$ ersetzt; das Zeichen φ kommt dann in der entstehenden Figur nicht mehr vor, vielmehr tritt als Funktionszeichen nur ψ, eventuell in mehrmaliger Überlagerung, auf, und die innersten Argumente sind Ziffern. Damit sind wir zu einem berechenbaren Ausdruck gelangt; denn ψ soll ja eine bereits bekannte Funktion sein. Diese Berechnung hat man nun von innen her auszuführen, und die dadurch gewonnene Ziffer soll der Ziffer $\mathfrak{m}$ zugeordnet werden.

Aus dem Inhalt dieser Anweisung ersehen wir zunächst, daß sie sich in jedem Falle einer vorgelegten Ziffer $\mathfrak{m}$ grundsätzlich erfüllen läßt

[26.1] From here on, the further development of elementary number theory is clear; only the method of *recursive definition* requires further discussion as a basic principle. Let us bring to mind what this method consists in: A new function symbol is introduced, say φ, and the definition of this function is given by two equations, which in the simplest case have the form:
$$\varphi(1) = \mathfrak{a},$$
$$\varphi(\mathfrak{n}+1) = \psi(\varphi(\mathfrak{n}), \mathfrak{n}).$$

Here, $\mathfrak{a}$ is a numeral, and ψ is a function composed of functions that are already known, such that $\psi(\mathfrak{b},\mathfrak{c})$ can be calculated for given numerals $\mathfrak{b}$, $\mathfrak{c}$, and again returns a numeral as its value.

For example, the function[26.2]
$$\varrho(\mathfrak{n}) = 1 \cdot 2 \cdot \ldots \cdot \mathfrak{n}$$
can be defined by the equations:
$$\varrho(1) = 1,$$
$$\varrho(\mathfrak{n}+1) = \varrho(\mathfrak{n}) \cdot (\mathfrak{n}+1).$$

It is not immediately obvious in which sense we are to understand this method of definition. To explain it, we first have to make the notion of a function more precise. Here, we understand a *function* to be an intuitive instruction by which a numeral is assigned to a given numeral, or a pair, a triple, ... of numerals. A pair of equations of the above sort — we call it a *"recursion"* — is to be viewed as an *abbreviated communication* of the following instruction:

Let $\mathfrak{m}$ be an arbitrary numeral. If $\mathfrak{m} = 1$, then the numeral $\mathfrak{a}$ is assigned to $\mathfrak{m}$. Otherwise, $\mathfrak{m}$ has the form $\mathfrak{b}+1$. Then let us first write down schematically:
$$\psi(\varphi(\mathfrak{b}), \mathfrak{b}).$$
Now, if $\mathfrak{b} = 1$, then $\varphi(\mathfrak{b})$ is replaced with $\mathfrak{a}$. Otherwise, $\mathfrak{b}$ has the form $\mathfrak{c}+1$, and then $\varphi(\mathfrak{b})$ is replaced with
$$\psi(\varphi(\mathfrak{c}), \mathfrak{c}).$$
Now again either $\mathfrak{c} = 1$, or $\mathfrak{c}$ is of the form $\mathfrak{d}+1$. In the first case, $\varphi(\mathfrak{c})$ is replaced with $\mathfrak{a}$; in the second case with
$$\psi(\varphi(\mathfrak{d}), \mathfrak{d}).$$
The continuation of this procedure always terminates: The numerals
$$\mathfrak{b}, \ \mathfrak{c}, \ \mathfrak{d}, \ \ldots,$$
which we obtain one after the other, arise from the *deconstruction of the numeral* $\mathfrak{m}$, which must terminate, just as the construction of $\mathfrak{m}$ terminates. When we have arrived at 1 in this process of deconstruction, then $\varphi(1)$ is replaced with $\mathfrak{a}$. The symbol φ does not occur any longer in the resulting figure; rather, only ψ occurs as a function symbol, possibly multiply nested in its own argument positions,[26.3] and the innermost arguments are numerals. Thus, we have arrived at a computable expression because ψ is supposed to be a function that is already known.[26.4] Now this computation is to be carried out from the inside out, and the numeral finally obtained is the value to be assigned to the numeral $\mathfrak{m}$.

[26.1] §2(a)(3) (as listed in the table of contents) starts with this paragraph.

[26.2] In the following line, we have added two dots "·" around "...".

[26.3] The original of "multiply nested in its own argument positions" reads "in mehrmaliger Überlagerung". More literally, this may be translated as "in multiple overlapping" or as "in multiple superposition", which, however, is confusing in English and is also in conflict with modern terminology in the area of term rewriting, which provides a formalization for this kind of recursion in the most general form, cf. [WIRTH, 2009].

[26.4] Note that any function is supposed to be computable here, because it is supposed to be an "instruction", as mentioned above.

und daß das Ergebnis eindeutig festgelegt ist. Zugleich ergibt sich aber auch, daß für jede gegebene Ziffer $\mathfrak{n}$ die Gleichung

$$\varphi(\mathfrak{n} + 1) = \psi(\varphi(\mathfrak{n}), \mathfrak{n})$$

erfüllt wird, wenn wir darin $\varphi(\mathfrak{n})$ und $\varphi(\mathfrak{n} + 1)$ durch die den Ziffern $\mathfrak{n}$ und $\mathfrak{n} + 1$ gemäß unserer Vorschrift zugeordneten Ziffern ersetzen und dann für die bekannte Funktion ψ ihre Definition substituieren.

Ganz entsprechend ist der etwas allgemeinere Fall zu behandeln, wo in der zu definierenden Funktion φ noch eine oder mehrere unbestimmte Ziffern als „*Parameter*" auftreten. Die Rekursionsgleichungen haben im Falle eines Parameters $\mathfrak{t}$ die Form

$$\varphi(\mathfrak{t}, 1) = \alpha(\mathfrak{t}),$$
$$\varphi(\mathfrak{t}, \mathfrak{n} + 1) = \psi(\varphi(\mathfrak{t}, \mathfrak{n}), \mathfrak{t}, \mathfrak{n}),$$

wobei α ebenso wie ψ eine bekannte Funktion ist. Z. B. wird durch die Rekursion
$$\varphi(\mathfrak{t}, 1) = \mathfrak{t}$$
$$\varphi(\mathfrak{t}, \mathfrak{n} + 1) = \varphi(\mathfrak{t}, \mathfrak{n}) \cdot \mathfrak{t},$$

die Funktion $\varphi(\mathfrak{t}, \mathfrak{n}) = \mathfrak{t}^{\mathfrak{n}}$ definiert.

Es handelt sich hier bei der Definition durch Rekursion wiederum nicht um ein selbständiges Definitionsprinzip, sondern die Rekursion hat im Rahmen der elementaren Zahlentheorie lediglich die Bedeutung einer Vereinbarung über eine abgekürzte Beschreibung gewisser Bildungsprozesse, durch die man aus einer oder mehreren gegebenen Ziffern wieder eine Ziffer erhält. —

Als ein Beispiel dafür, daß wir im Rahmen der anschaulichen Zahlentheorie auch *Unmöglichkeitsbeweise* führen können, werde der Satz genommen, welcher die Irrationalität von $\sqrt{2}$ zum Ausdruck bringt: Es kann nicht zwei Ziffern $\mathfrak{m}, \mathfrak{n}$ geben, so daß[1]

$$\mathfrak{m} \cdot \mathfrak{m} = 2 \cdot \mathfrak{n} \cdot \mathfrak{n}.$$

Der Beweis wird bekanntermaßen so geführt: Man zeigt zunächst, daß jede Ziffer entweder durch 2 teilbar oder von der Form $(2 \cdot \mathfrak{k}) + 1$ ist und daß daher $\mathfrak{a} \cdot \mathfrak{a}$ nur dann durch 2 teilbar sein kann, wenn $\mathfrak{a}$ durch 2 teilbar ist.

Wäre nun ein Zahlenpaar $\mathfrak{m}, \mathfrak{n}$ gegeben, das die obige Gleichung erfüllt, so könnten wir alle Zahlenpaare $\mathfrak{a}, \mathfrak{b}$, wo

$\mathfrak{a}$ der Reihe $1, \ldots, \mathfrak{m}$,

$\mathfrak{b}$ der Reihe $1, \ldots, \mathfrak{n}$

angehört, daraufhin ansehen, ob

$$\mathfrak{a} \cdot \mathfrak{a} = 2 \cdot \mathfrak{b} \cdot \mathfrak{b}$$

---

[1] Wir benutzen hier die übliche, zufolge des assoziativen Gesetzes der Multiplikation statthafte Schreibweise mehrgliedriger Produkte ohne Klammern.

From content of this instruction, we see at first that this instruction can be carried out for any given numeral m | and that the result is uniquely determined. Yet, at the same time, we find that for a given numeral n the equation

$$\varphi(n+1) = \psi(\varphi(n), n)$$

is satisfied when we replace $\varphi(n)$ and $\varphi(n+1)$ with the numerals assigned according to our instructions to n and n+1, respectively, and then substitute the known function ψ by its definition.

There is a slightly more general case of a recursive definition where one or several indefinite numerals may occur as *"parameters"*. This case can be dealt with in exactly the same way. In the case of one parameter t, the recursive equations have the form

$$\varphi(t, 1) = \alpha(t),$$
$$\varphi(t, n+1) = \psi(\varphi(t, n), t, n),$$

where both α and ψ are known functions. For example, the recursion

$$\varphi(t, 1) = t,$$
$$\varphi(t, n+1) = \varphi(t, n) \cdot t$$

defines the function $\varphi(t, n) = t^n$.

Just as above,[27.1] definition by recursion is not an independent principle of definition:[27.2] Rather, within the framework of elementary number theory, recursion merely constitutes a convention for abbreviating descriptions of certain formation processes[27.3] by which a numeral is obtained from one or several given numerals. —

[27.4] As an example for the fact that we can also carry out *"impossibility proofs"*[27.5] within intuitive number theory, we take the theorem expressing the irrationality of $\sqrt{2}$: There cannot be two numerals m, n, such that[1]

$$m \cdot m = 2 \cdot n \cdot n.$$

The proof proceeds in the familiar way: First one shows that every numeral is either divisible by 2, or of the form $(2 \cdot t) + 1$; therefore $a \cdot a$ is divisible by 2 only if a is divisible by 2.

Given a pair of numbers m, n satisfying the above equation, we could examine all pairs of numbers a, b such that

a is in the series 1, ..., m,

b is in the series 1, ..., n,

and determine whether

$$a \cdot a = 2 \cdot b \cdot b$$ |

---

[1] We use the familiar presentation of multiplication of several elements without parentheses here, which is permissible according to the associative law of multiplication.

[27.1] Cf. Page 23.

[27.2] As already remarked in Note 23.5, recursion (i.e. inductive function definition) is a principle separate from deductive logic. If functions are understood as "intuitive instructions" (cf. p. 26), however, then the principle is again already required for defining the notion of the relevant inductive data type. In Note 23.5, this data type are the numerals. In the present case, it is the data type of functions as intuitive instructions.

[27.3] The German original of our translation "formation processes" reads "Bildungsprozesse". The alternative "processes of composition" is not in line with our general intention to translate "Bildung" as "formation". If BERNAYS had meant "construction processes", however, he would have used "Aufbauprozesse" or "Konstruktionsprozesse" here; and if he had meant "instructions", he would have used the word "Vorschrift".

[27.4] § 2(a)(4) (as listed in the table of contents) starts with this paragraph.

[27.5] Cf. Page 19 and especially Note 19.2.

ist oder nicht. Unter den Wertepaaren, welche der Gleichung genügen, wählen wir ein solches, worin $\mathfrak{b}$ den kleinstmöglichen Wert hat. Es kann nur *ein* solches geben; dieses sei $\mathfrak{m}'$, $\mathfrak{n}'$. Aus der Gleichung

$$\mathfrak{m}' \cdot \mathfrak{m}' = 2 \cdot \mathfrak{n}' \cdot \mathfrak{n}'$$

folgt nun nach dem vorher Bemerkten, daß $\mathfrak{m}'$ durch 2 teilbar ist:

also erhalten wir
$$\mathfrak{m}' = 2 \cdot \mathfrak{l}',$$
$$2 \cdot \mathfrak{l}' \cdot 2 \cdot \mathfrak{l}' = 2 \cdot \mathfrak{n}' \cdot \mathfrak{n}',$$
$$2 \cdot \mathfrak{l}' \cdot \mathfrak{l}' = \mathfrak{n}' \cdot \mathfrak{n}'.$$

Hiernach wäre aber $\mathfrak{n}'$, $\mathfrak{l}'$ ein Zahlenpaar, das unserer Gleichung genügt, und zugleich wäre $\mathfrak{l}' < \mathfrak{n}'$. Dieses widerspricht aber der Bestimmung von $\mathfrak{n}'$.

Der hiermit bewiesene Satz läßt sich allerdings auch positiv aussprechen: Wenn $\mathfrak{m}$ und $\mathfrak{n}$ irgend zwei Ziffern sind, so ist $\mathfrak{m} \cdot \mathfrak{m}$ von $2 \cdot \mathfrak{n} \cdot \mathfrak{n}$ verschieden.

Soviel mag zur Charakterisierung der elementaren Behandlung der Zahlentheorie genügen. Diese haben wir als eine Theorie der Ziffern, also einer gewissen Art besonders einfacher Figuren, entwickelt. Die Bedeutung dieser Theorie für die Erkenntnis beruht auf der Beziehung der Ziffern zu dem eigentlichen *Anzahl-Begriff*. Diese Beziehung erhalten wir auf folgende Art:

Es sei eine konkrete (also jedenfalls endliche) Gesamtheit von Dingen vorgelegt. Man nehme nacheinander die Dinge der Gesamtheit vor und lege ihnen der Reihe nach die Ziffern 1, 11, 111, ... als Nummern bei. Wenn kein Ding mehr übrig ist, so sind wir zu einer gewissen Ziffer $\mathfrak{n}$ gelangt. Diese ist damit zunächst als *Ordinalzahl* für die Gesamtheit der Dinge in der gewählten Reihenfolge bestimmt.

Nun machen wir uns aber leicht klar, daß die resultierende Ziffer $\mathfrak{n}$ gar nicht von der gewählten Reihenfolge abhängt. Denn seien

$$a_1, a_2, \ldots, a_\mathfrak{n}$$

die Dinge der Gesamtheit in der gewählten Reihenfolge und

$$b_1, b_2, \ldots, b_\mathfrak{t}$$

die Dinge in einer anderen Reihenfolge. Dann können wir von der ersten Numerierung zu der zweiten folgendermaßen durch eine Reihe von Vertauschungen der Nummern übergehen: Falls a_1 von b_1 verschieden ist, so vertauschen wir zunächst die Nummer $\mathfrak{r}$, die das Ding b_1 in der ersten Numerierung hat, mit 1, d. h. wir legen dem Ding $a_\mathfrak{r}$ die Nummer 1, dem Ding a_1 die Nummer $\mathfrak{r}$ bei. In der hierdurch entstehenden Numerierung hat das Ding b_1 die Nummer 1; auf dieses folgt, mit der Nummer 2 versehen, entweder das Ding b_2, oder dieses Ding hat hier eine andere

| holds or not. From the pairs of values satisfying the equation we choose the one in which $\mathfrak{b}$ has the smallest value. There can only be *one* such pair; let it be designated by $\mathfrak{m}', \mathfrak{n}'$. From the equation
$$\mathfrak{m}' \cdot \mathfrak{m}' \;=\; 2 \cdot \mathfrak{n}' \cdot \mathfrak{n}',$$
it follows by our previous remarks that $\mathfrak{m}'$ is divisible by 2:
$$\mathfrak{m}' \;=\; 2 \cdot \mathfrak{f}',$$
and hence we get
$$2 \cdot \mathfrak{f}' \cdot 2 \cdot \mathfrak{f}' \;=\; 2 \cdot \mathfrak{n}' \cdot \mathfrak{n}',$$
$$2 \cdot \mathfrak{f}' \cdot \mathfrak{f}' \;=\; \mathfrak{n}' \cdot \mathfrak{n}'.$$

According to this, however, the pair of numbers $\mathfrak{n}', \mathfrak{f}'$ would satisfy our equation and at the same time it would be the case that $\mathfrak{f}' < \mathfrak{n}'$. But this contradicts the definition of $\mathfrak{n}'$.

On the other hand, the sentence just proved can also be stated positively:[28.1] If $\mathfrak{m}$ and $\mathfrak{n}$ are two arbitrary numerals, then $\mathfrak{m} \cdot \mathfrak{m}$ is different from $2 \cdot \mathfrak{n} \cdot \mathfrak{n}$.

[28.2] Let this much suffice as a characterization of the elementary treatment of number theory. We have developed this theory here as one of numerals, i.e. of a certain kind of particularly simple figures. The epistemic significance of this theory rests upon the relation of the numerals to the intrinsic *notion of a cardinal number*. We obtain this relation in the following way:

Imagine a concrete (thus, in any case, finite) totality of things. Consider the things in this totality one by one and assign the numerals 1, 11, 111, ... to them successively. When no thing[28.3] is left, we have arrived at a certain numeral $\mathfrak{n}$. Then this numeral is assigned — as an *ordinal number* at first — to the totality of the things in the chosen order.

Now we can easily see, however, that the resulting numeral $\mathfrak{n}$ does not at all depend on the chosen order. For let
$$a_1, \; a_2, \; \ldots, \; a_\mathfrak{n}$$
be the things of the totality in the chosen order and let
$$b_1, \; b_2, \; \ldots, \; b_\mathfrak{f}$$
be the same things in some other order. Then we can turn the first numeration into the second by swapping the numbers in the following way: If a_1 and b_1 are different, then we first swap the number $\mathfrak{r}$ of the thing b_1 in the first numeration with 1; i.e. we correlate the thing[28.4] $a_\mathfrak{r}$ with the number 1 and the thing a_1 with the number $\mathfrak{r}$. In the resulting numeration the thing b_1 has the number 1; following[28.5] b_1, and correlated with the number 2, is either the thing b_2, or this thing has some other number $\mathfrak{s}$, which is by all means different from 1; then we swap in this numeration the number $\mathfrak{s}$ with 2, whereby we get a numeration in which the thing b_1 has the number 1 and b_2 has the number 2; b_3 has either the number 3 or another number $\mathfrak{t}$, definitely different from 1 and 2, which we then swap again with 3.

[28.1] Note that switching between a negative statement of the form $\overline{(Em)}(En)(2m^2 = n^2)$ and the related positive statement $(m)(n)(2m^2 \neq n^2)$ is standard today. For PIERRE FERMAT (1607?–1665), however, this was still one of the hardest steps in his otherwise outstanding inductive proofs by *descente infinie*. Cf. [FERMAT, 1891ff., Vol. II, p. 432], [BUSSOTTI, 2006], and [WIRTH, 2010, § 2.4.3].

[28.2] §§ 2(b) and 2(b)(1) (as listed in the table of contents) start with this paragraph.

[28.3] Here and in the following, "no thing" is not a misprint, but the translation of "kein Ding".

[28.4] We have corrected the subsequent misprint of "$a_\mathfrak{r}$" of the German original to "a_r".

[28.5] From this "following", we see that what the proof calls a "numeration" is actually an array or a list, and the positions in the first list are swapped until it becomes identical to the second list. The data types of arrays and lists seem to have been non-standard in mathematics until they became popular in programming languages in the 1950s.

Nummer $\mathfrak{z}$, die jedenfalls auch von 1 verschieden ist; dann vertauschen wir in der Numerierung diese Nummer $\mathfrak{z}$ mit 2, so daß nun eine Numerierung entsteht, in der das Ding b_1 die Nummer 1, b_2 die Nummer 2 hat. b_3 hat hier entweder die Nummer 3 oder eine andere, jedenfalls von 1 und 2 verschiedene Nummer $\mathfrak{t}$; diese vertauschen wir dann wieder mit 3.

Mit diesem Verfahren müssen wir zu einem Abschluß gelangen; denn durch jede Vertauschung wird die Numerierung der betrachteten Gesamtheit mit der Numerierung

$$b_1, b_2, \ldots, b_{\mathfrak{f}}$$

vom Anfang aus um mindestens eine Stelle weiter zur Übereinstimmung gebracht, so daß man schließlich für b_1 die Nummer 1, für b_2 die Nummer 2, ..., für $b_{\mathfrak{f}}$ die Nummer $\mathfrak{f}$ bekommt, und dann ist kein weiteres Ding mehr übrig. Andererseits bleibt aber bei jeder der vorgenommenen Vertauschungen der Vorrat der verwendeten Ziffern ganz derselbe; es wird ja nur die Nummer eines Dinges gegen die eines anderen ausgewechselt. Es geht also die Numerierung jedesmal von 1 bis $\mathfrak{n}$, folglich ist auch

$$\mathfrak{f} = \mathfrak{n}.$$

Somit ist die Ziffer $\mathfrak{n}$ der betrachteten Gesamtheit unabhängig von der Reihenfolge zugeordnet, und wir können sie in diesem Sinne der Gesamtheit als ihre *Anzahl* beilegen[1]. Wir sagen, die Gesamtheit besteht aus $\mathfrak{n}$ Dingen.

Hat eine konkrete Gesamtheit mit einer anderen die Anzahl gemeinsam, so gewinnen wir, indem wir für jede eine Numerierung vornehmen, eine umkehrbar eindeutige Zuordnung der Dinge der einen Gesamtheit zu denen der anderen. Liegt andererseits eine solche Zuordnung zwischen zwei gegebenen Gesamtheiten von Dingen vor, so haben beide dieselbe Anzahl, wie ja unmittelbar aus unserer Definition der Anzahl folgt.

Von der Definition der Anzahl gelangen wir nun durch inhaltliche Überlegungen zu den Grundsätzen der *Anzahlenlehre* wie z. B. zu dem Satz, daß bei der Vereinigung zweier Gesamtheiten ohne gemeinsames Element, deren Anzahlen $\mathfrak{a}$ und $\mathfrak{b}$ sind, eine Gesamtheit von $\mathfrak{a} + \mathfrak{b}$ Dingen entsteht. —

Anschließend an die Darstellung der elementaren Zahlentheorie möge noch kurz der elementare inhaltliche Standpunkt in der *Algebra* gekennzeichnet werden. Es soll sich handeln um die elementare Theorie der ganzen rationalen Funktionen einer oder mehrerer Variablen mit ganzen Zahlen als Koeffizienten.

Als Objekte der Theorie haben wir hier wieder gewisse Figuren, die „Polynome", die aus einem bestimmten Vorrat von Buchstaben,

---

[1] Diese Überlegung ist von HELMHOLTZ in seinem Aufsatz „Zählen und Messen" (1887) durchgeführt worden. (HERMANN V. HELMHOLTZ, Schriften zur Erkenntnistheorie. Berlin: Julius Springer 1921. Siehe S. 80—82.)

This procedure must terminate; for with every swap the numeration of the considered totality is brought at least one step closer to a correspondence with the numeration

$$b_1, \; b_2, \; \ldots, \; b_{\mathfrak{f}}$$

such that for b_1 one gets the number 1, for b_2 the number 2, ..., for $b_{\mathfrak{f}}$ the number $\mathfrak{f}$, and then no further thing will remain. On the other hand, however, the stock of numerals remains exactly the same in each of the swaps; for all that happens is that the number of one thing is swapped with that of another. The numeration thus runs from 1 to $\mathfrak{n}$ in each case, and it follows that

$$\mathfrak{f} \; = \; \mathfrak{n}.$$

Thus, the numeral $\mathfrak{n}$ assigned to the considered totality is independent of the order, and, in this sense, we can attach it to the totality as its *cardinal number*.[1] We say that the totality consists of $\mathfrak{n}$ things.

If a concrete totality has its cardinal number in common with another totality, then, by doing a numeration for each of them, we get a one-to-one correspondence of the things in one totality onto those in the other. On the other hand, if such a correspondence between two given totalities exists, the two have the same cardinal number. Indeed, this follows immediately from our definition of a cardinal number.

Having started with the definition of a cardinal number, we now arrive on the basis of contentual considerations at the basic principles of the *theory of cardinal numbers*; such as the theorem that the union of two disjoint totalities with the cardinal numbers $\mathfrak{a}$ and $\mathfrak{b}$, respectively, results in a totality of $\mathfrak{a}+\mathfrak{b}$ things. —

[29.1]Following the presentation of elementary number theory, we will now briefly characterize the elementary contentual standpoint in *algebra* as well. We will deal with the elementary theory of polynomial functions[29.2] in one or more variables with integers as coefficients.

The objects of the theory are again certain figures, the "polynomials", | which are constructed with the help of the symbols $+, -, \cdot$ and parentheses, starting from numerals and from a given stock of letters $x, y, z, \ldots$, called "variables". Here the symbols $+$ and $\cdot$ are not to be viewed as symbols for communication (as was the case in elementary number theory), but belong to the objects of the theory.

---

[1]This consideration was made in [HELMHOLTZ, 1887]. ([HELMHOLTZ, 1921, pp. 80–82]).

[29.1]§ 2(b)(2) (as listed in the table of contents) starts with this paragraph.

[29.2]The German noun phrase "ganze rationale Funktion" ("integer rational function") of the German original is a now obsolete, traditional technical term. Today, we would simply say "polynomiale Funktion" ("polynomial function") instead.

$x, y, z, \ldots$, die „Variablen" heißen, und aus Ziffern mit Hilfe der Zeichen $+, -, \cdot$ und von Klammern zusammengesetzt sind. Die Zeichen $+, \cdot$ sind also hier nicht, wie in der elementaren Zahlentheorie, als Zeichen zur Mitteilung aufzufassen, sondern gehören zu den Objekten der Theorie.

Kleine deutsche Buchstaben benutzen wir wieder als Zeichen zur Mitteilung, aber nicht nur für Ziffern, sondern auch für irgendwelche Polynome.

Die Zusammensetzung der Polynome aus den genannten Zeichen geschieht nach folgenden Bildungsregeln:

Eine Variable sowie auch eine Ziffer kann für sich als Polynom genommen werden.

Aus zwei Polynomen $\mathfrak{a}, \mathfrak{b}$ können die Polynome

$$\mathfrak{a} + \mathfrak{b}, \quad \mathfrak{a} - \mathfrak{b}, \quad \mathfrak{a} \cdot \mathfrak{b}$$

gebildet werden, aus einem Polynom $\mathfrak{a}$ kann $(-\mathfrak{a})$ gebildet werden. Dabei gelten die üblichen Regeln für das Klammernsetzen. Als Zeichen zur Mitteilung werden noch eingeführt:

die Nummern $2, 3, \ldots$, so wie in der elementaren Zahlentheorie;

das Zeichen 0 für $1 - 1$;

die übliche Potenzbezeichnung: z. B. bedeutet: $x^\mathfrak{z}$, wenn $\mathfrak{z}$ eine Ziffer ist, dasjenige Polynom, das aus $\mathfrak{z}$ entsteht, indem statt jeder 1 die Variable x gesetzt und zwischen je zwei aufeinanderfolgende x das Zeichen „$\cdot$" gesetzt wird;

das Zeichen $=$ zur Mitteilung der gegenseitigen *Ersetzbarkeit* zweier Polynome.

Die Ersetzbarkeit wird durch folgende inhaltlichen Regeln bestimmt:

1. Die assoziativen und kommutativen Gesetze für „$+$" und „$\cdot$".
2. Das distributive Gesetz

$$\mathfrak{a} \cdot (\mathfrak{b} + \mathfrak{c}) = (\mathfrak{a} \cdot \mathfrak{b}) + (\mathfrak{a} \cdot \mathfrak{c}).$$

3. Die Regeln für „$-$":

$$\mathfrak{a} - \mathfrak{b} = \mathfrak{a} + (-\mathfrak{b}),$$

$$(\mathfrak{a} + \mathfrak{b}) - \mathfrak{b} = \mathfrak{a}.$$

4. $1 \cdot \mathfrak{a} = \mathfrak{a}$.
5. Sind zwei Polynome $\mathfrak{m}, \mathfrak{n}$ frei von Variablen und von „$-$" und besteht *im Sinne der Deutung der elementaren Zahlentheorie* die Gleichung $\mathfrak{m} = \mathfrak{n}$, so ist $\mathfrak{m}$ durch $\mathfrak{n}$ ersetzbar.

Diese Regeln der Ersetzbarkeit beziehen sich auch auf solche Polynome, die als *Bestandteile* von anderen Polynomen auftreten. Aus ihnen leiten sich die weiteren Sätze über die Ersetzbarkeit ab, welche die „Identitäten" und Theoreme der elementaren Algebra bilden. Als

Again, we use small German letters as our symbols for communication, not just for numerals, but for arbitrary polynomials as well.

The construction of polynomials from these symbols proceeds according to the following formation rules:[30.1]

(a) A variable as well as a numeral can be taken as a polynomial on its own.

(b) From two polynomials $\mathfrak{a}, \mathfrak{b}$ the polynomials
$$\mathfrak{a} + \mathfrak{b}, \qquad \mathfrak{a} - \mathfrak{b}, \qquad \mathfrak{a} \cdot \mathfrak{b}$$
can be constructed.

(c) From a polynomial $\mathfrak{a}$, the polynomial $(-\mathfrak{a})$ can be constructed.

The customary rules for parentheses apply here. Moreover, we introduce the following symbols for communication:[30.2]

- the numbers 2, 3, ..., as in elementary number theory;
- the symbol 0 for $1-1$;
- the customary symbols for powers: for instance, for a numeral $\mathfrak{z}$, the symbol $x^{\mathfrak{z}}$ denotes that polynomial which results from $\mathfrak{z}$ by putting the variable x in place of every 1 and the symbol "$\cdot$" between every two successive x;
- the symbol $=$ for communicating the mutual *replaceability* of two polynomials.

Replaceability is defined by the following contentual rules:

1. The associative and commutative laws for "+" and "$\cdot$".

2. The distributive law:
$$\mathfrak{a} \cdot (\mathfrak{b}+\mathfrak{c}) = (\mathfrak{a}\cdot\mathfrak{b}) + (\mathfrak{a}\cdot\mathfrak{c}).$$

3. The rules for "$-$":[30.1]
   (a) $\quad \mathfrak{a} - \mathfrak{b} = \mathfrak{a} + (-\mathfrak{b})$,
   (b) $\quad (\mathfrak{a}+\mathfrak{b}) - \mathfrak{b} = \mathfrak{a}$.

4. $1 \cdot \mathfrak{a} = \mathfrak{a}$.

5. If two polynomials $\mathfrak{m}, \mathfrak{n}$ are free of variables and do not contain "$-$", and if the equation $\mathfrak{m} = \mathfrak{n}$ holds in *the sense of the interpretation of elementary number theory*, then $\mathfrak{m}$ is replaceable with $\mathfrak{n}$.

These rules of replaceability also apply to polynomials occurring as *parts* of other polynomials. Moreover, from these rules, further sentences about replaceability can be derived, which constitute the "identities" and theorems of elementary algebra. Let us mention | some of the most simple provable identities:

[30.1] The labels (i.e. the letters in parentheses) of the items were introduced in the translation for facilitating further reference.

[30.1] The labels (i.e. the bullets "•") of the items were introduced in the translation for clarification.

einige der einfachsten beweisbaren Identitäten seien genannt:

$$\mathfrak{a} + 0 = \mathfrak{a} \qquad -(\mathfrak{a} - \mathfrak{b}) = \mathfrak{b} - \mathfrak{a},$$
$$\mathfrak{a} - \mathfrak{a} = 0 \qquad -(-\mathfrak{a}) = \mathfrak{a},$$
$$\mathfrak{a} \cdot 0 = 0 \qquad (-\mathfrak{a}) \cdot (-\mathfrak{b}) = \mathfrak{a} \cdot \mathfrak{b}.$$

Unter den Theoremen, welche durch inhaltliche Überlegung eingesehen werden, seien folgende grundlegenden Sätze erwähnt:

a) Sind $\mathfrak{a}$, $\mathfrak{b}$ zwei Polynome, die durch einander ersetzbar sind und von denen mindestens eines die Variable x enthält, und gehen aus $\mathfrak{a}$, $\mathfrak{b}$ die Polynome $\mathfrak{a}_1$, $\mathfrak{b}_1$ hervor, indem die Variable x überall, wo sie vorkommt, durch ein und dasselbe Polynom $\mathfrak{c}$ ersetzt wird, so ist auch $\mathfrak{a}_1$ durch $\mathfrak{b}_1$ ersetzbar.

b) Aus einer richtigen Gleichung zwischen Polynomen erhält man durch Einsetzung von Ziffern für die Variablen richtige Zahlengleichungen im Sinne der Zahlentheorie (vorausgesetzt, daß das Rechnen mit negativen Zahlen in die Zahlentheorie einbezogen wird). — Die Bedeutung dieses Satzes b) möge an einem einfachen Beispiel erläutert werden: Die Gleichung

$$(x + y) \cdot (x + y) = x^2 + 2 \cdot x \cdot y + y^2$$

besagt zunächst nichts anderes, als daß nach unseren Festsetzungen $(x + y) \cdot (x + y)$ durch $x^2 + 2 \cdot x \cdot y + y^2$ ersetzbar ist. Auf Grund des Satzes b) können wir aber hieraus folgern, daß, wenn $\mathfrak{m}$ und $\mathfrak{n}$ Zahlzeichen sind, $(\mathfrak{m} + \mathfrak{n}) \cdot (\mathfrak{m} + \mathfrak{n})$ im Sinne der Zahlentheorie mit $\mathfrak{m} \cdot \mathfrak{m} + 2 \cdot \mathfrak{m} \cdot \mathfrak{n} + \mathfrak{n} \cdot \mathfrak{n}$ übereinstimmt.

c) Jedes Polynom ist ersetzbar entweder durch 0 oder durch eine Summe verschiedener Potenzprodukte der Variablen — (als solches gilt hier auch das Polynom 1) —, deren jedes mit einem positiven oder negativen Zahlfaktor versehen ist.

An Hand dieser Normalform gewinnen wir ein Verfahren, um von zwei vorgelegten Polynomen zu entscheiden, ob sie durch einander ersetzbar sind oder nicht. Es gilt nämlich der Satz:

d) Ein Polynom, das aus einer Summe verschiedener Potenzprodukte mit Zahlfaktoren besteht, ist nicht durch 0 ersetzbar, und zwei solche Polynome sind nur dann durch einander ersetzbar, wenn sie, abgesehen von der Reihenfolge der Summanden sowie der Reihenfolge der Faktoren, in den Potenzprodukten und ihren Zahlfaktoren übereinstimmen.

Der zweite Teil dieses Satzes folgt aus dem ersten, und dieser kann mit Hilfe des Satzes b) durch Betrachtung geeigneter Einsetzungen von Ziffern bewiesen werden.

Als spezielle Folgerung aus d) ergibt sich der Satz:

e) Wenn eine Ziffer $\mathfrak{m}$, aufgefaßt als Polynom, durch eine Ziffer $\mathfrak{n}$ ersetzbar ist, so stimmt $\mathfrak{m}$ mit $\mathfrak{n}$ überein.

$$\mathfrak{a}+0 \;\;=\;^{31.1}\;\; \mathfrak{a} \qquad\qquad -(\mathfrak{a}-\mathfrak{b}) \;\;=\;^{31.5}\;\; \mathfrak{b}-\mathfrak{a}$$
$$\mathfrak{a}-\mathfrak{a} \;\;=\;^{31.2}\;\; 0 \qquad\qquad -(-\mathfrak{a}) \;\;=\;^{31.4}\;\; \mathfrak{a}$$
$$\mathfrak{a}\cdot 0 \;\;=\;^{31.3}\;\; 0 \qquad\qquad (-\mathfrak{a})\cdot(-\mathfrak{b}) \;\;=\;^{31.6}\;\; \mathfrak{a}\cdot\mathfrak{b}$$

Here are some of the basic theorems seen to be true via contentual considerations:

(a) If $\mathfrak{a}$ and $\mathfrak{b}$ are two polynomials such that they are replaceable with one another and at least one of them contains the variable x, and if the polynomials $\mathfrak{a}_1$, $\mathfrak{b}_1$ result from $\mathfrak{a}$, $\mathfrak{b}$ by substituting the same polynomial $\mathfrak{c}$ for the variable x at all places, then $\mathfrak{a}_1$ is replaceable with $\mathfrak{b}_1$.

(b) From a correct equation between polynomials one obtains a correct equation between numbers in the sense of number theory by substituting numerals for the variables (presupposing that calculations with negative numbers are incorporated into number theory).

The meaning of Theorem (b) may be illustrated by a simple example: The equation
$$(x+y)\cdot(x+y) \;\;=\;\; x^2 + 2\cdot x\cdot y + y^2$$
initially says nothing except that according to our definitions $(x+y)\cdot(x+y)$ is replaceable with $x^2+2\cdot x\cdot y+y^2$. By Theorem (b), however, we can infer from this that, if $\mathfrak{m}$ and $\mathfrak{n}$ are number symbols, then $(\mathfrak{m}+\mathfrak{n})\cdot(\mathfrak{m}+\mathfrak{n})$ coincides with $\mathfrak{m}\cdot\mathfrak{m}+2\cdot\mathfrak{m}\cdot\mathfrak{n}+\mathfrak{n}\cdot\mathfrak{n}$ in the sense of number theory.

(c) Every polynomial is replaceable with 0 or with a sum of different products of powers of variables — (the polynomial 1 also counts as such a product here) — each conjoined with a positive or negative numerical factor.

By means of this normal form and the following theorem, we obtain a procedure for deciding, for two given polynomials, whether they are replaceable with one another or not:

(d) A polynomial which is the sum of different products of powers with numerical factors is not replaceable with 0; and two such polynomials are replaceable with one another only if they coincide in their products of powers and their numerical factors, disregarding the order of the summands and the order of the factors.

The second part of this theorem follows from the first, which can be proved using (b) by considering suitable substitutions of numerals.

As a corollary to (d) we have the following theorem:

(e) If a numeral $\mathfrak{m}$ (taken as a polynomial)[31.7] is replaceable with a numeral $\mathfrak{n}$, then $\mathfrak{m}$ coincides with $\mathfrak{n}$.

[31.1] $\mathfrak{a}+0 \;=\; \mathfrak{a}+(1-1) \;=^{3a}\; \mathfrak{a}+(1+(-1)) \;=^1\; (\mathfrak{a}+1)+(-1) \;=^{3a}\; (\mathfrak{a}+1)-1 \;=^{3b}\; \mathfrak{a}$.

[31.2] $\mathfrak{a}-\mathfrak{a} \;=^{31.1}\; (\mathfrak{a}+0)-\mathfrak{a} \;=^1\; (0+\mathfrak{a})-\mathfrak{a} \;=^{3b}\; 0$.

[31.3] We could not find the following proofs without the help of the automatic theorem prover WALDMEISTER [HILLENBRAND & LÖCHNER, 2002]. We have (6): $-\mathfrak{b}+\mathfrak{b}+\mathfrak{a} \;=^1\; (\mathfrak{a}+\mathfrak{b})+(-\mathfrak{b}) \;=^{3a}\; (\mathfrak{a}+\mathfrak{b})-\mathfrak{b} \;=^{3b}\; \mathfrak{a}$ and (7): $\mathfrak{a}\cdot\mathfrak{c}+\mathfrak{a}\cdot 0 \;=^2\; \mathfrak{a}\cdot(\mathfrak{c}+0) \;=^{31.1}\; \mathfrak{a}\cdot\mathfrak{c}$ and then $\mathfrak{a}\cdot 0 \;=^{6,1,3a}\; (\mathfrak{a}\cdot\mathfrak{c}+\mathfrak{a}\cdot 0)-\mathfrak{a}\cdot\mathfrak{c} \;=^7\; \mathfrak{a}\cdot\mathfrak{c}-\mathfrak{a}\cdot\mathfrak{c} \;=^{31.2}\; 0$.

[31.4] $-(-\mathfrak{a}) \;=^{6,1}\; -(-\mathfrak{a})+(-\mathfrak{a})+\mathfrak{a} \;=^6\; \mathfrak{a}$.

[31.5] We have (8): $-(\mathfrak{a}+\mathfrak{b})+\mathfrak{a} \;=^{6,1}\; -(\mathfrak{a}+\mathfrak{b})+\mathfrak{a}+\mathfrak{b}+(-\mathfrak{b}) \;=^6\; -\mathfrak{b}$ and then $-(\mathfrak{a}-\mathfrak{b}) \;=^{3a,1}\; -(-\mathfrak{b}+\mathfrak{a}) \;=^{6,1}\; -(-(\mathfrak{a}+(-\mathfrak{a}))+\mathfrak{a}) \;=^8\; -(-(-\mathfrak{a}+\mathfrak{b})) \;=^{31.4}\; -\mathfrak{a}+\mathfrak{b} \;=^{1,3a}\; \mathfrak{b}-\mathfrak{a}$.

[31.6] We have (9): $-0 \;=\; -(1-1) \;=^{31.5}\; 1-1 \;=\; 0$ and (10): $-(\mathfrak{a}\cdot\mathfrak{b}) \;=^8\; -(\mathfrak{a}\cdot(-\mathfrak{b})+\mathfrak{a}\cdot\mathfrak{b})+\mathfrak{a}\cdot(-\mathfrak{b}) \;=^2\; -(\mathfrak{a}\cdot(-\mathfrak{b}+\mathfrak{b}))+\mathfrak{a}\cdot(-\mathfrak{b}) \;=^{31.1}\; -(\mathfrak{a}\cdot(-\mathfrak{b}+\mathfrak{b}+0))+\mathfrak{a}\cdot(-\mathfrak{b}) \;=^6\; -(\mathfrak{a}\cdot 0)+\mathfrak{a}\cdot(-\mathfrak{b}) \;=^{31.3}\; -0+\mathfrak{a}\cdot(-\mathfrak{b}) \;=^{9,1,31.1}\; \mathfrak{a}\cdot(-\mathfrak{b})$ and then $(-\mathfrak{a})\cdot(-\mathfrak{b}) \;=^{10}\; -((-\mathfrak{a})\cdot\mathfrak{b}) \;=^1\; -(\mathfrak{b}\cdot(-\mathfrak{a})) \;=^{10}\; -(-(\mathfrak{b}\cdot\mathfrak{a})) \;=^{31.4}\; \mathfrak{b}\cdot\mathfrak{a} \;=^1\; \mathfrak{a}\cdot\mathfrak{b}$; note that we also have (11): $(-1)\cdot\mathfrak{a} \;=^1\; \mathfrak{a}\cdot(-1) \;=^{10}\; -(\mathfrak{a}\cdot 1) \;=^1\; -(1\cdot\mathfrak{a}) \;=^4\; -\mathfrak{a}$.

[31.7] Here "taken as a polynomial" means that we should interpret the numeral as a polynomial without variables according to Item (a) on Page 30.

Methodisch sei zu diesen Sätzen noch bemerkt: Die in den Sätzen a), e) vorkommende Voraussetzung der Ersetzbarkeit von Polynomen ist so zu verstehen, daß wir annehmen, man habe die Ersetzbarkeit des einen Polynoms durch das andere gemäß den Regeln festgestellt. Bei dem Satz c) wird die Behauptung der Ersetzbarkeit näher bestimmt durch die Angabe eines Verfahrens, welches im Beweise des Satzes beschrieben wird.

Wir befinden uns also hier, ebenso wie in der elementaren Zahlentheorie, ganz im Bereich des elementaren inhaltlichen Schließens. Und das gilt auch für die weiteren Sätze und Beweise der elementaren Algebra. —

Die ausgeführte Betrachtung der Anfangsgründe von Zahlentheorie und Algebra diente dazu, uns das direkte inhaltliche, in Gedanken-Experimenten an anschaulich vorgestellten Objekten sich vollziehende und von axiomatischen Annahmen freie Schließen in seiner Anwendung und Handhabung vorzuführen. Diese Art des Schließens wollen wir, um einen kurzen Ausdruck zu haben, als das „*finite*" Schließen und ebenso auch die diesem Schließen zugrunde liegende methodische Einstellung als die „finite" Einstellung oder den „finiten" Standpunkt bezeichnen. Im gleichen Sinne wollen wir von finiten Begriffsbildungen und Behauptungen sprechen, indem wir allemal mit dem Worte „finit" zum Ausdruck bringen, daß die betreffende Überlegung, Behauptung oder Definition sich an die Grenzen der grundsätzlichen Vorstellbarkeit von Objekten sowie der grundsätzlichen Ausführbarkeit von Prozessen hält und sich somit im Rahmen konkreter Betrachtung vollzieht.

Zur Charakterisierung des finiten Standpunktes seien noch einige allgemeine Gesichtspunkte hervorgehoben, betreffend den Gebrauch der logischen Urteilsformen im finiten Denken, wobei wir zur Exemplifizierung Aussagen über *Ziffern* betrachten wollen.

Ein *allgemeines* Urteil über Ziffern kann finit nur im hypothetischen Sinn gedeutet werden, d. h. als eine Aussage über jedwede vorgelegte Ziffer. Ein solches Urteil spricht ein Gesetz aus, das sich an jedem vorliegenden Einzelfall verifizieren muß.

Ein *Existenzsatz* über Ziffern, also ein Satz von der Form „es gibt eine Ziffer $\mathfrak{n}$ von der Eigenschaft $\mathfrak{A}(\mathfrak{n})$", ist finit aufzufassen als ein „Partialurteil", d. h. als eine unvollständige Mitteilung einer genauer bestimmten Aussage, welche entweder in der direkten Angabe einer Ziffer von der Eigenschaft $\mathfrak{A}(\mathfrak{n})$ oder der Angabe eines Verfahrens zur Gewinnung einer solchen Ziffer besteht, — wobei zur Angabe eines Verfahrens gehört, daß für die Reihe der auszuführenden Handlungen eine bestimmte Grenze aufgewiesen wird.

In entsprechender Weise sind diejenigen Urteile finit zu interpretieren, in denen eine allgemeine Aussage mit einer Existenzbehauptung

| Concerning these theorems, let us make the following remarks:[32.1] In Theorems (a) and (e), respectively, the precondition regarding the replaceability of polynomials is to be understood as the assumption that the replaceability of the one polynomial with the other has been established according to our rules.[32.2] Moreover, in Theorem (c), the assertion of replaceability is made more precise by a procedure described in the proof of the theorem.[32.3]

All in all, just as in elementary number theory, we do not exceed the realm of elementary contentual inference here; and this holds for the further theorems and proofs in elementary algebra as well. —

[32.4]Our treatment of the basics of number theory and algebra was meant to demonstrate how to apply and implement direct contentual inference that takes place in thought experiments on intuitively conceived objects and is free of axiomatic assumptions. Let us call this kind of inference *"finitistic"* inference for short, and likewise the methodological attitude underlying this kind of inference as the "finitistic" attitude or the "finitistic" standpoint. In the same sense, we will speak of finitistic concept formations and assertions: With each use of the word "finitistic", we convey the idea that the relevant consideration, assertion, or definition is confined to[32.5]

- objects that are conceivable in principle, and
- processes that can be effectively executed in principle,

and thus it remains within the scope of a concrete treatment.

Yet, to characterize the finitistic standpoint, let us emphasize some general aspects of the usage of logical forms of judgment in finitistic thinking, and consider some propositions on *numerals* as examples.

A *universal* judgment about numerals can be given a finitistic interpretation only in a hypothetical sense, as a proposition about every arbitrary numeral that is effectively given. Such a judgment expresses a law which has to prove true in every particular case at hand.

In the finitistic sense, an *existential* sentence about numerals (i.e. a sentence of the form "there is a numeral $\mathfrak{n}$ with the property $\mathfrak{A}(\mathfrak{n})$") is to be understood as a "partial judgment". That is to say, it is to be understood as an incomplete communication of a more precisely determined proposition, which consists either in the direct presentation of a numeral with the property $\mathfrak{A}(\mathfrak{n})$, or in the presentation of a procedure for obtaining such a numeral. In the latter case, the presentation has to include a definite limit for the number of operational steps in the sequential execution of the procedure.

[32.1]This is one of the rare occurrences of the word "methodisch" in the German original that we completely omit in our translation. Cf. also Note 1.3.

[32.2]This is just to say that the replaceability of polynomials is the least fixed point of the congruence closure of the relation defined by Items 1 to 5 on Page 30.

[32.3]It is unclear to which proof of Theorem (c) this remark may refer. Maybe BERNAYS just failed to say "zu beschreiben sein wird" ("to be described") instead of "beschrieben wird" ("described").

[32.4]§§ 2(c) and 2(c)(1) (as listed in the table of contents) start with this paragraph.

[32.5]The itemization was introduced in the translation for facilitating readability.

verknüpft ist. So hat man z. B. einen Satz von der Form „zu jeder Ziffer $\mathfrak{k}$ von der Eigenschaft $\mathfrak{A}(\mathfrak{k})$ gibt es eine Ziffer $\mathfrak{l}$, für welche $\mathfrak{B}(\mathfrak{k}, \mathfrak{l})$ gilt", finit aufzufassen als unvollständige Mitteilung von einem Verfahren, welches gestattet, zu jeder vorgelegten Ziffer $\mathfrak{k}$ von der Eigenschaft $\mathfrak{A}(\mathfrak{k})$ eine Ziffer $\mathfrak{l}$ zu finden, welche zu $\mathfrak{k}$ in der Beziehung $\mathfrak{B}(\mathfrak{k}, \mathfrak{l})$ steht.

Besondere Achtsamkeit erfordert die Anwendung der *Negation*.

Die Verneinung ist unproblematisch bei „elementaren" Urteilen, welche eine Frage betreffen, über die sich durch eine direkte anschauliche Feststellung (einen „Befund") entscheiden läßt. Sind z. B. $\mathfrak{k}, \mathfrak{l}$ bestimmte Ziffern, so läßt sich direkt feststellen, ob

$$\mathfrak{k} + \mathfrak{k} = \mathfrak{l}$$

zutrifft oder nicht, d. h. ob $\mathfrak{k} + \mathfrak{k}$ mit $\mathfrak{l}$ übereinstimmt oder von $\mathfrak{l}$ verschieden ist.

Die Negation eines solchen elementaren Urteils besagt einfach, daß das Ergebnis der betreffenden anschaulichen Entscheidung von dem im Urteil behaupteten Sachverhalt abweicht; und es besteht für ein elementares Urteil ohne weiteres die Alternative, daß entweder dieses selbst oder seine Negation zutrifft.

Dagegen für ein allgemeines und ein existentiales Urteil ist es nicht ohne weiteres klar, was im finiten Sinne als seine Negation gelten soll.

Betrachten wir daraufhin zunächst die Existenzaussagen. Daß es eine Ziffer $\mathfrak{n}$ von einer Eigenschaft $\mathfrak{A}(\mathfrak{n})$ nicht gibt, kann in unscharfem Sinne gemeint sein, als die Feststellung, daß eine Ziffer von dieser Eigenschaft uns zur Angabe nicht zur Verfügung steht. Eine solche Feststellung hat aber wegen ihrer Bezogenheit auf einen zufälligen Erkenntniszustand keine objektive Bedeutung. Soll aber unabhängig vom Erkenntnisstande das Nichtvorhandensein einer Ziffer $\mathfrak{n}$ von der Eigenschaft $\mathfrak{A}(\mathfrak{n})$ behauptet werden, so kann das im finiten Sinne nur durch eine Unmöglichkeitsbehauptung geschehen, welche besagt, daß eine Ziffer $\mathfrak{n}$ nicht die Eigenschaft $\mathfrak{A}(\mathfrak{n})$ besitzen *kann*.

Wir kommen so auf eine *verschärfte* Negation; diese aber ist nicht das genau kontradiktorische Gegenteil der Existenzbehauptung, „es gibt eine Ziffer $\mathfrak{n}$ von der Eigenschaft $\mathfrak{A}(\mathfrak{n})$", die (als Partialurteil) auf eine bekannte Ziffer von jener Eigenschaft hinweist oder auf ein Verfahren, das wir zur Gewinnung einer solchen Ziffer besitzen.

Die Existenzaussage und ihre verschärfte Negation sind nicht, wie eine elementare Aussage und ihre Negation, Aussagen über die beiden allein in Betracht kommenden Ergebnisse *einer und derselben Entscheidung*, sondern sie entsprechen zwei getrennten Erkenntnismöglichkeiten, nämlich einerseits der Auffindung einer Ziffer von einer gegebenen Eigenschaft, andererseits der Einsicht in ein allgemeines Gesetz über Ziffern.

Daß eine von diesen beiden Möglichkeiten sich bieten muß, ist nicht logisch selbstverständlich. Wir können daher vom finiten Standpunkt

Judgments combining a universal proposition with an existential assertion are to be interpreted accordingly. | Thus, for example, a sentence of the form "for every numeral $\mathfrak{k}$ with the property $\mathfrak{A}(\mathfrak{k})$, there is a numeral $\mathfrak{l}$ such that $\mathfrak{B}(\mathfrak{k},\mathfrak{l})$ holds" is to be understood in the finitistic sense as an incomplete communication of a procedure which, for any given numeral $\mathfrak{k}$ with the property $\mathfrak{A}(\mathfrak{k})$, permits us to find a numeral $\mathfrak{l}$ such that the relation $\mathfrak{B}(\mathfrak{k},\mathfrak{l})$ holds.

Applying *negation* requires particular care.

Negating is unproblematic for "elementary" judgments concerning a question decidable by a direct intuitive observation (a "finding"). For example, if $\mathfrak{k}$ and $\mathfrak{l}$ are definite numerals, then it can be directly observed whether

$$\mathfrak{k}+\mathfrak{k} = \mathfrak{l}$$

is the case or not, i.e. whether $\mathfrak{k}+\mathfrak{k}$ coincides with $\mathfrak{l}$ or is different from $\mathfrak{l}$.

The negation of such an elementary judgment simply says that the result of the related intuitive decision deviates from the state of affairs expressed in the judgment; and for an elementary judgment it may be taken for granted that either the judgment itself or its negation holds.

For a universal or existential judgment, however, it is not immediately obvious what should count as its negation in the finitistic sense.

Let us first consider existential assertions. That there is no numeral $\mathfrak{n}$ with the property $\mathfrak{A}(\mathfrak{n})$ may be meant in a vague sense: as the statement that we do not have a numeral with this property at our disposal. Such a statement, however, has no objective significance because it is relative to an accidental state of knowledge. However, if we wish to assert the unavailability of a numeral $\mathfrak{n}$ with the property $\mathfrak{A}(\mathfrak{n})$ independently of our state of knowledge, then one can express this finitistically only by an assertion of impossibility, which says that a numeral $\mathfrak{n}$ *cannot* have the property $\mathfrak{A}(\mathfrak{n})$.

Thus, we arrive at a *sharpened* notion of negation. This notion, however, is not the exact contradictory opposite of an existential assertion such as "there is a numeral $\mathfrak{n}$ with the property $\mathfrak{A}(\mathfrak{n})$", which (as a partial judgment) indicates either a known numeral with this property, or a procedure that we have at our disposal for obtaining such a numeral.

Unlike an elementary proposition and its negation, the existential proposition and its sharpened negation are not propositions about the only two possible results of *one and the same decision*; rather, they correspond to two distinct epistemic possibilities: Namely, on one hand, the discovery of a numeral with the given property, and, on the other hand, the insight into a general law about numerals.[33.1]

It is by no means logically self-evident that one of these two possibilities has to arise. Thus, from the finitistic standpoint, | we cannot make use of the alternative[33.2] that either there is a numeral $\mathfrak{n}$ for which $\mathfrak{A}(\mathfrak{n})$ holds, or that it is excluded that $\mathfrak{A}(\mathfrak{n})$ holds for some numeral.

[33.1] Thus, for an elementary judgment $\mathfrak{A}(x)$, the sharpened negation of the finitistic proposition $(Ex)\,\mathfrak{A}(x)$ is logically equivalent to the finitistic proposition $(x)\,\overline{\mathfrak{A}(x)}$. This seems to be just the same as in classical logic, but note that, in general, $(x)\,\overline{\mathfrak{A}(x)} \lor (Ex)\,\mathfrak{A}(x)$ does not hold in the finitistic sense.

[33.2] That we cannot make use of this alternative simply means that a case analysis based on it would be incomplete in a finitistic sense.

nicht die **Alternative** benutzen, daß es entweder eine Ziffer $\mathfrak{n}$ gibt, für die $\mathfrak{A}(\mathfrak{n})$ zutrifft, oder daß das Zutreffen von $\mathfrak{A}(\mathfrak{n})$ auf eine Ziffer ausgeschlossen ist.

Ähnlich wie bei dem Existentialurteil verhält es sich bei einem allgemeinen Urteil von der Form „für jede Ziffer $\mathfrak{n}$ gilt $\mathfrak{A}(\mathfrak{n})$" betreffs der finiten Negation. Die Verneinung der Gültigkeit eines solchen Urteils ergibt ohne weiteres noch keinen finiten Sinn; wird sie aber zu der Behauptung verschärft, daß die Allgemeingültigkeit von $\mathfrak{A}(\mathfrak{n})$ sich durch ein Gegenbeispiel widerlegen läßt, dann bildet diese verschärfte Negation nicht das kontradiktorische Gegenteil des allgemeinen Urteils; nämlich es ist dann wiederum nicht logisch selbstverständlich, daß entweder das allgemeine Urteil oder die verschärfte Negation zutreffen muß, daß also entweder $\mathfrak{A}(\mathfrak{n})$ für jede vorgelegte Ziffer $\mathfrak{n}$ zutrifft oder daß sich eine Ziffer angeben läßt, für welche $\mathfrak{A}(\mathfrak{n})$ unzutreffend ist.

Allerdings ist zu bemerken, daß die Auffindung eines Gegenbeispiels nicht die einzige Möglichkeit bildet, ein allgemeines Urteil zu widerlegen. Es kann auch in anderer Weise die Verfolgung der Konsequenzen des allgemeinen Urteils auf einen Widerspruch führen. Dieser Umstand hebt jedoch die Schwierigkeit nicht auf, vielmehr wird dadurch die Komplikation noch erhöht. Nämlich es ist weder die Alternative logisch ersichtlich, daß ein allgemeines Urteil über Ziffern entweder zutreffen oder in seinen Konsequenzen auf einen Widerspruch führen, also widerlegbar sein müsse, noch auch ist es selbstverständlich, daß ein solches Urteil, wenn es widerlegbar ist, dann auch durch ein Gegenbeispiel widerlegbar ist.

Die komplizierte Situation, die wir hier in betreff der Verneinung von Urteilen beim finiten Standpunkt vorfinden, entspricht der These BROUWERS von der Ungültigkeit des Satzes vom ausgeschlossenen Dritten für unendliche Gesamtheiten. Diese Ungültigkeit besteht beim finiten Standpunkt in der Tat insofern, als es hier für das existentiale sowie für das allgemeine Urteil nicht gelingt, eine dem Satz vom ausgeschlossenen Dritten genügende Negation finiten Inhalts zu finden.

Diese Darlegungen mögen zur Kennzeichnung des finiten Standpunktes genügen. Sehen wir uns nun die Arithmetik in ihrer üblichen Behandlung daraufhin an, ob sie diesem methodischen Standpunkt entspricht, so bemerken wir, daß dieses nicht der Fall ist, daß vielmehr durch die arithmetischen Schlußweisen und Begriffsbildungen mannigfach die Grenzen der finiten Betrachtung überschritten werden.

Die Überschreitung des finiten Standpunktes findet bereits in den Schlußweisen der Zahlentheorie statt, indem hier Existenzaussagen über ganze Zahlen — wir sprechen in der üblichen Mathematik von „ganzen Zahlen" (genauer „positiven ganzen Zahlen" und kurz auch „Zahlen") anstatt von „Ziffern" — zugelassen werden ohne Rücksicht auf die Möglichkeit einer tatsächlichen Bestimmung der betreffenden Zahl, und indem man Gebrauch macht von der Alternative, daß eine

With respect to finitistic negation, the case of a universal judgment of the form "$\mathfrak{A}(\mathfrak{n})$ holds for every numeral $\mathfrak{n}$" is similar to that of an existential judgment. Negating the validity of such a judgment does not yet make finitistic sense: In its sharpened form, namely the assertion that the validity of $\mathfrak{A}(\mathfrak{n})$ can be refuted with a counterexample, this sharpened negation no longer constitutes the contradictory opposite of the universal judgment: Again, it is not logically self-evident that either the universal judgment or the sharpened negation must hold; i.e. that either $\mathfrak{A}(\mathfrak{n})$ holds for every given numeral $\mathfrak{n}$ or that some numeral can be presented for that $\mathfrak{A}(\mathfrak{n})$ does not hold.

It is to be noted, however, that finding a counterexample does not constitute the only possibility to refute a universal judgment. It may well be possible that pursuing the consequences of a universal judgment leads to a contradiction in some other way. This possibility, however, does not eliminate the difficulty but rather complicates matters. Namely, it is neither logically apparent that a universal judgment on numerals must either hold or else lead to a contradiction in its consequences (i.e. be refutable),[34.1] nor is it even self-evident that such a judgment — if it is refutable — is also refutable with a counterexample.[34.2]

The complicated situation we find here with respect to negating judgments from the finitistic standpoint corresponds to BROUWER's thesis of the invalidity of the law of the excluded middle for infinite totalities. Indeed, this invalidity holds for the finitistic standpoint in that we cannot find a finitistic negation of existential and universal judgments satisfying the law of the excluded middle.[34.3]

[34.4]These explanations may suffice as a characterization of the finitistic standpoint. If we now look at the customary treatment of arithmetic and see whether it corresponds to this methodological standpoint, then we notice that this is not the case: Rather, in arithmetic, concept formations and modes of inference often transgress the limits of finitistic treatment in multiple ways.

The transgression of the finitistic standpoint occurs already in the modes of inference in number theory; namely in[34.5]

- the admission of existential propositions about integers, regardless of the actual possibility to determine the integer in question; and in
- applications of the alternative that an | assertion about integers either holds for all integers or that there is an integer for that it does not hold.

Note that in customary mathematics, instead of "numerals", we speak of "integers" (or, more precisely, of "positive integers", or "numbers" for short).

[34.1]For example, in an *incomplete* theory, such as the contentual treatment of number theory just presented, or any attempt to axiomatize it formally, there will be some universal judgements that are neither provable nor refutable.

[34.2]Note that all this may already occur in a classical axiomatic theory: In the case of $\vdash \overline{(x)\,\mathfrak{A}(x)}$, it may not only be the case that we cannot present a numeral $\mathfrak{n}$ such that $\vdash \overline{\mathfrak{A}(\mathfrak{n})}$, but it may even be the case that we do not have $\vdash \overline{\mathfrak{A}(\mathfrak{n})}$ for any numeral $\mathfrak{n}$. Moreover, this may even be the case if the theory is complete: This means that we may have $\vdash \overline{(x)\,\mathfrak{A}(x)}$ and $\vdash \mathfrak{A}(\mathfrak{n})$ for each numeral $\mathfrak{n}$. In such a case, the theory is not ω-*complete*, and its model must be non-standard in the sense that it includes other objects in its domain beside those that are denoted by the numerals.

[34.3]Note that from the axiom schema $\mathfrak{B} \vee \overline{\mathfrak{B}}$ ("law of the excluded middle"), we can conclude

$$(x)\,\mathfrak{A}(x) \;\vee\; (Ex)\,\overline{\mathfrak{A}(x)} \quad (\text{"tertium non datur for integers"}) \text{ (cf. Page 34ff.)},$$

if we assume $\overline{(x)\,\mathfrak{A}(x)} \to (Ex)\,\overline{\mathfrak{A}(x)}$, which — in intuitionistic logic — is not valid in general, but is again implied (in general: properly) by the *tertium non datur for integers*. Cf. Note 35.1.

[34.4]§ 2(c)(2) (as listed in the table of contents) starts with this paragraph.

[34.5]The itemization was introduced in the translation for facilitating readability.

Aussage über ganze Zahlen entweder für alle Zahlen zutrifft oder daß es eine Zahl gibt, für die sie unzutreffend ist.

Diese Alternative, das „tertium non datur" für ganze Zahlen, kommt implizite auch zur Anwendung bei dem „Prinzip der kleinsten Zahl", welches besagt: „Wenn eine Aussage über ganze Zahlen für mindestens eine Zahl zutrifft, so gibt es eine kleinste Zahl, für die sie zutrifft."

Das Prinzip der kleinsten Zahl hat in seinen *elementaren* Anwendungen finiten Charakter. In der Tat, sei $\mathfrak{A}(n)$ die betreffende Aussage über eine Zahl n, und sei $\mathfrak{m}$ eine bestimmte Zahl, für welche $\mathfrak{A}(\mathfrak{m})$ zutrifft, so gehe man die Zahlen von 1 bis $\mathfrak{m}$ durch; man muß dann einmal zuerst zu einer Zahl $\mathfrak{k}$ gelangen, für die $\mathfrak{A}(\mathfrak{k})$ richtig ist, da ja spätestens $\mathfrak{m}$ eine solche Zahl ist. Diese Zahl $\mathfrak{k}$ ist dann die kleinste Zahl von der Eigenschaft $\mathfrak{A}$.

Diese Überlegung beruht aber auf zwei Voraussetzungen, die bei den nichtelementaren Anwendungen des Prinzips der kleinsten Zahl nicht immer erfüllt sind. Erstens wird vorausgesetzt, daß das Zutreffen der Aussage $\mathfrak{A}$ auf eine Zahl in dem Sinne statthat, daß uns eine Zahl $\mathfrak{m}$ von der Eigenschaft $\mathfrak{A}(\mathfrak{m})$ wirklich gegeben ist, während bei den Anwendungen die Existenz einer Zahl von der Eigenschaft $\mathfrak{A}$ vielfach nur mittels des „tertium non datur" erschlossen ist, ohne daß wir dadurch zur wirklichen Bestimmung einer solchen Zahl gelangen. Die zweite Voraussetzung ist, daß sich für eine jede Zahl $\mathfrak{k}$ aus der Reihe der Zahlen von 1 bis $\mathfrak{m}$ entscheiden läßt, ob $\mathfrak{A}(\mathfrak{k})$ zutrifft oder nicht; diese Entscheidungsmöglichkeit besteht allerdings für elementare Aussagen $\mathfrak{A}(n)$ ohne weiteres; dagegen kann bei einer nichtelementaren Aussage $\mathfrak{A}(n)$ die Frage, ob sie für eine gegebene Zahl $\mathfrak{k}$ zutrifft, ein ungelöstes Problem bilden.

Sei z. B. $\psi(a)$ eine Funktion, die durch eine Aufeinanderfolge von Rekursionen und Einsetzungen definiert ist, so wie wir sie in der finiten Zahlentheorie zulassen, und $\mathfrak{A}(n)$ bedeute die Aussage, daß es eine Zahl a gibt, für welche $\psi(a) = n$ ist. Dann ist für eine vorgelegte Zahl $\mathfrak{k}$ die Frage, ob $\mathfrak{A}(\mathfrak{k})$ zutrifft, im allgemeinen (d. h. wenn die Funktion ψ nicht besonders einfach ist) nicht durch direktes Zusehen entscheidbar, vielmehr hat sie den Charakter eines mathematischen Problems. Denn die Rekursionen, welche in die Definition von ψ eingehen, liefern ja die Werte der Funktion nur *für vorgelegte Argumentwerte*, während es sich bei der Frage, ob es eine Zahl a gibt, für die $\psi(a)$ den Wert $\mathfrak{k}$ hat, um den gesamten Wertverlauf der Funktion ψ handelt.

In allen den Fällen nun, wo die genannten Voraussetzungen für die finite Begründung des Prinzips der kleinsten Zahl nicht erfüllt sind, muß zur Begründung dieses Prinzips das „tertium non datur" für die ganzen Zahlen herangezogen werden[1].

---

[1] Wir werden den Beweis des Prinzips der kleinsten Zahl später im Rahmen des Formalismus vorführen. Siehe § 6 S. 284—285.

This alternative — i.e. the "tertium non datur" for integers[35.1] — is also implicitly applied in connection with the *least-number principle*, which says: "If a proposition about integers holds for at least one number, then there is a least number for which it holds."

In its *elementary* applications, the least-number principle has a finitistic character. Indeed, let $\mathfrak{A}(n)$ be the proposition in question about a number n, and let $\mathfrak{m}$ be a definite number for which $\mathfrak{A}(\mathfrak{m})$ holds, then one may go through the numbers from 1 to $\mathfrak{m}$: At some time, one must reach a first number $\mathfrak{f}$ for which $\mathfrak{A}(\mathfrak{f})$ is correct; because, at the latest, $\mathfrak{m}$ is obviously such a number. Then this number $\mathfrak{f}$ is the least number with property $\mathfrak{A}$.

This consideration rests on two presuppositions, however, which are not always satisfied in non-elementary applications of the least-number principle:[35.2]

- The first presupposition is that "$\mathfrak{A}$ holds for some number" means that some number $\mathfrak{m}$ with the property $\mathfrak{A}(\mathfrak{m})$ is actually given. In many applications of the least-number principle, however, the existence of a number with the property $\mathfrak{A}$ is inferred only by means of the "tertium non datur"; without succeeding thereby in an actual determination of such a number.[35.3]

- The second presupposition is that we can decide, for any number $\mathfrak{f}$ in the series from 1 to $\mathfrak{m}$, whether $\mathfrak{A}(\mathfrak{f})$ holds or not. Of course, this possibility of deciding can be taken for granted for elementary propositions $\mathfrak{A}(n)$. For a non-elementary proposition $\mathfrak{A}(n)$, however, the question whether it holds for a given number $\mathfrak{f}$ may present itself as an unsolved problem.

For example, let $\psi(a)$ be a function defined by a succession of recursions and substitutions, as we admit them[35.4] in finitistic number theory. Let $\mathfrak{A}(n)$ stand for the proposition that there is a number a for which $\psi(a) = n$. Then, in general (i.e. unless the function ψ is particularly simple), for a given number $\mathfrak{f}$, the question whether $\mathfrak{A}(\mathfrak{f})$ holds is not decidable by direct inspection, but has the character of a mathematical problem: The recursions entering into the definition of ψ give the values of the function only *for given values of the arguments*, while the question whether there is a number a for which $\psi(a)$ has the value $\mathfrak{f}$ involves the whole graph of the function ψ.

Now in all cases where the mentioned presuppositions for the finitistic justification of the least-number principle are not satisfied, the foundational justification of this principle requires reference to the "tertium non datur" for integers.[1]

---

[1] We will give the proof of the least-number principle later within the framework of the formalism. Cf. §6, pp. 284–285.

[35.1] Note that here, in the tradition of [HILBERT, 1922b, p.176], the *tertium non datur for integers* is
$$(x)\,\mathfrak{A}(x) \;\lor\; (Ex)\,\overline{\mathfrak{A}(x)}.$$
Thus, it is *not at all* an instance of the *law of the excluded middle* (formally: $\mathfrak{B} \lor \overline{\mathfrak{B}}$), which is often called "tertium non datur" as well. Moreover, it is not intuitionistically implied by the instance $(Ex)\,\overline{\mathfrak{A}(x)} \;\lor\; \overline{(Ex)\,\mathfrak{A}(x)}$; cf. Note 34.3. Finally, note that the axiom $(A \to B) \to \{(\overline{A} \to B) \to B\}$ is also called "Prinzip des tertium non datur" ("principle of the tertium non datur") in [HILBERT, 1923a, p.153].

[35.2] The itemization was introduced in the translation for facilitating readability.

[35.3] For example, in mathematical induction and especially in FERMAT's *descente infinie*, the argumentation is often via assuming the existence of a counterexample and then using the least number principle to derive a contradiction. In general, this is an argumentation that uses the *tertium non datur* in the strongest form. Cf. the proof of the irrationality of $\sqrt{2}$ on Page 27 for an example of the assumption of such a counterexample. Cf. e.g. [BUSSOTTI, 2006], [WIRTH, 2004; 2010] for a deeper historical, logical, and mathematical treatment of this subject.

[35.4] In the German original, the pronoun "sie" may actually be a third person singular (English: "she"; or, after changing gender in the translation: "it") and refer to "a function" instead of "recursions and substitutions", but the third person plural ("them") is the preferred reading.

§ 2. Die elementare Zahlentheorie. — Das finite Schließen.

Es seien einige Beispiele von zahlentheoretischen Alternativen aufgeführt, welche sich mittels des tertium non datur für ganze Zahlen ergeben, dagegen auf finitem Wege vom heutigen Stande unserer Kenntnis nicht zu erweisen sind:

„Entweder ist jede gerade Zahl, die >2 ist, als Summe zweier Primzahlen darstellbar, oder es gibt eine gerade Zahl, die >2 und nicht als Summe zweier Primzahlen darstellbar ist."

„Entweder ist jede ganze Zahl der Form $2^{(2^f)} + 1$ für $f > 4$ in zwei Faktoren >1 zerlegbar, oder es gibt eine Primzahl der Form $2^{(2^f)} + 1$ mit $f > 4$."

„Entweder ist jede genügend große ganze Zahl als Summe von weniger als 8 dritten Potenzen darstellbar, oder es gibt zu jeder ganzen Zahl n eine größere ganze Zahl m, welche nicht als Summe von weniger als 8 dritten Potenzen darstellbar ist."

„Entweder gibt es beliebig große Primzahlen p von der Eigenschaft, daß $p + 2$ ebenfalls eine Primzahl ist, oder es gibt eine größte Primzahl von dieser Eigenschaft."

„Entweder besteht für jede ganze Zahl $n > 2$ und beliebige positive ganze Zahlen a, b, c die Ungleichung $a^n + b^n \neq c^n$, oder es gibt eine kleinste solche ganze Zahl $n > 2$, für welche die Gleichung $a^n + b^n = c^n$ mit positiven ganzen Zahlen a, b, c lösbar ist."

Derartige Beispiele der Zahlentheorie sind geeignet, um uns die einfachsten Formen nichtfiniter Argumentationen zu verdeutlichen. Dagegen wird uns in der Zahlentheorie das Erfordernis zur Überschreitung des finiten Standpunktes nicht wirklich fühlbar; denn es gibt wohl kaum einen zahlentheoretisch geführten Beweis, bei dem sich die etwa benutzten nichtfiniten Schlußweisen nicht durch ziemlich leichte Modifikationen umgehen ließen.

Ganz anders steht es damit in der Analysis (Infinitesimalrechnung); hier gehört die nichtfinite Art der Begriffsbildung und der Beweisführung geradezu zur Methode der Theorie.

Wir wollen uns den Grundbegriff der Analysis, den Begriff der reellen Zahl, kurz vergegenwärtigen. Man definiert die reelle Zahl entweder als eine Folge beständig wachsender rationaler Zahlen

$$r_1 < r_2 < r_3 < \ldots,$$

welche alle unter einer gemeinsamen Schranke liegen, („Fundamentalreihe") oder als einen unendlichen Dezimalbruch bzw. Dualbruch, oder als eine Einteilung der rationalen Zahlen in zwei Klassen, bei welcher jede Zahl der ersten Klasse kleiner ist als jede Zahl der zweiten Klasse („Dedekindscher Schnitt").

Dabei liegt die Auffassung zugrunde, daß die rationalen Zahlen eine festumgrenzte Gesamtheit bilden, die als ein *Individuenbereich*

Let us mention a few examples of number-theoretical alternatives that we obtain by means of the "tertium non datur" for integers, but which cannot be proved finitistically according to the current state of our knowledge:[36.1]

- "Either every even number > 2 is representable as the sum of two prime numbers,[36.2] or there is an even number > 2 not representable as the sum of two prime numbers."
- "Either every integer of the form $2^{(2^{\mathfrak{f}})}+1$ with $\mathfrak{f} > 4$ is divisible into two factors > 1, or there is a prime number of the form $2^{(2^{\mathfrak{f}})}+1$ with $\mathfrak{f} > 4$." [36.3]
- "Either every sufficiently large integer is representable as the sum of less than 8 cubes, or for every integer n there is an integer m greater than n which is not representable as the sum of less than 8 cubes."
- "Either there are arbitrarily great prime numbers p with the property that $p+2$ is also a prime number, or there is a greatest prime number with this property."
- "Either for every integer $n > 2$ and arbitrary positive integers a, b, c, the inequation $a^n + b^n \neq c^n$ holds,[36.4] or there is a least such integer $n > 2$ for which the equation $a^n + b^n = c^n$ has a solution in the positive integers."

Such examples from number theory are appropriate for making the simplest forms of non-finitistic argumentations clear. Yet, we do not really feel the need to go beyond the finitistic standpoint in number theory: There is hardly any number-theoretical proof in which the non-finitistic inferences may not be circumvented by fairly easy modifications.

This is completely different in analysis (infinitesimal calculus), where the non-finitistic mode of concept formation and reasoning is a constituent part of the method of the theory.

[36.5]Let us briefly recall the fundamental notion of analysis, namely the notion of a real number. The real number is defined as one of the following: a strictly increasing sequence of rational numbers $$r_1 < r_2 < r_3 < \ldots,$$
which are all less than a common bound ("fundamental sequence"[36.6]); an infinite decimal or binary fraction; a partition of the rational numbers into two classes such that every number in the first class is smaller than every number of the second ("DEDEKIND cut").

[36.1]The labels (i.e. the bullets "•") of the items were introduced in the translation for clarification.

[36.2]This is GOLDBACH's Conjecture, open since CHRISTIAN GOLDBACH (1690–1764) proposed it to LEONHARD EULER (1707–1783) in 1742.

[36.3]In his letter for CHRISTIAAN HUYGENS (1629–1695), PIERRE FERMAT (1607?–1665) claimed to have proved $(n \in \mathbf{N})\left(2^{2^n}+1 \text{ is prime} \right)$, which, however, has 5 as its least counterexample because $2^7 \cdot 5 + 1$ divides $2^{2^5}+1$. Cf. e.g. [MAHONEY, 1994], [WIRTH, 2010, § 7].

[36.4]This is FERMAT's Last Theorem, for which a proof (neither direct, nor finitistic, nor even number theoretical) was published only at the end of the 20th century; cf. [WILES, 1995], [TAYLOR & WILES, 1995].

[36.5]§§ 2(d) and 2(d)(1) (as listed in the table of contents) start on this page, probably with this paragraph. In [HILBERT & BERNAYS, 2001], however, the beginning of §§ 2(d) and 2(d)(1) is indicated two paragraphs before.

[36.6]Let us consider a sequence $(r_i)_{i \in \mathbf{N}}$. It is a "CANTORsche Fundamentalreihe" [CANTOR, 1872, p.123f.] if for all $\varepsilon > 0$ and all $m \in \mathbf{N}$ there is an $n \in \mathbf{N}$ such that, for all $j \in \mathbf{N}$ with $j \geq n$, we have $|r_{j+m} - r_j| < \varepsilon$. It is a "Fundamentalfolge" (or "CAUCHY-Folge" ("CAUCHY sequence")) if for all $\varepsilon > 0$ there is an $n \in \mathbf{N}$ such that, for all $i, j \in \mathbf{N}$ with $i \geq n$ and $j \geq n$, we have $|r_i - r_j| < \varepsilon$. It is a "Fundamentalreihe" ("fundamental sequence") as defined in the text here if there is some b such that, for all $i \in \mathbf{N}$, we have $r_i < r_{i+1} < b$. Thus, the notion of a "CAUCHY-Folge" seems to be slightly stronger than the notion of a "CANTORsche Fundamentalreihe", and the notion of a "Fundamentalreihe" as defined in the text here seems to be incomparable to both, unless we restrict a "CANTORsche Fundamentalreihe" to be increasing and apply non-trivial reasoning in the theory of real numbers.

betrachtet werden kann. Auch die Gesamtheit der möglichen Folgen von rationalen Zahlen, bzw. der möglichen Einteilungen aller rationalen Zahlen wird in der Analysis als ein Individuenbereich gedacht.

Allerdings genügt es, an Stelle der Gesamtheit der rationalen Zahlen die Gesamtheit der ganzen Zahlen zugrunde zu legen und an Stelle der Einteilungen aller rationalen Zahlen diejenigen aller ganzen Zahlen zu betrachten. In der Tat ist ja jede positive rationale Zahl gegeben durch ein Zahlenpaar m, n, und jede rationale Zahl überhaupt läßt sich darstellen als Differenz zweier positiven rationalen Zahlen, d. h. als ein Paar von Zahlenpaaren $(m, n; p, q)$. Auch läßt sich ja jeder Dualbruch von der Form
$$0, a_1 a_2 a_3 \ldots,$$
worin $a_1, a_2, a_3, \ldots$ alle entweder $= 0$ oder $= 1$ sind, als eine Einteilung aller ganzen Zahlen deuten, nämlich als die Einteilung in solche Zahlen k, für welche $a_k = 0$ ist, und solche, für welche $a_k = 1$ ist. Jeder Einteilung der positiven ganzen Zahlen entspricht auf diese Weise umkehrbar eindeutig ein Dualbruch der obigen Form, und andrerseits läßt sich jede reelle Zahl darstellen als Summe einer ganzen Zahl und eines Dualbruchs dieser Form.

Statt der Einteilungen können wir auch *Mengen* von ganzen Zahlen betrachten; denn jede Menge von ganzen Zahlen bestimmt die Einteilung in solche Zahlen, die zur Menge gehören und solche, die nicht dazu gehören, und ist auch umgekehrt durch diese Einteilung vollständig bestimmt. — Dieselbe Bemerkung gilt auch für den DEDEKINDschen Schnitt, der ebenfalls durch eine *Menge* von rationalen Zahlen, nämlich die Menge der kleineren rationalen Zahlen vertreten werden kann. Eine solche Menge ist durch folgende Eigenschaften charakterisiert: 1. sie enthält mindestens eine und nicht alle rationalen Zahlen; 2. zugleich mit einer rationalen Zahl enthält sie alle kleineren rationalen Zahlen und mindestens eine größere.

Durch solche Umformungen wird aber die existentiale Voraussetzung, die wir für die Analysis zugrunde legen müssen, nur unwesentlich abgeschwächt. Es bleibt das Erfordernis, die Mannigfaltigkeit der ganzen Zahlen und auch die der Mengen von ganzen Zahlen als festen Individuenbereich aufzufassen, für den das „tertium non datur" Gültigkeit hat und mit Bezug auf den eine Aussage über die Existenz einer ganzen Zahl bzw. einer Zahlenmenge mit einer Eigenschaft $\mathfrak{E}$ unabhängig von ihrer Deutbarkeit als Partialurteil sinnvoll ist. Während also das Unendlich-Große und das Unendlich-Kleine durch diese Theorie der reellen Zahlen im eigentlichen Sinne ausgeschaltet wird und nur noch im Sinne einer Redeweise bestehen bleibt, wird das *Unendliche als Gesamtheit* beibehalten. Ja, man kann sagen, daß die Vorstellung von den unendlichen Gesamtheiten systematisch erst hier in der strengen Grundlegung der Analysis eingeführt und zur Geltung gebracht wurde.

The underlying point of view is that the rational numbers form a well-determined and fixed totality which can be seen as a *domain of individuals*. | In analysis, the totalities that are thought of as domains of individuals include also the possible sequences of rational numbers and the possible partitions of all rational numbers.

As a basis, however, it suffices to take the totality of integers instead of the totality of rational numbers, and the totality of the partitions of all integers instead of that of the partitions of all rational numbers. Indeed, every positive rational number is given by a pair of numbers m, n, and every rational number can be represented as the difference of two positive rational numbers, i.e. as a pair of number pairs $(m, n; p, q)$. Also every binary fraction of the form
$$0 . a_1 a_2 a_3 \ldots,$$
where all $a_1, a_2, a_3, \ldots$ are either $= 0$ or $= 1$, can be interpreted as a partition of all integers, namely the partition into those numbers k for which $a_k = 0$, and those for which $a_k = 1$. In this way there is a one-to-one correspondence of the partitions of the positive integers onto the binary fractions of the above form; and, on the other hand, every real number can be represented as the sum of an integer and a binary fraction of this form.

Instead of partitions, we can also consider *sets* of integers; this is because every set of integers determines the partition of those numbers that belong to the set and those that do not; and conversely, every set of integers is again completely determined by such a partition. — The same remark holds for the DEDEKIND cut, which can likewise be represented as a *set* of rational numbers, namely by the set of rational numbers that are smaller. Such a set is characterized by the following properties:[37.1]

1. it contains at least one rational number, but not all;

2. for every rational number it contains, it also contains all the smaller rational numbers and at least one rational number that is greater.

The existential[37.2] presupposition that we need for the grounding of analysis, however, is weakened by these conversions only non-essentially. There still remains the requirement to construe the manifold of the integers and also the manifold of the sets of integers as fixed domains of individuals for which the "tertium non datur" holds and with reference to which a proposition about the existence[37.3] of an integer or a set of integers with a property $\mathfrak{E}$ is meaningful, even without the possibility of interpreting this proposition as a partial judgment. Thus — in this theory of real numbers[37.4] — the *infinite as a totality* is retained; whereas the infinitely great and the infinitely small in the proper sense are excluded and remain just as a manner of speaking. Moreover, we can indeed say that here, in the rigorous grounding of analysis, the conception of infinite totalities was for the first time systematically introduced and brought to bear. |

[37.1]The vertical listing of the items was introduced in the translation.

[37.2]For this notion of "existence", cf. the discussion on Pages 1.a and 2, and especially in Note 1.6 on Page 1.b.

[37.3]Note that the meaning of "existence" is definitely different here from the other form of existence (or existential form) of Note 37.2; namely because it relates to the discussion of existential sentences on Page 32f. The line of thinking developed here may actually be the historical reason for calling the other form simply "existence", although "existence of the (infinite) extension of a notion" or "modality of existence of the actual infinite" may actually be more appropriate designations here. See also the discussions in Note 1.6 on Page 1.b, and in Note 20.3 on Page 20.

[37.4]The theory of real numbers to which the text refers here is the rational reconstruction of analysis as well as the reduction of the real numbers to natural numbers and set theory, which originated from KARL WEIERSTRASZ (1815–1897) as well as RICHARD DEDEKIND (1831–1916) and GOTTLOB FREGE (1848–1925). These subjects and their relation to the infinite are discussed in detail in [HILBERT, 1926].

§ 2. Die elementare Zahlentheorie. — Das finite Schließen.

Um uns wirklich davon zu überzeugen, daß die Voraussetzung der Totalität des Bereiches der ganzen Zahlen bzw. der rationalen Zahlen und ferner des Bereiches der Mengen (Einteilungen) von ganzen bzw. von rationalen Zahlen wesentlich in der Begründung der Analysis zur Anwendung kommt, brauchen wir uns nur einige von den grundlegenden Begriffsbildungen und Überlegungen vorzuführen.

Wird die reelle Zahl durch eine Folge von wachsenden rationalen Zahlen
$$r_1 < r_2 < r_3 < \ldots$$
definiert, so ist schon der Begriff der Gleichheit von reellen Zahlen nicht finit. Denn ob zwei solche Folgen von rationalen Zahlen dieselbe reelle Zahl definieren, hängt davon ab, ob es zu jeder Zahl in der einen Folge eine größere in der anderen Folge und umgekehrt gibt. Ein allgemeines Verfahren zur Entscheidung hierüber besitzen wir aber nicht.

Geht man andererseits aus von der Definition der reellen Zahl durch einen DEDEKINDschen Schnitt, so hat man zu beweisen, daß jede beschränkte Folge wachsender rationaler Zahlen einen Schnitt erzeugt, welcher die obere Grenze der Folge darstellt. Diesen Schnitt erhält man als die Einteilung der rationalen Zahlen in solche, die von mindestens einer Zahl aus der Folge übertroffen werden, und solche, die nicht übertroffen werden. Das heißt: eine rationale Zahl r wird zur ersten oder zur zweiten Klasse gerechnet, je nachdem es unter den Zahlen der Folge eine solche gibt, die $> r$ ist, oder alle Zahlen der Folge $\leq r$ sind. Dies ist wiederum keine finite Unterscheidung.

Ähnlich verhält es sich, wenn man die reellen Zahlen durch unendliche Dezimalbrüche oder durch Dualbrüche definiert. Es muß wiederum gezeigt werden, daß eine beschränkte Folge rationaler Zahlen
$$r_1 < r_2 < \ldots$$
einen Dezimalbruch bzw. Dualbruch bestimmt. Nehmen wir der Einfachheit halber an, es handle sich um eine Folge positiver echter Brüche:
$$0 < b_1 < b_2 < \cdots < 1,$$
und es soll der Dualbruch
$$0, a_1 a_2 a_3 \ldots$$
bestimmt werden, der die obere Grenze der Folge von Brüchen darstellt. Diese Bestimmung geschieht folgendermaßen:

a_1 ist $= 0$ oder $= 1$, je nachdem alle Brüche $b_n < \frac{1}{2}$ sind oder nicht;
a_{m+1} ist $= 0$ oder $= 1$, je nachdem alle Brüche b_n kleiner sind als
$$\frac{a_1}{2} + \frac{a_2}{4} + \cdots + \frac{a_m}{2^m} + \frac{1}{2^{m+1}}$$
oder nicht.

In allen diesen Fällen hat man es mit Alternativen zu tun, bei denen es sich darum handelt, ob alle rationalen Zahlen einer gegebenen Folge
$$r_1, r_2, r_3, \ldots$$

$^{38.1}$In the grounding of analysis, the presuppositions of the totalities of the domains of integers and rational numbers are essential; moreover, the same holds for the domains of sets (partitions) of integers and rational numbers. To become absolutely convinced of this essentiality, we only need to present some of the basic considerations and concept formations of analysis.

If the real number is defined by a sequence of increasing rational numbers

$$r_1 < r_2 < r_3 < \ldots,$$

then already the notion of equality of real numbers is not finitistic: Whether two sequences of rational numbers define the same real number depends on whether, for every number in one of the two sequences, there is a greater one in the other sequence, and conversely; but we do not have a general procedure at our disposal for deciding this.

On the other hand, if we define the real number by a DEDEKIND cut, then we have to prove that every bounded sequence of increasing rational numbers gives rise to a cut, which represents the least upper bound of the sequence. The cut is obtained by partitioning the rational numbers into those which are less than at least one number of the sequence, and those which are not. That is to say, a rational number r is said to be in the first or second class, respectively, depending on whether there is a number $> r$ among the numbers of the sequence, or whether all numbers in the sequence are $\leq r$. Again, this is not a finitistic distinction.

Things are similar if we define the real numbers by infinite decimal or binary fractions. Again, it must be shown that a bounded sequence of rational numbers

$$r_1 < r_2 < \ldots$$

determines a decimal or binary fraction. For the sake of simplicity, let us suppose that we are given a sequence of positive proper fractions

$$0 < b_1 < b_2 < \cdots < 1,$$

and our task is to determine the binary fraction

$$0 \, . \, a_1 \, a_2 \, a_3 \, \ldots$$

that represents the least upper bound of the sequence of fractions. This determination proceeds as follows:$^{38.2}$

- a_1 is $= 0$ or $= 1$, depending on whether $b_n < \frac{1}{2}$ holds for all fractions or not;

- a_{m+1} is $= 0$ or $= 1$, depending on whether or not all fractions b_n are less than
$$\frac{a_1}{2} + \frac{a_2}{4} + \cdots + \frac{a_m}{2^m} + \frac{1}{2^{m+1}} \, .$$

In all these cases, we are dealing with alternatives of the following kind: "All rational numbers in a given sequence

$$r_1, \, r_2, \, r_3, \, \ldots$$

satisfy a certain inequation, or there is at least one exception." The truth of such an alternative relies on the "tertium non datur" for integers, because it depends on the following presupposition: Either, for every integer n as an index, the rational number r_n satisfies the inequation in question, or there is at least one integer n for which r_n violates the inequation.

$^{38.1}$§ 2(d)(2) (as listed in the table of contents) starts with this paragraph.
$^{38.2}$The labels (i.e. the bullets "•") of the items were introduced in the translation for clarification.

eine gewisse Ungleichung erfüllen, oder ob diese mindestens einmal eine Ausnahme erleidet. Eine solche Alternative benutzt das „tertium non datur" für die ganzen Zahlen; denn es wird dabei vorausgesetzt, daß entweder für alle ganzen Zahlen n als Index die rationale Zahl r_n der betreffenden Ungleichung genügt, oder daß es eine ganze Zahl n gibt, für welche r_n gegen die Ungleichung verstößt.

Mit dieser Inanspruchnahme der *Gesamtheit der ganzen Zahlen* als Individuenbereich kommen wir jedoch für die Analysis nicht aus, sondern wir brauchen überdies die *Gesamtheit der rellen Zahlen* als Individuenbereich. Wie wir sahen, ist diese Gesamtheit im wesentlichen gleichzusetzen mit derjenigen der Mengen von ganzen Zahlen.

Die Erforderlichkeit des Individuenbereiches der reellen Zahlen macht sich schon beim Beweise des Satzes von der oberen Grenze einer beschränkten Menge von reellen Zahlen geltend. Um die Existenz der oberen Grenze für eine beschränkte, also etwa im Intervall von 0 bis 1 gelegene Menge von reellen Zahlen auf Grund der DEDEKINDschen Definition der reellen Zahl zu beweisen, betrachtet man die Einteilung der rationalen Zahlen in solche, die von einer reellen Zahl aus der Menge übertroffen werden, und solche, die nicht übertroffen werden. Man rechnet also eine rationale Zahl r zur ersten Klasse, dann und nur dann, wenn es in der Menge eine reelle Zahl $a > r$ gibt.

Nun muß man sich klarmachen, daß eine Menge uns in der Analysis im allgemeinen nur durch eine definierende Eigenschaft gegeben ist, d. h. die Menge wird eingeführt als die Gesamtheit derjenigen reellen Zahlen, welche eine gewisse Bedingung $\mathfrak{B}$ erfüllen. Die Frage, ob es in einer betrachteten Menge eine reelle Zahl $a > r$ gibt, kommt also darauf hinaus, ob es eine reelle Zahl gibt, welche größer als r ist und zugleich eine gewisse Bedingung $\mathfrak{B}$ erfüllt. In dieser Fassung wird es deutlich, daß wir die Gesamtheit der reellen Zahlen als einen Individuenbereich zugrunde legen[1].

Es sei noch bemerkt, daß der beschriebene Prozeß zur Gewinnung der oberen Grenze auf die Bildung einer *Vereinigungsmenge* hinauskommt. In der Tat ist ja jede reelle Zahl definiert durch eine Einteilung der rationalen Zahlen in kleinere und größere bzw. durch die Menge der kleineren rationalen Zahlen. Die gegebene Menge von reellen Zahlen stellt sich hiernach dar als eine Menge $\mathfrak{M}$ von Mengen von rationalen Zahlen. Und die obere Grenze der Menge $\mathfrak{M}$ wird gebildet von der Menge derjenigen rationalen Zahlen, welche mindestens einer der Mengen aus $\mathfrak{M}$ angehören. Die Gesamtheit dieser rationalen Zahlen ist aber gerade die Vereinigungsmenge von $\mathfrak{M}$.

Es gelingt auch nicht etwa, die Heranziehung des Individuenbereiches der reellen Zahlen dadurch zu umgehen, daß man anstatt der DEDEKIND-

---

[1] Auf den hier vorliegenden Sachverhalt hat WEYL in seiner Schrift „Das Kontinuum" (Leipzig 1918) besonders nachdrücklich hingewiesen.

The presupposition of the *totality of the integers* as a domain of individuals, however, is not yet sufficient for analysis: We need the *totality of the real numbers* as a domain of individuals as well. As we saw above,[39.1] this totality is essentially equivalent to that of the sets of integers.

It becomes necessary to have the real numbers as a domain of individuals already in the proof of the theorem on the least upper bound of a bounded set of real numbers, such as the reals in the interval from 0 to 1. To prove the existence of this least upper bound on the basis of DEDEKIND's definition of the real number, one considers the partition of the rational numbers into those which are exceeded by a real number in the bounded set, and those which are not. A rational number r is thus assigned to the first class if and only if there is a real number a in the set with $a > r$.

Now, we have to realize that, in analysis, a set is given in general only by its defining property; i.e. the set is introduced as the totality of those real numbers which satisfy a certain condition $\mathfrak{B}$. Therefore, the question whether there is a real number a with $a > r$ in a given set amounts to the question whether there is a real number which is greater than r and simultaneously satisfies a certain condition $\mathfrak{B}$. In this formulation, it becomes obvious that we depend here on the totality of the real numbers as a domain of individuals.[1]

Moreover, let us remark that the process just described for determining the least upper bound amounts to forming a *union*. Indeed, every real number is defined by a partition of the rational numbers in into smaller and greater ones, or just by the set of the smaller rational numbers. The given set of real numbers is thus represented as a set $\mathfrak{M}$ of sets of rational numbers. And the least upper bound of the set $\mathfrak{M}$ is given as the set of those rational numbers which belong to at least one of the sets in $\mathfrak{M}$. The totality of those rational numbers, however, is just the union of $\mathfrak{M}$.

---

[1] HERMANN WEYL has pointed out this state of affairs with particular insistence in [WEYL, 1918].

[39.1] Cf. Page 37.

schen Definition der reellen Zahlen die Definition durch eine Fundamentalreihe oder durch einen Dualbruch benutzt. Vielmehr wird hierdurch der Prozeß nur noch komplizierter, indem noch ein rekursives Verfahren hinzutritt. Es sei dies kurz für den Fall der Definition der reellen Zahlen durch Dualbrüche angegeben. Wir haben es dann zu tun mit einer Menge von Dualbrüchen

$$0, a_1 a_2 a_3 \ldots,$$

die wiederum durch eine gewisse Bedingung $\mathfrak{B}$ bestimmt ist; und die obere Grenze stellt sich dar durch einen Dualbruch

$$0, b_1 b_2 \ldots,$$

der folgendermaßen definiert ist:

$b_1 = 0$, wenn bei allen Dualbrüchen, welche der Bedingung $\mathfrak{B}$ genügen, an erster Dualstelle 0 steht, sonst ist $b_1 = 1$;

$b_{n+1} = 0$, wenn bei allen Dualbrüchen, welche der Bedingung $\mathfrak{B}$ genügen, und deren erste n Dualziffern bzw. mit $b_1, b_2, \ldots, b_n$ übereinstimmen, an der $(n + 1)$ ten Stelle 0 steht, sonst ist $b_{n+1} = 1$.

Hier tritt die Gesamtheit der reellen Zahlen auf als die Gesamtheit aller Dualbrüche, und wir machen Gebrauch von der Voraussetzung, daß für die aus Nullen und Einsen gebildeten unendlichen Folgen das „tertium non datur" gilt. —

Nun ist aber auch diese Voraussetzung der Gesamtheit aller reellen Zahlen bzw. aller Dualbrüche als eines Individuenbereiches noch nicht ausreichend. Dies zeigt sich an folgendem einfachen Fall: Es sei a die obere Grenze einer Menge von reellen Zahlen. Wir wollen zeigen, daß es eine Folge von reellen Zahlen *aus der Menge* gibt, welche gegen a konvergiert. Hierzu schließen wir folgendermaßen:

Aus der Eigenschaft der oberen Grenze folgt, daß es für jede ganze Zahl n eine Zahl c_n in der Menge gibt, so daß

$$a - \frac{1}{n} < c_n \leqq a,$$

also

$$|a - c_n| < \frac{1}{n}$$

ist. Die Zahlen c_n bilden somit eine gegen a konvergente Folge, und sie gehören alle der betrachteten Menge an.

Wenn wir so argumentieren, so verdecken wir durch die Ausdrucksweise einen prinzipiellen Beweispunkt. Indem wir nämlich die Schreibweise c_n anwenden, denken wir uns für jede Zahl n unter denjenigen reellen Zahlen c, welche zu der betrachteten Menge gehören, und die Ungleichung

$$a - \frac{1}{n} < c \leqq a$$

erfüllen, eine bestimmte ausgezeichnet.

The presupposition of the totality of the real numbers as a domain of individuals cannot be bypassed by using the definition of the real numbers as fundamental sequences or as binary fractions instead of DEDEKIND's definition. | Rather, that only complicates the process even more, namely by adding a recursive procedure. Let us briefly indicate this, for the case of defining the real numbers by binary fractions. In this case, we are dealing with a set of binary fractions
$$0.a_1 a_2 a_3 \ldots,$$
which is determined by a certain condition $\mathfrak{B}$ again; and its least upper bound is given as a binary fraction
$$0.b_1 b_2 b_3 \ldots,$$
which is defined as follows:[40.1]

$$b_1 = \begin{cases} 0 & \text{if 0 occupies the first binary position in all binary fractions satisfying } \mathfrak{B}; \\ 1 & \text{otherwise}; \end{cases}$$

$$b_{n+1} = \begin{cases} 0 & \text{if 0 occupies the } (n+1)^{\text{st}} \text{ position in all those binary fractions} \\ & \quad - \text{ which satisfy condition } \mathfrak{B} \text{ and} \\ & \quad - \text{ in which the first } n \text{ binary positions coincide with } b_1, b_2, \ldots, b_n, \\ & \quad \text{respectively}; \\ 1 & \text{otherwise.} \end{cases}$$

Here the totality of the real numbers appears as the totality of all binary fractions and we make use of the assumption that the "tertium non datur" holds for infinite sequences formed of zeros and ones. —

[40.2]Now even this presupposition of the totality of all real numbers (binary fractions) as a domain of individuals is still not sufficient. This can be seen from the following simple instance: Let a be the least upper bound of a set of real numbers. We want to show that there is a sequence of real numbers *from this set* that converges to a. For this, we reason as follows:

From the property of a least upper bound, it follows that for every integer n there is a number c_n in the set such that
$$a - \tfrac{1}{n} < c_n \leq a$$
and thus
$$|a - c_n| < \tfrac{1}{n}$$
Thus, the numbers c_n constitute a sequence converging to a, and they all belong to the set under consideration.

Arguing in this way, a principal point of the proof is concealed with our mode of expression. Namely, by using the notation c_n, we suppose that, for every number n, among those real numbers c in the set under consideration which satisfy the inequation
$$a - \tfrac{1}{n} < c \leq a,$$
a certain c_n is chosen. |

[40.1]The two big left braces ({) in the following two lines and the itemization were introduced in the translation.

[40.2]§ 2(d)(3) (as listed in the table of contents) starts with this paragraph.

Hierin liegt eine Voraussetzung. Was wir unmittelbar schließen können, ist nur, daß es zu jeder Zahl n eine Teilmenge $\mathfrak{M}_n$ unserer betrachteten Menge gibt, welche aus den Zahlen besteht, die der obigen Ungleichung genügen, und daß für jedes n diese Teilmenge mindestens ein Element enthält. Nun wird vorausgesetzt, daß wir in jeder dieser Mengen

$$\mathfrak{M}_1, \mathfrak{M}_2, \mathfrak{M}_3, \ldots$$

je ein Element, c_1 in $\mathfrak{M}_1$, c_2 in $\mathfrak{M}_2$, ... c_n in $\mathfrak{M}_n$, auszeichnen können, so daß wir eine bestimmte unendliche Folge von reellen Zahlen erhalten.

Wir haben hier einen Spezialfall des *Auswahlprinzips* vor uns, welches allgemein folgendes besagt: „Wenn es zu jedem Ding x einer Gattung $\mathfrak{G}_1$ mindestens ein Ding y der Gattung $\mathfrak{G}_2$ gibt, welches zu x in der Beziehung $\mathfrak{B}(x, y)$ steht, so gibt es eine Funktion φ, welche jedem Ding x der Gattung $\mathfrak{G}_1$ eindeutig ein solches Ding $\varphi(x)$ der Gattung $\mathfrak{G}_2$ zuordnet, welches zu x in der Beziehung $\mathfrak{B}(x, \varphi(x))$ steht."

In dem vorliegenden Fall ist die Gattung $\mathfrak{G}_1$ die der positiven ganzen Zahlen, $\mathfrak{G}_2$ die der reellen Zahlen, die Beziehung $\mathfrak{B}(x, y)$ besteht in der Ungleichung

$$a - \frac{1}{x} < y \leq a,$$

und die Funktion φ, deren Existenz aus dem Auswahlaxiom entnommen wird, ist die Zuordnung der reellen Zahl c_x zu ihrer Nummer x.

In der Anwendung des Auswahlprinzips, welches zuerst von ZERMELO als eine besondere Voraussetzung erkannt und in mengentheoretischer Fassung formuliert worden ist, liegt eine weitere Art der Überschreitung des finiten Standpunktes vor, die über die Anwendung des „tertium non datur" noch hinausgeht. Die angestellte Betrachtung von methodischen Beispielen lehrt uns, daß die Begründung der Infinitesimalrechnung, wie sie seit der Entdeckung der strengen Methoden gegeben wird, nicht im Sinne einer Zurückführung auf das *finite* zahlentheoretische Denken erfolgt. Die hier vollzogene *Arithmetisierung* der Größenlehre ist insofern *keine restlose*, als gewisse systematische Grundvorstellungen eingeführt werden, die nicht dem Bereich des anschaulichen arithmetischen Denkens angehören. Die Einsicht, welche uns die strenge Begründung der Analysis gebracht hat, besteht darin, daß diese wenigen Grundannahmen schon genügen, um die Größenlehre als Theorie der Zahlenmengen aufzubauen.

Von den Methoden der Analysis werden große Gebiete der Mathematik beherrscht, so die Funktionentheorie, die Differentialgeometrie, die Topologie (Analysis situs). Der weitgehendste Gebrauch von nichtfiniten Annahmen, noch weit über die Voraussetzungen der Analysis hinaus, wird in der allgemeinen *Mengenlehre* gemacht, deren Methoden auch in die neuere abstrakte Algebra und in die Topologie eingreifen.

There is a presupposition involved here. All we can infer immediately is that for every number n there is a subset $\mathfrak{M}_n$ of our set under consideration which consists of those numbers satisfying the above inequation, and that, for every n, this subset has at least one element. Now it is assumed that in each of these sets

$$\mathfrak{M}_1, \ \mathfrak{M}_2, \ \mathfrak{M}_3, \ \ldots$$

we can choose an element c_1 in $\mathfrak{M}_1$, c_2 in $\mathfrak{M}_2$, $\ldots$, c_n in $\mathfrak{M}_n$, such that we obtain a definite infinite sequence of real numbers.

This is a special case of the *principle of choice*,[41.1] which generally says the following: "If for every thing x of a genus[41.2] $\mathfrak{G}_1$, there is at least one thing y of a genus $\mathfrak{G}_2$ which stands to x in the relation $\mathfrak{B}(x,y)$, then there is a function φ that assigns to every thing x of the genus $\mathfrak{G}_1$ a unique thing $\varphi(x)$ of the genus $\mathfrak{G}_2$ which stands to x in the relation $\mathfrak{B}(x, \varphi(x))$."

In the present case, the genus $\mathfrak{G}_1$ is that of the positive integers and $\mathfrak{G}_2$ that of the real numbers,[41.3] the relation $\mathfrak{B}(x,y)$ is defined by the inequation

$$a - \tfrac{1}{x} \ < \ y \ \leq \ a,$$

and the function φ, whose existence is derived from the Axiom of Choice, assigns the real number c_x to its number x.

The principle of choice was first recognized as a special presupposition by ERNST ZERMELO, who formulated a set-theoretical version of it. Its application involves a further transgression of the finitistic standpoint, which goes even beyond the application of the "tertium non datur". From our treatment of the methods of the infinitesimal calculus, we learn that its grounding (as it has been given since the discovery of the rigorous methods) is not accomplished in the sense of a reduction to *finitistic* number-theoretical thinking. The *arithmetization* of the theory of magnitudes that is accomplished here is *not a complete one* because certain systematic fundamental conceptions are introduced which do not belong to the realm of intuitive arithmetical thinking. The insight that the rigorous grounding of analysis has brought us is that those few fundamental assumptions already suffice to construct the theory of magnitudes as a theory of sets of numbers.

The methods of analysis dominate large areas of mathematics, such as function theory, differential geometry, and topology (*analysis situs*). The most extensive use of non-finitistic assumptions — going far beyond the presuppositions of analysis — is made in general *set theory*,[41.4] whose methods also play a major rôle in modern abstract algebra and in topology.

[41.1] Principles of choice are typically formalized as forms of the Axiom of Choice [RUBIN & RUBIN, 1985] (such as the Well-Ordering Theorem and Zorn's Lemma) or as weak forms of the Axiom of Choice [HOWARD & RUBIN, 1998] (such as the Principle of Dependent Choice and König's Lemma).

[41.2] Cf. Note 2.1 on Page 2 for translating "Gattung" as "genus".

[41.3] Actually, we have to restrict either $\mathfrak{G}_2$ or the second argument of $\mathfrak{B}$ to the real numbers *of our set under consideration* here.

[41.4] We translate the noun phrase "allgemeine Mengenlehre" as "general set theory". Note that this name is meant here only in the most general sense, and not in the sense of the technical term denoting many special set theories from ADOLF ABRAHAM FRAENKEL (1891–1965) to GEORGE BOOLOS (1940–1996), including the one of [BERNAYS, 1942b; 1943].

Die Arithmetik in ihrer üblichen Behandlung entspricht somit keineswegs dem finiten Standpunkt, sondern beruht wesentlich auf hinzutretenden Prinzipien des Schließens. Wir sehen uns daher, wenn wir die Arithmetik in ihrer vorhandenen Form beibehalten wollen und andererseits die Anforderungen des finiten Standpunktes unter dem Gesichtspunkt der Evidenz anerkennen, vor die Aufgabe gestellt, die Anwendung jener Prinzipien, mit denen wir über das finite Denken hinausgehen, durch einen Nachweis ihrer Widerspruchsfreiheit zu rechtfertigen. Wenn ein solcher Nachweis für die Widerspruchsfreiheit der üblichen arithmetischen Schlußweisen gelingt, so haben wir damit auch die Gewähr, daß die Ergebnisse dieser Schlußweisen niemals durch eine finite Feststellung oder eine finite Überlegung widerlegt werden können; denn die finiten Methoden sind ja in der üblichen Arithmetik inbegriffen, und eine finite Widerlegung eines mit den üblichen Mitteln der Arithmetik bewiesenen Satzes würde daher einen Widerspruch innerhalb der üblichen Arithmetik bedeuten.

Wir kommen so auf das im § 1 gestellte Problem zurück. Nun bleibt aber noch die Frage zu beantworten, von welcher die Betrachtungen dieses Paragraphen ausgingen: ob wir nicht, anstatt einen Unmöglichkeitsbeweis für das Auftreten eines Widerspruchs in der Arithmetik an Hand der Formalisierung der Schlüsse zu führen, einfacher direkt ohne zusätzliche Annahmen die ganze Arithmetik begründen und damit jenen Unmöglichkeitsbeweis entbehrlich machen können.

Die Antwort hierauf ist zum einen Teil bejahend, zum anderen verneinend. Nämlich was die Möglichkeit einer direkten finiten Begründung der Arithmetik, in einem für die praktischen Anwendungen ausreichenden Umfange, betrifft, so ist diese durch die Untersuchungen von KRONECKER und von BROUWER aufgezeigt worden.

KRONECKER, der als erster die Anforderungen des finiten Standpunktes geltend gemacht hat, ging darauf aus, die nichtfiniten Schlußweisen allenthalben aus der Mathematik auszuschalten. In der Theorie der algebraischen Zahlen und Zahlenkörper ist er damit zum Ziel gekommen[1]. Hier gelingt auch die Einhaltung des finiten Standpunktes noch in solcher Weise, daß man von den Sätzen und Beweismethoden nichts Wesentliches aufzugeben braucht.

Nachdem die Problemstellung KRONECKERs lange Zeit hindurch gänzlich abgelehnt worden war, hat in neuerer Zeit BROUWER sich an die Aufgabe gemacht, die Arithmetik unabhängig von dem Satz vom ausgeschlossenen Dritten zu begründen, und hat im Sinne dieses Programms erhebliche Teile der Analysis und Mengenlehre entwickelt[2].

---

[1] Die Ergebnisse dieser Untersuchungen wurden von KRONECKER nicht systematisch publiziert, sondern nur in Vorlesungen mitgeteilt.

[2] Ein ausführliches Verzeichnis der Veröffentlichungen BROUWERs über diesen Gegenstand findet sich in dem Lehrbuch von A. FRAENKEL: „Einleitung in die Mengenlehre", *Dritte* Auflage. Berlin: Julius Springer 1928.

| [42.1]Arithmetic[42.2] as customarily treated is thus by no means in accordance with the finitistic standpoint, but rests essentially on additional principles of inference. Thus, if we want to keep arithmetic in its present form, and if we also want to acknowledge the demands of the finitistic standpoint with its emphasis on self-evidence, then we are confronted with the following task: We have to justify the application of those principles which go beyond[42.3] the scope of finitistic thinking by proving their consistency. If we succeed in proving the consistency of the customary modes of inference in arithmetic, then we also have the guarantee that the results of these modes can never be refuted by a finitistic observation or consideration; this is because the finitistic methods are included in customary arithmetic, and a finitistic refutation of a sentence proved by customary means of arithmetic would therefore mean a contradiction within customary arithmetic itself.

This means that we return to the issue posed in §1. The questions that initiated the treatment in the current section and still remain to be answered are: Do we really have to give an impossibility proof of the occurrence of a contradiction[42.4] in arithmetic by means of a complete formalization of the inferences? Can we not ground all of arithmetic directly, by simpler means and without additional assumptions (so that we can dispense with the impossibility proof)?

The answers to these questions are partly affirmative and partly negative. Specifically, in the investigations of KRONECKER and BROUWER, possibilities of a direct finitistic grounding of arithmetic have been demonstrated to an extent that is sufficient for practical applications.

KRONECKER was the first to enforce the requirements of the finitistic standpoint; he aimed at eliminating the non-finitistic modes of inference from mathematics altogether. He achieved his goal for the theory of algebraic numbers and number fields.[1] In these areas, we still succeed in observing the finitistic standpoint without sacrificing any essential theorems or methods of proof.

More recently, after KRONECKER's posing of the problem had been completely rejected for a long time, BROUWER set himself the task of a grounding of arithmetic that is independent of the law of the excluded middle,[42.5] and developed substantial portions of analysis and set theory in terms of this program.[2] | With his approach, however, essential theorems have to be abandoned and considerable complications regarding concept formation have to be accepted.

---

[1]The results of these investigations were not published systematically by KRONECKER, but communicated only in lectures.

[2]A detailed list of BROUWER's publications in this area is found in the textbook of FRAENKEL [1928].

[42.1]§ 2(e) (as listed in the table of contents) starts on this page, probably with this paragraph.

[42.2]Note that, in the whole text, "arithmetic" is to be understood as the arithmetic not only of natural numbers (number theory), but of real numbers as well.

[42.3]At a first glance, it seems to be appropriate here to translate "über das finite Denken hinausgehen" as "to transcend finitistic thinking", because something like KANT's notion of transcendence from phenomena to noumena may be intended here (according to the *Critique of Pure Reason* [KANT, 1787]). On a deeper inspection of the German text, however, such an intention becomes unlikely: Otherwise, instead of the verb "hinausgehen", BERNAYS would probably have chosen the more specific German verb "überschreiten" (as on Page 3, translated as "transcend"), or even the formal German verb "transzendieren" (or "trancendieren" as in [KANT, 1787]).

[42.4]The noun phrase "an impossibility proof of the occurrence of a contradiction" is synonymous here with "a proof of the impossibility of the derivation of a contradiction".

[42.5]Cf. Page 34 and Note 34.3 for the law of the excluded middle.

Allerdings müssen bei diesem Verfahren wesentliche Sätze preisgegeben und beträchtliche Komplikationen der Begriffsbildung in Kauf genommen werden.

Der methodische Standpunkt des „Intuitionismus", den BROUWER zugrunde legt, bildet eine gewisse *Erweiterung der finiten Einstellung* insofern, als BROUWER zuläßt, daß eine Annahme über das Vorliegen einer Folgerung bzw. eines Beweises eingeführt wird, ohne daß die Folgerung bzw. der Beweis nach anschaulicher Beschaffenheit bestimmt ist. So sind z. B. vom Standpunkt BROUWERS Sätze zugelassen von der Form „wenn unter der Voraussetzung 𝔄 der Satz 𝔅 gilt, so gilt auch ℭ" oder auch „die Annahme, daß 𝔄 widerlegbar sei, führt auf einen Widerspruch" bzw. nach BROUWERS Ausdrucksweise: „die Absurdität von 𝔄 ist absurd".

Eine derartige weitere Fassung des finiten Standpunktes, welche erkenntnistheoretisch darauf hinauskommt, daß man zu den anschaulichen Einsichten noch Überlegungen allgemein logischen Charakters hinzunimmt, erweist sich als erforderlich, wenn man mittels der finiten Betrachtungen über einen gewissen elementaren Bereich hinaus gelangen will. Wir werden in einem vorgerückten Stadium unserer Betrachtungen auf dieses Erfordernis hingeführt werden.

Wenngleich nun durch die genannten Untersuchungen ein Weg gewiesen ist, wie man sich in der Mathematik weitgehend ohne die nichtfiniten Schlußweisen behelfen kann, so wird doch damit ein Nachweis für die Widerspruchsfreiheit der üblichen Methoden der Arithmetik keineswegs entbehrlich gemacht. Denn die Vermeidung der nichtfiniten Methoden des Schließens erfolgt nicht im Sinne einer vollen Ersetzung dieser Methoden durch andere Überlegungen, vielmehr gelingt sie in der Analysis und den an sie anschließenden Gebieten der Mathematik nur um den Preis einer wesentlichen Einbuße an Systematik und Beweistechnik.

Dem Mathematiker kann aber nicht zugemutet werden, eine solche Einbuße ohne zwingenden Grund hinzunehmen. Die Methoden der Analysis sind in einem Ausmaß erprobt, wie wohl sonst kaum eine wissenschaftliche Voraussetzung, und sie haben sich aufs glänzendste bewährt. Wenn wir diese Methoden unter dem Gesichtspunkt der Evidenz kritisieren, so entsteht für uns die Aufgabe, den Grund für ihre Anwendbarkeit aufzuspüren, so wie wir es überall in der Mathematik tun, wo ein erfolgreiches Verfahren auf Grund von Vorstellungen geübt wird, die an Evidenz zu wünschen übriglassen.

Es ist also, sofern wir den finiten Standpunkt einnehmen, ein nicht abzuweisendes Problem, uns betreffs der Anwendbarkeit der nichtfiniten Methoden eine klare Einsicht zu verschaffen, und sofern uns unser Vertrauen auf diese Methoden nicht täuscht, kann diese Einsicht nur darin bestehen, daß wir Gewißheit darüber er-

The methodological standpoint of "intuitionism" underlying BROUWER's approach constitutes a certain *extension of the finitistic attitude*[43.1] insofar as assumptions on the existence of derivations or proofs may be introduced even if their intuitive nature is not determined. For instance, from BROUWER's standpoint, sentences of the following form are allowed: "If the sentence $\mathfrak{B}$ holds under the assumption $\mathfrak{A}$, then $\mathfrak{C}$ holds as well." And also: "The assumption that $\mathfrak{A}$ is refutable leads to a contradiction." Or in BROUWER's way of speaking: "The absurdity of $\mathfrak{A}$ is absurd".

Such an extension of the finitistic standpoint, which epistemologically amounts to the addition of considerations of general logical character to the intuitive insights, turns out to be necessary if — by means of finitistic treatment — we want to go beyond a certain elementary area. We will be led to this demand at an advanced stage of our treatment.

And although the aforementioned investigations indicate a path in mathematics that is free from non-finitistic modes of inference to a large extent, a consistency proof of the customary methods of arithmetic is by no means made dispensable. This is because the avoidance of non-finitistic methods of inference is not achieved in the sense of a complete replacement of these methods by other considerations; rather, this avoidance succeeds only at the cost of a substantial loss of systematics and proof techniques in analysis and the associated advanced areas of mathematics.

A mathematician, however, cannot be expected to accept such a loss without compelling reasons. The methods of analysis have arguably been put to test to a greater extent than almost any other scientific presuppositions; and they have proved themselves most splendidly. If we criticize these methods on the basis of their self-evidence, then we are confronted with the task of tracking the reasons of their applicability (just as we do everywhere in mathematics where a successful approach is practiced on the basis of conceptions which leave something to be desired with regard to their self-evidence).

Thus, insofar as we take the finitistic standpoint, we cannot escape the problem of obtaining a clear insight into the applicability of non-finitistic methods. And if our trust in these methods does not deceive us, this insight can only consist in gaining certainty that these | customary methods of arithmetic can never lead to a demonstrably false result; or, more precisely, that the results of their application are consistent with each other as well as with any fact evident from the finitistic standpoint.

[43.1]This must not be understood in the sense that the intuitionism of L. E. J. BROUWER (1881–1966) would have been, as a matter of historical fact, built on top of HILBERT's finitistic standpoint. BROUWER's intuitionism, as documented in [BROUWER, 1925a; 1925b; 1926], was first described in preliminary form (but not under this name) in [BROUWER, 1907]. Thus, it precedes HILBERT's finitistic standpoint in time by a decade at least. Moreover, note that in the years around 1930, when this volume was written and first published, the label "intuitionism" was used to denote HILBERT's finitistic standpoint as well as BROUWER's intuitionism; cf. e.g. [WIRTH &AL., 2009, § 3.2].

halten, daß diese üblichen arithmetischen Methoden niemals zu einem nachweislich falschen Ergebnis führen können, genauer gesagt, daß die Ergebnisse ihrer Anwendung sowohl miteinander wie auch mit jeder vom finiten Standpunkt ersichtlichen Tatsache im Einklang stehen.

Dieses Problem ist aber kein anderes als das eines Nachweises der Widerspruchsfreiheit unserer üblichen Arithmetik.

Zur Behandlung dieses Problems haben wir im § 1 bereits die in der symbolischen Logik ausgebildete Methode der Formalisierung des logischen Schließens in Aussicht genommen[1]. Diese Methode erfüllt jedenfalls die Bedingung, daß durch sie die Aufgabe des geforderten Nachweises der Widerspruchsfreiheit — sofern die vollständige Formalisierung der üblichen Arithmetik gelingt — zu einem *finiten* Problem gemacht wird. Denn wenn die übliche Arithmetik formalisiert ist, d. h. ihre Voraussetzungen und Schlußweisen in Ausgangsformeln und Regeln der Ableitung übersetzt sind, so stellt sich ein arithmetischer Beweis als eine anschaulich überblickbare Aufeinanderfolge von Prozessen dar, deren jeder einem von vornherein angegebenen Bestande von in Betracht kommenden Handlungen angehört. Wir haben also dann grundsätzlich dieselbe methodische Sachlage wie in der elementaren Zahlentheorie, und so wie es dort gelingt, Unmöglichkeitsbeweise in finitem Sinne zu führen, z. B. dafür, daß es nicht zwei Ziffern $\mathfrak{m}$, $\mathfrak{n}$ geben kann derart, daß

$$\mathfrak{m} \cdot \mathfrak{m} = 2 \cdot \mathfrak{n} \cdot \mathfrak{n},$$

so ist es auch ein finites Problem, zu zeigen, daß es in der formalisierten Arithmetik nicht zwei Beweise geben kann derart, daß die Endformel des einen mit der Negation der Endformel des anderen übereinstimmt.

Von einer Lösung dieses Problems sind wir allerdings noch weit entfernt. Doch sind in der Verfolgung dieses Zieles bereits mannigfache lohnende Ergebnisse gewonnen worden, und es hat sich auf diesem Wege ein neues Feld der Forschung eröffnet, indem die Formalisierung des logischen Schließens zu einer systematischen *Beweistheorie* verwertet wurde, welche die Frage nach der Tragweite der logischen Schlußweisen, die von der traditionellen Logik nur in einer sehr speziellen Form gestellt und gelöst wurde, in systematischer Allgemeinheit behandelt und durch deren Untersuchungsmethode die Probleme der Grundlagen der Mathematik mit den logischen Problemen in unmittelbaren Zusammenhang treten.

Diese Beweistheorie, auch *„Metamathematik"* genannt, soll im folgenden entwickelt werden. Wir beginnen mit der Formalisierung des Schließens, die wir zunächst unabhängig von der Anwendung auf die Beweistheorie darlegen wollen.

---

[1] Vgl. S. 18.

This problem, however, is none other than that of proving the consistency of our customary arithmetic.

Toward the treatment of this problem, in §1 we have already envisaged the method of formalizing logical inference as developed in symbolic logic.[1] In any case — provided that a complete formalization of customary arithmetic succeeds — this method satisfies the condition of turning the task of a demanded proof of consistency into a *finitistic* problem: When customary arithmetic is formalized (i.e. when its presuppositions and modes of inference are translated into initial formulas[44.1] and rules of deduction), then an arithmetical proof presents itself as a succession of intuitively comprehensible processes, each of which belongs to a stock of relevant operations fixed in advance. Then, in principle, we have the same situation as in our treatment of elementary number theory.[44.2] And just as it is possible to carry out impossibility proofs in a finitistic sense there, for example that there are no two numerals $\mathfrak{m}$, $\mathfrak{n}$ such that

$$\mathfrak{m} \cdot \mathfrak{m} \;=\; 2 \cdot \mathfrak{n} \cdot \mathfrak{n},$$

it is also a finitistic problem to show that there are no two proofs in formalized arithmetic such that the end formula[44.3] of the first coincides with the negation of the end formula of the second.

We are still far away, however, from a solution to this problem. Yet, in pursuing this goal, a variety of rewarding results has already been obtained. Moreover, in this pursuit, a new field of research has been opened up, in which the formalization of logical inference has been turned into a systematic *proof theory*. This proof theory treats the question of the scope of logical modes of inference — a question which was posed and solved within traditional logic only in a very special form — with systematic generality. And via the investigational methods of this proof theory, the problems of the foundations of mathematics become immediately connected to the problems of logic.

This proof theory, also called *"metamathematics"*, will be developed in what follows. We will start with the formalization of inference, and we will initially present this formalization independent of its application to proof theory.

---

[1] Cf. p. 18.

[44.1] We translate the German term "Ausgangsformel" as "initial formula". An initial formula is a premise in a proof or a leaf in a proof tree.

[44.2] This is one of the rare cases where we translate "methodisch" as "our treatment of". Cf. also Notes 1.3 and 20.5.

[44.3] We translate the German term "Endformel" as "end formula". An end formula is the result of a proof or the root of a proof tree.

Appendix

The bilingual part is ended here.

The appendix consists of

- our new bibliographical references,
- our new overall index,
- BERNAYS' original Namenverzeichnis (index of persons),
- BERNAYS' original Sachverzeichnis (subject index),
- our glossary German–English, and
- our guidelines for the typesetting of the English translation.

References

[ABRAMSKY &AL., 1992] S. Abramsky, Dov Gabbay, and T. S. E. Maibaum, editors. *Handbook of Logic in Computer Sci.*. Clarendon Press, 1992.

[ABRUSCI, 1983] V. Michele Abrusci. PAUL HERTZ's logical works — contents and relevance. 1983. In [ANON, 1983, pp. 369–374].

[ACKERMANN, 1925a] Wilhelm Ackermann. Begründung des „tertium non datur" mittels der HILBERT-schen Theorie der Widerspruchsfreiheit. *Mathematische Annalen*, 93:1–36, 1925. Received March 30, 1924. Inauguraldissertation, Göttingen 1924.

[ACKERMANN, 1925b] Wilhelm Ackermann. Die Widerspruchsfreiheit des Auswahlaxioms (Vorläufige Mitteilung). *Nachrichten von der Gesellschaft der Wissenschaften zu Göttingen, Mathematisch-Physikalische Klasse, Weidmannsche Buchhandlung, Berlin*, 1924:247–250, 1925. Presented by DAVID HILBERT at the meeting of Jan. 16, 1925.

[ACKERMANN, 1928a] Wilhelm Ackermann. Über die Erfüllbarkeit gewisser Zählausdrücke. *Mathematische Annalen*, 100:638–649, 1928. Received Feb. 9, 1928.

[ACKERMANN, 1928b] Wilhelm Ackermann. Zum HILBERTschen Aufbau der reellen Zahlen. *Mathematische Annalen*, 99:118–133, 1928. Received Jan. 20, 1927.

[ACKERMANN, 1934] Wilhelm Ackermann. Review of [HILBERT & BERNAYS, 1934]. *Jahrbuch über die Fortschritte der Mathematik*, 60:17–19, 1934.

[ACKERMANN, 1935a] Wilhelm Ackermann. Untersuchungen über das Eliminationsproblem der mathematischen Logik. *Mathematische Annalen*, 110:390–413, 1935. Received Jan. 13, 1934.

[ACKERMANN, 1935b] Wilhelm Ackermann. Zum Eliminationsproblem der mathematischen Logik. *Mathematische Annalen*, 111:61–63, 1935. Received Nov. 9, 1934.

[ACKERMANN, 1936] Wilhelm Ackermann. Beiträge zum Entscheidungsproblem der mathematischen Logik. *Mathematische Annalen*, 112:419–432, 1936. Received July 25, 1935.

[ACKERMANN, 1937] Wilhelm Ackermann. Die Widerspruchsfreiheit der allgemeinen Mengenlehre. *Mathematische Annalen*, 114:305–315, 1937. Received Nov. 15, 1936.

[ACKERMANN, 1938] Wilhelm Ackermann. Mengentheoretische Begründung der Logik. *Mathematische Annalen*, 115:1–22, 1938. Received April 23, 1937.

[ACKERMANN, 1940] Wilhelm Ackermann. Zur Widerspruchsfreiheit der Zahlentheorie. *Mathematische Annalen*, 117:163–194, 1940. Received Aug. 15, 1939.

[ACKERMANN, 1954] Wilhelm Ackermann. *Solvable Cases of the Decision Problem*. North-Holland (Elsevier), 1954.

[ANON, 1887] Anon, editor. *Philosophische Aufsätze. EDUARD ZELLER zu seinem fünfzigjährigen Doctor-Jubiläum gewidmet*. Verlag Fues, Leipzig, 1887. Reprinted as facsimilie by Zentral-Antiquariat der Deutschen Demokratischen Republik, Leipzig, 1962.

[ANON, 1899] Anon, editor. *Festschrift zur Feier der Enthüllung des GAUSZ-WEBER-Denkmals in Göttingen, herausgegeben von dem Fest-Comitee*. Verlag von B. G. Teubner, Leipzig, 1899.

[ANON, 1905] Anon, editor. *Verhandlungen des Dritten Internationalen Mathematiker-Kongresses, Heidelberg, Aug. 8–13, 1904*. Verlag von B. G. Teubner, Leipzig, 1905.

[ANON, 1931] Anon. Minutes of the meeting of the vienna circle on 15 Jan 1931. CARNAP Archives of the University of Pittsburgh. See also [SIEG, 1988, Note 11] and [MANCOSU, 1999b, pp. 36–37], 1931.

[ANON, 1941] Anon, editor. *Der Große Duden. Rechtschreibung der deutschen Sprache und der Fremdwörter. Zwölfte, neubearbeitete und erweiterte Auflage*. Bibliographisches Institut, Leipzig, 1941.

[ANON, 1950] Anon, editor. *Études de Philosophie des Sciences. En hommage à FERDINAND GONSETH à l'occasion de son soixantième anniversaire*. Number 20 in Bibliothéque Scientifique. Éditions du Griffon, Neuchâtel (Suisse), 1950.

References

[ANON, 1972] Anon, editor. *Das Kontinuum und andere Monographien*. Chelsea, New York, 1972. Reprint of 4 monographies by HERMANN WEYL ([WEYL, 1918]), EDMUND G. H. LANDAU, and BERNHARD RIEMANN.

[ANON, 1983] Anon, editor. *Atti del Convegno Internazionale di Storia della Logica, Le teorie delle modalità. 4–8 December 1982, San Gimignano*. CLUEB, Bologna, 1983.

[AVIGAD & FEFERMAN, 1998] Jeremy Avigad and Sol(omon) Feferman. GÖDEL's functional ("dialectica") interpretation. 1998. In [BUSS, 1998, pp. 337–405].

[AVIGAD, 2006] Jeremy Avigad. Methodology and metaphysics in the development of DEDEKIND's theory of ideals. 2006. In [FERREIRÓS & GRAY, 2006, pp. 159–186].

[BARWISE, 1977] Jon Barwise, editor. *Handbook of mathematical logic*. North-Holland (Elsevier), 1977.

[BASLER, 1934] Otto Basler, editor. *Der Große Duden. Rechtschreibung der deutschen Sprache und der Fremdwörter. Elfte, neubearbeitete und erweiterte Auflage. Erster verbesserter Neudruck*. Bibliographisches Institut, Leipzig, 1934.

[BEHMANN, 1922] Heinrich Behmann. Beiträge zur Algebra der Logik, insbesondere zum Entscheidungsproblem. *Mathematische Annalen*, 86:163–229, 1922. Received July 16, 1921. Corrected in [BEHMANN, 1923].

[BEHMANN, 1923] Heinrich Behmann. Druckfehlerberichtigung zu dem Aufsatz [BEHMANN, 1922]. *Mathematische Annalen*, 88:168, 1923.

[BEHNKE, 1976] Heinrich Behnke. Die goldenen Jahre des Mathematischen Seminars der Universität Hamburg. *Mitteilungen der Mathematischen Gesellschaft in Hamburg*, X:225–240, 1976.

[BERKA & KREISER, 1973] Karel Berka and Lothar Kreiser, editors. *Logik-Texte – Kommentierte Auswahl zur Geschichte der modernen Logik*. Akademie-Verlag, Berlin, 1973. 2nd rev. ed. (1st ed. 1971; 4th rev. rev. ed. 1986).

[BERNAYS & SCHÖNFINKEL, 1928] Paul Bernays and Moses Schönfinkel. Zum Entscheidungsproblem der mathematischen Logik. *Mathematische Annalen*, 99:342–372, 1928. Received March 24, 1927.

[BERNAYS, 1922] Paul Bernays. Über HILBERTs Gedanken zur Grundlegung der Mathematik. *Jahresbericht der Deutschen Mathematiker-Vereinigung*, 31:10–19, 1922.

[BERNAYS, 1927] Paul Bernays. Probleme der theoretischen Logik. *Unterrichtsblätter für Mathematik und Naturwissenschaften*, 33:369–377, 1927. Talk given at the 56. Versammlung deutscher Philologen und Schulmänner in Göttingen. Also in [BERNAYS, 1976, pp. 1–16].

[BERNAYS, 1928] Paul Bernays. Zusatz zu HILBERTs Vortrag „Die Grundlagen der Mathematik". *Abhandlungen aus dem mathematischen Seminar der Univ. Hamburg*, 6:89–92, 1928. English translation *On the Consistency of Arithmetic* in [HEIJENOORT, 1971, pp. 485–489].

[BERNAYS, 1930/31] Paul Bernays. Die Philosophie der Mathematik und die HILBERTsche Beweistheorie. *Blätter für Deutsche Philosophie*, 4:326–367, 1930/31. Also in [BERNAYS, 1976, pp. 17–61].

[BERNAYS, 1935a] Paul Bernays. HILBERT's Untersuchungen über die Grundlagen der Arithmetik. 1935. In [HILBERT, 1932ff., Vol. 3, pp. 196–216].

[BERNAYS, 1935b] Paul Bernays. Sur le Platonisme dans les Mathématiques. *L'Enseignement Mathématique*, 34:52–69, 1935. German translation by PETER BERNATH in [BERNAYS, 1976, pp. 62–78].

[BERNAYS, 1937] Paul Bernays. A System of Axiomatic Set Theory — Part I. *J. Symbolic Logic*, 2:65–77, 1937. Received Sept. 29, 1936.

[BERNAYS, 1941] Paul Bernays. A System of Axiomatic Set Theory — Part II. *J. Symbolic Logic*, 6:1–17, 1941. Received June 12, 1940.

[BERNAYS, 1942a] Paul Bernays. A System of Axiomatic Set Theory: Part III. Infinity and Enumerability. Analysis. *J. Symbolic Logic*, 7:65–89, 1942. Received Oct. 7, 1940.

[BERNAYS, 1942b] Paul Bernays. A System of Axiomatic Set Theory: Part IV. General Set Theory. *J. Symbolic Logic*, 7:133–145, 1942. Received April 25, 1940.

[BERNAYS, 1943] Paul Bernays. A System of Axiomatic Set Theory: Part V. General Set Theory (continued). *J. Symbolic Logic*, 8:89–106, 1943. Received June 9, 1941.

References

[BERNAYS, 1948] Paul Bernays. A System of Axiomatic Set Theory — Part VI. *J. Symbolic Logic*, 13:65–79, 1948. Received April 28, 1947.

[BERNAYS, 1950] Paul Bernays. Mathematische Existenz und Widerspruchsfreiheit. 1950. In [ANON, 1950, pp. 11–25]. Also in [BERNAYS, 1976, pp. 92–106].

[BERNAYS, 1954a] Paul Bernays. A System of Axiomatic Set Theory — Part VII. *J. Symbolic Logic*, 19:81–96, 1954. Received March 30, 1953.

[BERNAYS, 1954b] Paul Bernays. Zur Beurteilung der Situation in der beweistheoretischen Forschung + Discussion. *Revue internationale de philosophie*, 8:9–13+15–21, 1954.

[BERNAYS, 1967] Paul Bernays. HILBERT, DAVID. 1967. In [EDWARDS, 1967, pp. 496–504].

[BERNAYS, 1969] Paul Bernays. HERTZ, PAUL. *Neue Deutsche Biographie*, 8:711f., 1969. http://www.deutsche-biographie.de/pnd11675446X.html.

[BERNAYS, 1976] Paul Bernays. *Abhandlungen zur Philosophie der Mathematik*. Wissenschaftliche Buchgesellschaft, Darmstadt, 1976.

[BERNSTEIN, 1919] Felix Bernstein. Die Mengenlehre GEORG CANTORs und der Finitismus. *Jahresbericht der Deutschen Mathematiker-Vereinigung*, 28:63–78, 1919.

[BORN, 1904] Max Born. Zahlbegriff und Quadratur des Kreises. Notes of [HILBERT, 1904]. HILBERT Nachlass, Georg August Universität Göttingen, Mathematisches Inst., Lesesaal, 1904.

[BORN, 1905] Max Born. Logische Prinzipien des mathematischen Denkens. Notes of [HILBERT, 1905b], 188 bound pp.. HILBERT Nachlass, Niedersächsische Staats- und Universitätsbibliothek Göttingen, Handschriftenabteilung, Cod. Ms. D. Hilbert 558a, 1905.

[BOURBAKI, 1939ff.] Nicolas Bourbaki. *Théorie des Ensembles*. Éléments des Mathématique. Hermann, Paris, 1939ff..

[BROUWER, 1907] L. E. J. Brouwer. *Over de Grondslagen der Wiskunde*. PhD thesis, University of Amsterdam, Department of Physics and Mathematics, 1907.

[BROUWER, 1925a] L. E. J. Brouwer. Zur Begründung der intutionistischen Mathematik I. *Mathematische Annalen*, 93:244–257, 1925.

[BROUWER, 1925b] L. E. J. Brouwer. Zur Begründung der intutionistischen Mathematik II. *Mathematische Annalen*, 95:453–472, 1925.

[BROUWER, 1926] L. E. J. Brouwer. Zur Begründung der intutionistischen Mathematik III. *Mathematische Annalen*, 96:451–488, 1926.

[BUCHHOLZ &AL., 1981] Wilfried Buchholz, Sol(omon) Feferman, Wolfram Pohlers, and Wilfried Sieg. *Iterated inductive definitions and subsystems of analysis: Recent proof-theoretical studies*. Number 897 in Lecture Notes in Mathematics. Springer, 1981.

[BULDT & WINSLOW, 2008] Bernd Buldt and Peter Winslow. Foundations of Mathematics, Vol. I, 1934, §§ 1–2. English translation of [HILBERT & BERNAYS, 1968, §§ 1–2]. Draft, 2008.

[BURRIS & SANKAPPANAVAR, 1981] Stanley Burris and H. P. Sankappanavar. *A Course in Universal Algebra*. Springer, 1981.

[BUSS, 1998] Samuel R. Buss, editor. *Handbook of proof theory*. Elsevier, 1998.

[BUSSOTTI, 2006] Paolo Bussotti. *From FERMAT to GAUSZ: indefinite descent and methods of reduction in number theory*. Number 55 in Algorismus. Dr. Erwin Rauner Verlag, Augsburg, 2006.

[BUTTERFIELD &AL., 2006] Jeremy Butterfield, John Earman, Dov Gabbay, Paul Thagard, and John Woods, editors. *Handbook of the Philosophy of Science: Philosophy of Physics, Part A*. North-Holland (Elsevier), 2006.

[CANTOR, 1872] Georg Cantor. Über die Ausdehnung eines Satzes aus der Theorie der trigonometrischen Reihen. *Mathematische Annalen*, 5:123–132, 1872. Received Nov. 8, 1871.

[CARNOT, 1803] Lazare Nicolas Carnot. *Géométrie de position à l'usage de ceux qui se destinent á mesurer les terrains*. Duprat, Paris, 1803.

References

[CHOLAK, 2000] Peter Cholak, editor. *Computability Theory and its Applications — Current Trends and Open Problems*. American Mathematical Society, 2000.

[CHURCH, 1936] Alonzo Church. A note on the Entscheidungsproblem. *J. Symbolic Logic*, 1:40–41,101–102, 1936.

[CHURCH, 1944] Alonzo Church. *Introduction to Mathematical Logic*. Number PMS-13 in Annals of Mathematics Studies. Princeton Univ. Press, 1944.

[CHURCH, 1956] Alonzo Church. *Introduction to Mathematical Logic*. Princeton Univ. Press, 1956. Rev. and extd. ed. of [CHURCH, 1944].

[CORRY, 2004] Leo Corry. *Modern algebra and the rise of mathematical structures*. Birkhäuser (Springer), 2004. 2nd rev. ed. (1st ed. 1996).

[COSTA &AL., 1991] Newton C. A. da Costa, F. A. Doria, and N. Papavero. MEINONG's Theory of Objects and HILBERT's ε-symbol. *Reports on Mathematical Logic*, 25:119–132, 1991. ISSN 0137-2904, Jagiellonian University, Krakow.

[DEDEKIND, 1872] Richard Dedekind. *Stetigkeit und irrationale Zahlen*. Vieweg, Braunschweig, 1872. Also in [DEDEKIND, 1930–32, Vol. 3, pp. 315–334]. Also in [DEDEKIND, 1969].

[DEDEKIND, 1888] Richard Dedekind. *Was sind und was sollen die Zahlen*. Vieweg, Braunschweig, 1888. Also in [DEDEKIND, 1930–32, Vol. 3, pp. 335–391]. Also in [DEDEKIND, 1969].

[DEDEKIND, 1889] Richard Dedekind. Letter to KEFERSTEIN. 1889. In [HEIJENOORT, 1971, pp. 98–103].

[DEDEKIND, 1930–32] Richard Dedekind. *Gesammelte mathematische Werke*. Vieweg, Braunschweig, 1930–32. Ed. by ROBERT FRICKE, EMMY NOETHER, and ÖYSTEIN ORE.

[DEDEKIND, 1969] Richard Dedekind. *Was sind und was sollen die Zahlen? Stetigkeit und irrationale Zahlen*. Vieweg, Braunschweig, 1969.

[DELZELL, 1996] Charles N. Delzell. KREISEL's unwinding of ARTIN's proof. 1996. In [ODIFREDDI, 1996, pp. 113–246].

[DERSHOWITZ & JOUANNAUD, 1990] Nachum Dershowitz and Jean-Pierre Jouannaud. Rewrite systems. 1990. In [LEEUWEN, 1990, Vol. B, pp. 243–320].

[DROSDOWSKI, 1979] Günther Drosdowski, editor. *Duden. Das große Wörterbuch der deutschen Sprache. In 6 Bänden*. Bibliographisches Institut AG, Mannheim, 1979. Revised reprint of the edition of 1977.

[DUGAC, 1976] P. Dugac. RICHARD DEDEKIND *et les fondements des mathématiques*. Vrin, Paris, 1976.

[DUPRÉ, 1998] Lyn Dupré. *BUGS in Writing — A Guide to Debugging Your Prose*. Addison-Wesley, 1998. Rev. ed..

[EBBINGHAUS, 2007] Heinz-Dieter Ebbinghaus. ERNST ZERMELO*: An approach to his life and work*. Springer, 2007.

[EDWARDS, 1967] Paul Edwards. *Encyclopedia of Philosophy*. MacMillan, New York and London, 1967.

[EUCLID, ca. 300 B.C.] Euclid of Alexandria. *Elements*. ca. 300 B.C.. Web version without the figures: http://www.perseus.tufts.edu/hopper/text?doc=Perseus:text:1999.01.0085. English translation: THOMAS L. HEATH (ed.). *The Thirteen Books of* EUCLID*'s Elements*. Cambridge Univ. Press, 1908; web version without the figures: http://www.perseus.tufts.edu/hopper/text?doc=Perseus:text:1999.01.0086. English web version (incl. figures): D. E. JOYCE (ed.). EUCLID*'s Elements*. http://aleph0.clarku.edu/~djoyce/java/elements/elements.html, Dept. Math. & Comp. Sci., Clark Univ., Worcester (MA).

[EWALD, 1996] William Ewald, editor. *From* KANT *to* HILBERT *— A source book in the foundations of mathematics*. Oxford Univ. Press, 1996.

[FEFERMAN, 1981] Sol(omon) Feferman. How we got from there to here. 1981. In [BUCHHOLZ &AL., 1981, pp. 1–15].

[FEFERMAN, 1988] Sol(omon) Feferman. HILBERT's Program relativized: proof-theoretical and foundational reductions. *J. Symbolic Logic*, 53:364–384, 1988.

[FEFERMAN, 1996] Sol(omon) Feferman. KREISEL's "Unwinding" Program. 1996. In [ODIFREDDI, 1996, pp. 247–273].

[FEFERMAN, 1998] Sol(omon) Feferman. *In the light of logic.* Oxford Univ. Press, 1998.

[FERMAT, 1891ff.] Pierre Fermat. *Œuvres de* FERMAT. Gauthier-Villars, Paris, 1891ff.. Ed. by PAUL TANNERY, CHARLES HENRY.

[FERREIRA & FERREIRA, 2006] Fernando Ferreira and Gilda Ferreira. Counting as integration in feasible analysis. *Math. Logic Quart.*, 52:315–320, 2006.

[FERREIRA, 1994] Fernando Ferreira. A feasible theory for analysis. *J. Symbolic Logic*, 59:1001–1011, 1994.

[FERREIRÓS & GRAY, 2006] José Ferreirós and Jeremy J. Gray, editors. *The architecture of modern mathematics — Essays in history and philosophy.* Oxford Univ. Press, 2006.

[FERREIRÓS, 2007] José Ferreirós. *Labyrinth of Thought — A history of set theory and its role in modern mathematics.* Birkhäuser (Springer), 2007. 2nd rev. ed. (1st ed. 1999).

[FINE, 1985] Kit Fine. *Reasoning with Arbitrary Objects.* Number 3 in Aristotelian Society Series. Basil Blackwell, Oxford, 1985.

[FRAASSEN, 1980] Bas C. van Fraassen. *The Scientific Image.* Clarendon Press, 1980.

[FRAENKEL, 1919] Adolf Abraham Fraenkel. *Einleitung in die Mengenlehre — Eine allgemeinverständliche Einführung in das Reich der unendlichen Größen.* Springer, 1919.

[FRAENKEL, 1923] Adolf Abraham Fraenkel. *Einleitung in die Mengenlehre — Eine elementare Einführung in das Reich des Unendlichgroßen.* Springer, 1923. 2nd rev. extd. ed. of [FRAENKEL, 1919].

[FRAENKEL, 1928] Adolf Abraham Fraenkel. *Einleitung in die Mengenlehre.* Springer, 1928. 3rd rev. extd. ed. of [FRAENKEL, 1919], thoroughly revised and majorly extended edition of [FRAENKEL, 1923].

[FREGE, 1879] Gottlob Frege. *Begriffsschrift, eine der arithmetischen nachgebildete Formelsprache des reinen Denkens.* Verlag von L. Nebert, Halle an der Saale, 1879. Corrected facsimile in [FREGE, 1964b]. Reprint of pp. III–VIII and pp. 1–54 in [BERKA & KREISER, 1973, pp. 48–106]. English translation in [HEIJENOORT, 1971, pp. 1–82].

[FREGE, 1884] Gottlob Frege. *Die Grundlagen der Arithmetik.* Verlag von M. & H. Marcus, Breslau, 1884. Reprinted 1934. Reprinted as facsimile: Georg Olms Verlag, Hildesheim, 1990.

[FREGE, 1893/1903] Gottlob Frege. *Grundgesetze der Arithmetik — Begriffsschriftlich abgeleitet.* Verlag von Hermann Pohle, Jena, 1893/1903. As facsimile with corrigenda by Christian Thiel: Georg Olms Verlag, Hildesheim, 1998. English translation: [FREGE, 1964a].

[FREGE, 1964a] Gottlob Frege. *The Basic Laws of Arithmetic.* Univ. of California Press, 1964. English translation of [FREGE, 1893/1903], with an introduction, by MONTGOMERY FURTH.

[FREGE, 1964b] Gottlob Frege. *Begriffsschrift und andere Aufsätze.* Wissenschaftliche Buchgesellschaft, Darmstadt, 1964. Zweite Auflage, mit EDMUND HUSSERLS und HEINRICH SCHOLZ' Anmerkungen, herausgegeben von IGNACIO ANGELELLI.

[FREGE, 1980] Gottlob Frege. *Translations from the philosophical writings of* GOTTLOB FREGE. Basil Blackwell, Oxford, 1980. Ed. by PETER GEACH and MAX BLACK. 3rd ed. (1st ed. 1966).

[FRIEDMAN & SIMPSON, 2000] Harvey Friedman and Stephen G. Simpson. Issues and problems in reverse mathematics. 2000. In [CHOLAK, 2000, pp. 137–144].

[GABBAY & WOODS, 2004ff.] Dov Gabbay and John Woods, editors. *Handbook of the History of Logic.* North-Holland (Elsevier), 2004ff..

[GENTZEN, 1935] Gerhard Gentzen. Untersuchungen über das logische Schließen. *Mathematische Zeitschrift*, 39:176–210,405–431, 1935. Also in [BERKA & KREISER, 1973, pp. 192–253]. English translation in [GENTZEN, 1969].

[GENTZEN, 1936] Gerhard Gentzen. Die Widerspruchsfreiheit der reinen Zahlentheorie. *Mathematische Annalen*, 112:493–565, 1936. English translation in [GENTZEN, 1969].

[GENTZEN, 1938] Gerhard Gentzen. Die gegenwärtige Lage in der mathematischen Grundlagenforschung – Neue Fassung des Widerspruchsfreiheitsbeweises für die reine Zahlentheorie. *Forschungen zur Logik und zur Grundlegung der exakten Wissenschaften, Neue Folge*, 4:3–44, 1938. Reprinted as facsimile by Wissenschaftliche Buchgesellschaft, Darmstadt, Verlag von S. Hirzel, Stuttgart. English translation in [GENTZEN, 1969].

References

[GENTZEN, 1943] Gerhard Gentzen. Beweisbarkeit und Unbeweisbarkeit von Anfangsfällen der transfiniten Induktion in der reinen Zahlentheorie. *Mathematische Annalen*, 119:140–161, 1943. Received July 9, 1942. English translation in [GENTZEN, 1969].

[GENTZEN, 1969] Gerhard Gentzen. *The Collected Papers of* GERHARD GENTZEN. North-Holland (Elsevier), 1969. Ed. by MANFRED E. SZABO.

[GEORGE, 1994] Alexander George. *Mathematics and Mind*. Oxford Univ. Press, 1994.

[GILLIES, 1992] Donald Gillies, editor. *Revolutions in mathematics*. Oxford Univ. Press, 1992.

[GILLMAN, 1987] Leonard Gillman. *Writing Mathematics Well*. The Mathematical Association of America, 1987.

[GOEB, 1917] Margarethe Goeb. Mengenlehre. Notes of [HILBERT, 1917b]. HILBERT Nachlass, Georg August Universität Göttingen, Mathematisches Inst., Lesesaal, 1917.

[GÖDEL, 1930] Kurt Gödel. Die Vollständigkeit der Axiome des logischen Funktionenkalküls. *Monatshefte für Mathematik und Physik*, 37:349–360, 1930. With English translation also in [GÖDEL, 1986ff., Vol. I, pp. 102–123].

[GÖDEL, 1931] Kurt Gödel. Über formal unentscheidbare Sätze der Principia Mathematica und verwandter Systeme I. *Monatshefte für Mathematik und Physik*, 38:173–198, 1931. With English translation also in [GÖDEL, 1986ff., Vol. I, pp. 145–195]. English translation also in [HEIJENOORT, 1971, pp. 596–616] and in [GÖDEL, 1962].

[GÖDEL, 1932] Kurt Gödel. Ein Spezialfall des Entscheidungsproblems der theoretischen Logik. *Ergebnisse eines mathematischen Kolloquiums*, 2:27–28, 1932. With English translation also in [GÖDEL, 1986ff., Vol. I, pp. 230–235].

[GÖDEL, 1933a] Kurt Gödel. The present situation in the foundations of mathematics. 1933. With English translation in [GÖDEL, 1986ff., Vol. III, pp. 36–53].

[GÖDEL, 1933b] Kurt Gödel. Zum Entscheidungsproblem des logischen Funktionenkalküls. *Monatshefte für Mathematik und Physik*, 40:433–443, 1933. With English translation also in [GÖDEL, 1986ff., Vol. I, pp. 306–327].

[GÖDEL, 1962] Kurt Gödel. *On formally undecidable propositions of Principia Mathematica and related systems*. Basic Books, New York, 1962. English translation of [GÖDEL, 1931] by B. MELTZER. With an introduction by R. B. BRAITHWAITE. 2nd ed. by Dover Publications, 1992.

[GÖDEL, 1986ff.] Kurt Gödel. *Collected Works*. Oxford Univ. Press, 1986ff. Ed. by SOL(OMON) FEFERMAN, JOHN W. DAWSON JR., WARREN GOLDFARB, JEAN VAN HEIJENOORT, STEPHEN C. KLEENE, CHARLES PARSONS, WILFRIED SIEG, &AL..

[GOETZ, 1986] Philip W. Goetz, editor. *The New Encyclopædia Britannica*. Encyclopædia Britannica, Inc., Chicago, 1986. 15th ed., first published in 1974.

[GOVE, 1993] Philipp Babcock Gove, editor. *Webster's Third New International Dictionary*. Merriam-Webster, Springfield (MA), 1993. Principal copyright 1961.

[GRAY, 1992] Jeremy J. Gray. A 19th century revolution in mathematical ontology. 1992. In [GILLIES, 1992, pp. 226–248].

[HALLETT, 1994] Michael Hallett. HILBERT's axiomatic method and the laws of thought. 1994. In [GEORGE, 1994, pp. 71–117].

[HAMACHER-HERMES, 1994] Adelheid Hamacher-Hermes. *Inhalts- oder Umfangslogik? Die Kontroverse zwischen* EDMUND HUSSERL *und* ANDREAS HEINRICH VOIGT. Verlag Karl Alber, Freiburg (Breisgau), München, 1994.

[HASENJAEGER, 1950] Gisbert Hasenjaeger. Über eine Art von Unvollständigkeit des Prädikatenkalküls erster Stufe. *J. Symbolic Logic*, 15:273–276, 1950. Received Dec. 30, 1949.

[HEIJENOORT, 1971] Jean van Heijenoort. *From* FREGE *to* GÖDEL: *A Source Book in Mathematical Logic, 1879–1931*. Harvard Univ. Press, 1971. 2nd rev. ed. (1st ed. 1967).

[HELLINGER, 1905] Ernst Hellinger. Logische Prinzipien des mathematischen Denkens. Notes of [HILBERT, 1905b]. HILBERT Nachlass, Georg August Universität Göttingen, Mathematisches Inst., Lesesaal, 1905.

References

[HELMHOLTZ, 1879] Hermann von Helmholtz. *Die Thatsachen in der Wahrnehmung*. Verlag von August Hirschwald, Berlin, 1879. Rede, gehalten zur Stiftungsfeier der Friedrich-Wilhelms-Universität zu Berlin am 3. August, 1978; mit 3 Beilagen. Also in [HELMHOLTZ, 1959, pp. 8–74].

[HELMHOLTZ, 1887] Hermann von Helmholtz. Zählen und Messen — erkenntnistheoretisch betrachtet. 1887. In [ANON, 1887, pp. 15–52]. With commentary by PAUL HERTZ also in [HELMHOLTZ, 1921, Chapter III]. Also in [HELMHOLTZ, 1959, pp. 75–112].

[HELMHOLTZ, 1921] Hermann von Helmholtz. *Schriften zur Erkenntnistheorie*. Springer, 1921. Ed. by MORITZ SCHLICK und PAUL HERTZ.

[HELMHOLTZ, 1959] Hermann von Helmholtz. *Die Tatsachen in der Wahrnehmung, Zählen und Messen erkenntnistheoretisch betrachtet*. Wissenschaftliche Buchgesellschaft, Darmstadt, 1959. Facsimilie reprint of [HELMHOLTZ, 1879] and [HELMHOLTZ, 1887] with new page numbering and new title pages.

[HENDRICKS &AL., 2000] Vincent F. Hendricks, Stig Andur Pedersen, and Klaus Vrowin Jørgensen, editors. *Proof Theory — History and Philosophical Significance*. Number 292 in Synthese Library. Kluwer (Springer), 2000.

[HENDRICKS &AL., 2006] Vincent F. Hendricks, Klaus Vrowin Jørgensen, Jesper Lützen, and Stig Andur Pedersen, editors. *Mathematics, Physics and Philosophy, 1860–1930*. Number 251 in Boston Studies in the Philosophy of Science. Springer, 2006.

[HERBRAND, 1930] Jacques Herbrand. *Recherches sur la théorie de la démonstration*. PhD thesis, Université de Paris, 1930. Thèses présentées à la faculté des Sciences de Paris pour obtenir le grade de docteur ès sciences mathématiques — 1re thèse: Recherches sur la théorie de la démonstration — 2me thèse: Propositions données par la faculté, Les équations de Fredholm — Soutenues le 1930 devant la commission d'examen — Président: M. VESSIOT, Examinateurs: MM. DENJOY, FRECHET — Vu et approuvé, Paris, le 20 Juin 1929, Le doyen de la faculté des Sciences, C. MAURAIN — Vu et permis d'imprimer, Paris, le 20 Juin 1929, Le recteur de l'Academie de Paris, S. CHARLETY — No. d'ordre 2121, Série A, No. de Série 1252 — Imprimerie J. Dziewulski, Varsovie — Univ. de Paris. Also in Prace Towarzystwa Naukowego Warszawskiego, Wydział III Nauk Matematyczno-Fizychnych, Nr. 33, Warszawa. Also in [HERBRAND, 1968, pp. 35–153]. Annotated English translation *Investigations in Proof Theory* by WARREN GOLDFARB (Chapters 1–4) and BURTON DREBEN and JEAN VAN HEIJENOORT (Chapter 5) with a brief introduction by GOLDFARB and extended notes by GOLDFARB (Notes A–C, K–M, O), DREBEN (Notes F–I), DREBEN and GOLDFARB (Notes D, J, and N), and DREBEN, GEORGE HUFF, and THEODORE HAILPERIN (Note E) in [HERBRAND, 1971, pp. 44–202]. English translation of § 5 with a different introduction by HEIJENOORT and some additional extended notes by DREBEN also in [HEIJENOORT, 1971, pp. 525–581]. (HERBRAND's *PhD thesis, his cardinal work, dated April 14, 1929; submitted at the Univ. of Paris; defended at the Sorbonne June 11, 1930; printed in Warsaw, 1930.*)

[HERBRAND, 1931] Jacques Herbrand. Sur le problème fondamental de la logique mathématique. *Revue de Métaphysique et de Morale*, 24:12–56, 1931. Also in [HERBRAND, 1968, pp. 167–207]. Annotated English translation *On the Fundamental Problem of Mathematical Logic* in [HERBRAND, 1971, pp. 215–271].

[HERBRAND, 1932] Jacques Herbrand. Sur la non-contradiction de l'Arithmetique. *J. für die reine und angewandte Mathematik* (CRELLEsches *J.*), 166:1–8, 1932. Received July 27, 1931. Without HASSE's obituary also in [HERBRAND, 1968, pp. 221–232]. Annotated English translation *On the Consistency of Arithmetic* in [HEIJENOORT, 1971, pp. 618–628]. Annotated English translation also in [HERBRAND, 1971, pp. 282–298]. (*Consistency of arithmetic, foreshadowing of (total) recursive functions, discussion of* [GÖDEL, 1931]).

[HERBRAND, 1968] Jacques Herbrand. *Écrits Logiques*. Presses Universitaires de France, Paris, 1968. Ed. by Jean van Heijenoort. English translation is [HERBRAND, 1971]. (HERBRAND's *logical writings, faultily retyped*).

[HERBRAND, 1971] Jacques Herbrand. *Logical Writings*. Harvard Univ. Press, 1971. Ed. by WARREN GOLDFARB. Translation of [HERBRAND, 1968] with additional annotations, brief introductions, and extended notes by GOLDFARB, BURTON DREBEN, and JEAN VAN HEIJENOORT. (*Still the best source on* HERBRAND's *logical writings today*).

[HERTZ, 1894] Heinrich Hertz. *Die Prinzipien der Mechanik (Gesammelte Werke, Band III)*. Barth, Leipzig, 1894.

References

[HERTZ, 1929a] Paul Hertz. Über Axiomensysteme beliebiger Satzsysteme. Vortrag, gehalten auf der Versammlung der Deutschen Mathematiker-Vereinigung in Hamburg, Sept. 1928. *Erkenntnis (Annalen der Philosophie (und Philosophischen Kritik))*, 8:178–204, 1929.

[HERTZ, 1929b] Paul Hertz. Über Axiomensysteme für beliebige Satzsysteme. *Mathematische Annalen*, 101:457–514, 1929.

[HERTZ, 1931] Paul Hertz. Vom Wesen des Logischen, insbesondere der Bedeutung des modus barbara. *Erkenntnis (Annalen der Philosophie (und Philosophischen Kritik))*, 2:369–392, 1931.

[HEYTING, 1930] Arend Heyting. Die formalen Regeln der intuitionistischen Logik. *Sitzungsberichte der Preußischen Akademie der Wissenschaften, Berlin, Physikalisch-mathematische Klasse II*, 1930:42–56, 1930. Short version also in [BERKA & KREISER, 1973], pp. 173–178.

[HILBERT & ACKERMANN, 1928] David Hilbert and Wilhelm Ackermann. *Grundzüge der theoretischen Logik*. Springer, 1928. 1st ed., the final version in a serious of three thorough revisions is [HILBERT & ACKERMANN, 1959].

[HILBERT & ACKERMANN, 1959] David Hilbert and Wilhelm Ackermann. *Grundzüge der theoretischen Logik*. Springer, 1959. 4th ed., most thoroughly revised edition of [HILBERT & ACKERMANN, 1928].

[HILBERT & BERNAYS, 1934] David Hilbert and Paul Bernays. *Die Grundlagen der Mathematik — Erster Band*. Number XL in Die Grundlehren der Mathematischen Wissenschaften in Einzeldarstellungen. Springer, 1934. 1st ed. (2nd ed. is [HILBERT & BERNAYS, 1968]).

[HILBERT & BERNAYS, 1939] David Hilbert and Paul Bernays. *Die Grundlagen der Mathematik — Zweiter Band*. Number L in Die Grundlehren der Mathematischen Wissenschaften in Einzeldarstellungen. Springer, 1939. 1st ed. (2nd ed. is [HILBERT & BERNAYS, 1970]).

[HILBERT & BERNAYS, 1968] David Hilbert and Paul Bernays. *Die Grundlagen der Mathematik I*. Number 40 in Die Grundlehren der Mathematischen Wissenschaften in Einzeldarstellungen. Springer, 1968. 2nd rev. ed. of [HILBERT & BERNAYS, 1934].

[HILBERT & BERNAYS, 1970] David Hilbert and Paul Bernays. *Die Grundlagen der Mathematik II*. Number 50 in Die Grundlehren der Mathematischen Wissenschaften in Einzeldarstellungen. Springer, 1970. 2nd rev. ed. of [HILBERT & BERNAYS, 1939].

[HILBERT & BERNAYS, 1979] David Hilbert and Paul Bernays. *Osnovaniya Matematiki 1*. Nauka Publishing House, Moscow, 1979. Russian Translation of [HILBERT & BERNAYS, 1968] by NICOLAI MAKAROVICH NAGORNYI.

[HILBERT & BERNAYS, 1982] David Hilbert and Paul Bernays. *Osnovaniya Matematiki 2*. Nauka Publishing House, Moscow, 1982. Russian Translation of [HILBERT & BERNAYS, 1970] by NICOLAI MAKAROVICH NAGORNYI.

[HILBERT & BERNAYS, 2001] David Hilbert and Paul Bernays. *Fondements des Mathématiques 1*. L'Harmattan, Paris, 2001. French Translation of [HILBERT & BERNAYS, 1968], incl. the parallel passages of [HILBERT & BERNAYS, 1934], by FRANÇOIS GAILLARD and MARCEL GUILLAUME, ISBN 2747515184.

[HILBERT & BERNAYS, 2003] David Hilbert and Paul Bernays. *Fondements des Mathématiques 2*. L'Harmattan, Paris, 2003. French Translation of [HILBERT & BERNAYS, 1970], incl. the parallel passages of [HILBERT & BERNAYS, 1939], by FRANÇOIS GAILLARD and MARCEL GUILLAUME, ISBN 2747515192.

[HILBERT, 1889/90] David Hilbert. Einführung in das Studium der Mathematik (Zahlbegriff, Höhere Analysis, Analytische Geometrie, Differential- und Integralrechnung). Lectures winter term 1889/90, Königsberg, announced as "Einleitung in die höhere Analysis, 2st.", notes by HILBERT, 59 bound pp.. HILBERT Nachlass, Handschriftenabteilung of the Staats- und Universitätsbibliothek of the Georg August Universität Göttingen, Cod. Ms. D. Hilbert 530, 1889/90.

[HILBERT, 1891] David Hilbert. Projektive Geometrie. Lectures summer term 1891, Königsberg, announced as "Geometrie der Lage, 2st.", notes by HILBERT, 111 bound + 1 loose pp.. HILBERT Nachlass, Handschriftenabteilung of the Staats- und Universitätsbibliothek of the Georg August Universität Göttingen, Cod. Ms. D. Hilbert 535. Introduction, §§ 1–4, and last part of § 8, with an introduction by RALF HAUBRICH, also in [HILBERT, 2004, pp. 15–64], 1891.

References

[HILBERT, 1894] David Hilbert. Die Grundlagen der Geometrie. Lectures summer term 1894, Königsberg, planned (but canceled) for summer term 1993, announced in 1894 as "Über die Axiome der Geometrie, 2st."; notes by HILBERT, 99 bound + 1 loose pp.. HILBERT Nachlass, Niedersächsische Staats- und Universitätsbibliothek Göttingen, Handschriftenabteilung, Cod. Ms. D. Hilbert 541. With an introduction by ULRICH MAJER also in [HILBERT, 2004, pp. 65–144], 1894.

[HILBERT, 1897/98] David Hilbert. Zahlbegriff und Quadratur des Kreises. Lectures winter term 1897/98, Göttingen, announced as "Über den Begriff der Irrationalzahl und das Problem der Quadratur des Kreises, 2st."; notes by HILBERT with additions from later years, such as notes of the lecture with the same title of winter term 1899/1900, 55 bound + 1 loose pp.. HILBERT Nachlass, Niedersächsische Staats- und Universitätsbibliothek Göttingen, Handschriftenabteilung, Cod. Ms. D. Hilbert 549, 1897/98.

[HILBERT, 1898/99] David Hilbert. Grundlagen der EUKLIDischen Geometrie. Lectures winter term 1898/99, Göttingen, announced as "Elemente der EUKLIDischen Geometrie, 2st."; notes by HILBERT, 116 bound pp.. HILBERT Nachlass, Niedersächsische Staats- und Universitätsbibliothek Göttingen, Handschriftenabteilung, Cod. Ms. D. Hilbert 551. With an introduction by MICHAEL HALLETT and together with [SCHAPER, 1898/99] also in [HILBERT, 2004, pp. 185–407], 1898/99.

[HILBERT, 1899] David Hilbert. Grundlagen der Geometrie. 1899. In [ANON, 1899, pp. 1–92]. 1st ed. without appendixes. Reprinted in [HILBERT, 2004, pp. 436–525]. *(Last edition of "Grundlagen der Geometrie" by HILBERT is [HILBERT, 1930a], which is also most complete regarding the appendixes. Last three editions by PAUL BERNAYS are [HILBERT, 1962; 1968; 1972], which are also most complete regarding supplements and figures. Its first appearance as a separate book was the French translation [HILBERT, 1900b]. Two substantially different English translations are [HILBERT, 1902] and [HILBERT, 1971]).*

[HILBERT, 1900a] David Hilbert. Über den Zahlbegriff. *Jahresbericht der Deutschen Mathematiker-Vereinigung*, 8:180–184, 1900. Received Dec. 1899. Reprinted as Appendix VI of [HILBERT, 1909; 1913; 1922a; 1923b; 1930a].

[HILBERT, 1900b] David Hilbert. Les principes fondamentaux de la géométrie. *Annales Scientifiques de l'École Normale Supérieure*, Série 3, 17:103–209, 1900. French translation by LÉONCE LAUGEL of special version of [HILBERT, 1899], revised and authorized by HILBERT. Also in published as a separate book by the same publisher (Gauthier-Villars, Paris).

[HILBERT, 1901] David Hilbert. Mathematische Probleme. Vortrag, gehalten auf dem internationalen Mathematiker-Kongress zu Paris 1900. *Archiv der Mathematik und Physik*, 3rd series, 1:44–63+213–237, 1901. Part of the essay is translated in [EWALD, 1996, pp. 1096–1105].

[HILBERT, 1901/02] David Hilbert. Zahlbegriff und Quadratur des Kreises. Lectures winter term 1901/02, Göttingen, announced as "Zahlbegriff und Quadratur des Kreises, 2st.". Only HILBERT's "Disposition" of the lecture is available in [HILBERT, 1897/98, pp. 52–54], 1901/02.

[HILBERT, 1902] David Hilbert. *The Foundations of Geometry*. Open Court, Chicago, 1902. English translation by E. J. TOWNSEND of special version of [HILBERT, 1899], revised and authorized by HILBERT, http://www.gutenberg.org/etext/17384.

[HILBERT, 1903] David Hilbert. *Grundlagen der Geometrie. — Zweite, durch Zusätze vermehrte und mit fünf Anhängen versehene Auflage. Mit zahlreichen in den Text gedruckten Figuren*. Druck und Verlag von B. G. Teubner, Leipzig, 1903. 2nd rev. extd. ed. of [HILBERT, 1899], rev. and extd. with five appendixes, newly added figures, and an index of notion names.

[HILBERT, 1904] David Hilbert. Zahlbegriff und Quadratur des Kreises. Lectures summer term 1904, Göttingen, announced as "Zahlbegriff und Quadratur des Kreises, 2st.". [BORN, 1904] is the only available set of notes, 1904.

[HILBERT, 1905a] David Hilbert. Über die Grundlagen der Logik und der Arithmetik. 1905. In [ANON, 1905, 174–185]. Reprinted as Appendix VII of [HILBERT, 1909; 1913; 1922a; 1923b; 1930a]. English translation *On the foundations of logic and arithmetic* by BEVERLY WOODWARD with an introduction by JEAN VAN HEIJENOORT in [HEIJENOORT, 1971, pp. 129–138].

[HILBERT, 1905b] David Hilbert. Logische Prinzipien des mathematischen Denkens. Lectures summer term 1905, Göttingen, announced as "Logische Prinzipien des mathematischen Denkens, 2st.". [BORN, 1905] and [HELLINGER, 1905] are the only available sets of notes, 1905.

References

[HILBERT, 1909] David Hilbert. *Grundlagen der Geometrie.* — *Dritte, durch Zusätze und Literaturhinweise von neuem vermehrte und mit sieben Anhängen versehene Auflage. Mit zahlreichen in den Text gedruckten Figuren.* Number VII in Wissenschaft und Hypothese. Druck und Verlag von B. G. Teubner, Leipzig, Berlin, 1909. 3rd rev. extd. ed. of [HILBERT, 1899], rev. ed. of [HILBERT, 1903], extd. with a bibliography and two additional appendixes (now seven in total) (Appendix VI: [HILBERT, 1900a]) (Appendix VII: [HILBERT, 1905a]).

[HILBERT, 1913] David Hilbert. *Grundlagen der Geometrie.* — *Vierte, durch Zusätze und Literaturhinweise von neuem vermehrte und mit sieben Anhängen versehene Auflage. Mit zahlreichen in den Text gedruckten Figuren.* Druck und Verlag von B. G. Teubner, Leipzig, Berlin, 1913. 4th rev. extd. ed. of [HILBERT, 1899], rev. ed. of [HILBERT, 1909].

[HILBERT, 1917a] David Hilbert. Axiomatisches Denken. *Mathematische Annalen*, 78:405–415, 1917. Talk given at the Swiss Mathematical Society in Zürich on Sept. 11, 1917. Reprinted in [HILBERT, 1932ff., Vol. 3, pp. 146–156]. English translation in [EWALD, 1996, pp. 1105–1115].

[HILBERT, 1917b] David Hilbert. Mengenlehre. Lectures summer term 1917, Göttingen, announced as "Mengenlehre, 4st.". [GOEB, 1917] is the only available set of notes, 1917.

[HILBERT, 1917/18] David Hilbert. Prinzipien der Mathematik. Lectures winter term 1917/18, Göttingen, announced as "Prinzipien der Mathematik, 2st.", notes by BERNAYS. HILBERT Nachlass, Georg August Universität Göttingen, Mathematisches Inst., Lesesaal, 1917/18.

[HILBERT, 1920] David Hilbert. Probleme der Mathematischen Logik. Lectures summer term 1920, Göttingen, announced as "Probleme der Mathematischen Logik, 1st.", notes by SCHÖNFINKEL and BERNAYS. HILBERT Nachlass, Georg August Universität Göttingen, Mathematisches Inst., Lesesaal, 1920.

[HILBERT, 1921] David Hilbert. Natur und mathematisches Erkennen. Manuscript of a talk given in København, March 14, 1921; 32 pp.. HILBERT Nachlass, Niedersächsische Staats- und Universitätsbibliothek Göttingen, Handschriftenabteilung, Cod. Ms. D. Hilbert 589, 1921.

[HILBERT, 1921/22] David Hilbert. Grundlagen der Mathematik. Lectures winter term 1921/22, Göttingen, announced as "Grundlegung der Mathematik, 4st.", notes by PAUL BERNAYS. (Cf. also [KNESER, 1921/22].) HILBERT Nachlass, Georg August Universität Göttingen, Mathematisches Inst., Lesesaal, 1921/22.

[HILBERT, 1922a] David Hilbert. *Grundlagen der Geometrie.* — *Fünfte, durch Zusätze und Literaturhinweise von neuem vermehrte und mit sieben Anhängen versehene Auflage. Mit zahlreichen in den Text gedruckten Figuren.* Verlag und Druck von B. G. Teubner, Leipzig, Berlin, 1922. 5th extd. ed. of [HILBERT, 1899]. Contrary to what the sub-title may suggest, this is an anastatic reprint of [HILBERT, 1913], extended with a very short preface on the changes w.r.t. [HILBERT, 1913], and with augmentations to Appendix II, Appendix III, and Chapter IV, § 21.

[HILBERT, 1922b] David Hilbert. Neubegründung der Mathematik (Erste Mitteilung). *Abhandlungen aus dem mathematischen Seminar der Univ. Hamburg*, 1:157–177, 1922. Reprinted with additional notes in [HILBERT, 1932ff., Vol. 3, pp. 157–177]. English translation in [EWALD, 1996, pp. 1115–1134].

[HILBERT, 1922/23] David Hilbert. Logische Grundlagen der Mathematik. Lectures winter term 1922/23, Göttingen, announced as "Grundlagen der Arithmetik, 2st.", fragmentary set of notes by BERNAYS, 33 pp.. (Cf. also [KNESER, 1922/23].) HILBERT Nachlass, Niedersächsische Staats- und Universitätsbibliothek Göttingen, Handschriftenabteilung, Cod. Ms. D. Hilbert 567, 1922/23.

[HILBERT, 1923a] David Hilbert. Die logischen Grundlagen der Mathematik. *Mathematische Annalen*, 88:151–165, 1923. Received Sept. 29, 1922. Talk given at the Deutsche Naturforschergesellschaft in Leipzig, Sept. 1922. English translation in [EWALD, 1996, pp. 1134–1148].

[HILBERT, 1923b] David Hilbert. *Grundlagen der Geometrie.* — *Sechste unveränderte Auflage. Anastatischer Nachdruck. Mit zahlreichen in den Text gedruckten Figuren.* Verlag und Druck von B. G. Teubner, Leipzig, Berlin, 1923. 6th rev. extd. ed. of [HILBERT, 1899], anastatic reprint of [HILBERT, 1922a].

[HILBERT, 1926] David Hilbert. Über das Unendliche — Vortrag, gehalten am 4. Juni 1925 gelegentlich einer zur Ehrung des Andenkens an WEIERSTRASZ von der Westfälischen Math. Ges. veranstalteten Mathematiker-Zusammenkunft in Münster i. W. *Mathematische Annalen*, 95:161–190, 1926. Received June 24, 1925. Reprinted as Appendix VIII of [HILBERT, 1930a]. English translation *On the infinite* by STEFAN BAUER-MENGELBERG with an introduction by JEAN VAN HEIJENOORT in [HEIJENOORT, 1971, pp. 367–392].

References

[HILBERT, 1928] David Hilbert. Die Grundlagen der Mathematik — Vortrag, gehalten auf Einladung des Mathematischen Seminars im Juli 1927 in Hamburg. *Abhandlungen aus dem mathematischen Seminar der Univ. Hamburg*, 6:65–85, 1928. Reprinted as Appendix IX of [HILBERT, 1930a]. English translation *The foundations of mathematics* by STEFAN BAUER-MENGELBERG and DAGFINN FØLLESDAL with a short introduction by JEAN VAN HEIJENOORT in [HEIJENOORT, 1971, pp. 464–479].

[HILBERT, 1930a] David Hilbert. *Grundlagen der Geometrie. — Siebente umgearbeitete und vermehrte Auflage. Mit 100 in den Text gedruckten Figuren.* Verlag und Druck von B. G. Teubner, Leipzig, Berlin, 1930. 7th rev. extd. ed. of [HILBERT, 1899], thoroughly revised edition of [HILBERT, 1923b], extd. with three new appendixes (now ten in total) (Appendix VIII: [HILBERT, 1926]) (Appendix IX: [HILBERT, 1928]) (Appendix X: [HILBERT, 1930b]).

[HILBERT, 1930b] David Hilbert. Probleme der Grundlegung der Mathematik. *Mathematische Annalen*, 102:1–9, 1930. Vortrag gehalten auf dem Internationalen Mathematiker-Kongreß in Bologna, Sept. 3, 1928. Received March 25, 1929. Reprinted as Appendix X of [HILBERT, 1930a]. Short version in *Atti del congresso internationale dei matematici, Bologna, 3–10 settembre 1928*, Vol. 1, pp. 135–141, Bologna, 1929.

[HILBERT, 1932ff.] David Hilbert. *Gesammelte Abhandlungen.* Springer, 1932ff..

[HILBERT, 1956] David Hilbert. *Grundlagen der Geometrie. — Achte Auflage, mit Revisionen und Ergänzungen von Dr. PAUL BERNAYS. Mit 124 Abbildungen.* B. G. Teubner Verlagsgesellschaft, Stuttgart, 1956. 8th rev. extd. ed. of [HILBERT, 1899], rev. ed. of [HILBERT, 1930a], omitting appendixes VI–X, extd. by PAUL BERNAYS, now with 24 additional figures and 3 additional supplements.

[HILBERT, 1962] David Hilbert. *Grundlagen der Geometrie. — Neunte Auflage, revidiert und ergänzt von Dr. PAUL BERNAYS. Mit 129 Abbildungen.* B. G. Teubner Verlagsgesellschaft, Stuttgart, 1962. 9th rev. extd. ed. of [HILBERT, 1899], rev. ed. of [HILBERT, 1956], extd. by PAUL BERNAYS, now with 129 figures, 5 appendixes, and 8 supplements (I 1, I 2, II, III, IV 1, IV 2, V 1, V 2).

[HILBERT, 1968] David Hilbert. *Grundlagen der Geometrie. — Zehnte Auflage, revidiert und ergänzt von Dr. PAUL BERNAYS. Mit 124 Abbildungen.* B. G. Teubner Verlagsgesellschaft, Stuttgart, 1968. 10th rev. extd. ed. of [HILBERT, 1899], rev. ed. of [HILBERT, 1962] by PAUL BERNAYS.

[HILBERT, 1971] David Hilbert. *The Foundations of Geometry.* Open Court, Chicago and La Salle (IL), 1971. Newly translated and fundamentally different 2nd ed. of [HILBERT, 1902], actually an English translation of [HILBERT, 1968] by LEO UNGER.

[HILBERT, 1972] David Hilbert. *Grundlagen der Geometrie. — 11. Auflage. Mit Supplementen von Dr. PAUL BERNAYS.* B. G. Teubner Verlagsgesellschaft, Stuttgart, 1972. 11th rev. extd. ed. of [HILBERT, 1899], rev. ed. of [HILBERT, 1968] by PAUL BERNAYS.

[HILBERT, 1999] David Hilbert. *Grundlagen der Geometrie. — 14. Auflage. Mit Supplementen von PAUL BERNAYS.* B. G. Teubner, Stuttgart, Leipzig, 1999. 14th rev. extd. ed. of [HILBERT, 1899], with contributions by MICHAEL TOEPELL (ed.), ALEXANDER KREUZER, and HEINRICH WEFELSCHEID.

[HILBERT, 2004] David Hilbert. *DAVID HILBERT's Lectures on the Foundations of Geometry, 1891–1902.* Springer, 2004. Ed. by MICHAEL HALLETT and ULRICH MAJER.

[HILLENBRAND & LÖCHNER, 2002] Thomas Hillenbrand and Bernd Löchner. The next WALDMEISTER loop. 2002. In [VORONKOV, 2002, pp. 486–500]. http://www.waldmeister.org.

[HOWARD & RUBIN, 1998] Paul Howard and Jean E. Rubin. *Consequences of the Axiom of Choice.* American Math. Society, 1998.

[HUNTINGTON, 1924] Edward V. Huntington. A new set of postulates for betweenness with proof of complete independence. *Transactions of the American Math. Soc.*, 26:257–282, 1924.

[JAMMER, 1974] Max Jammer. *The Philosophy of Quantum Mechanics: the interpretations of quantum mechanics in historical perspective.* John Wiley & Sons, New York, 1974.

[JAŚKOWSKI, 1934] Stanisław Jaśkowski. On the rules of suppositions in formal logic. *Studia Logica (Warszawa)*, 1:1–32, 1934.

[KALMÁR, 1933] László Kalmár. Über die Erfüllbarkeit derjenigen Zählausdrücke, welche in der Normalform zwei benachbarte Allzeichen enthalten. *Mathematische Annalen*, 108:466–484, 1933. Received Sept. 3, 1932.

References

[KANT, 1781] Immanuel Kant. *Critik der reinen Vernunft*. Johann Friedrich Hartknoch, Riga, 1781. 1st ed..

[KANT, 1787] Immanuel Kant. *Critik der reinen Vernunft*. Johann Friedrich Hartknoch, Riga, 1787. 2nd thoroughly rev. ed. of [KANT, 1781]. English translation: [KANT, 2008].

[KANT, 1800] Immanuel Kant. *Logik — ein Handbuch zu Vorlesungen*. Friedrich Nicolovius, Königsberg, 1800. Ed. by GOTTLOB BENJAMIN JÄSCHE.

[KANT, 2008] Immanuel Kant. *Critique of Pure Reason*. Penguin Classics, 2008. English translation of [KANT, 1787], including the differences of the first edition ([KANT, 1781]), by MAX MÜLLER and MARCUS WEIGELT, 2nd rev. ed. (1st ed. 2007).

[KERNIGHAN & RITCHIE, 1977] Brian W. Kernighan and Dennis M. Ritchie. *The C Programming Language*. Prentice–Hall, Inc., 1977.

[KLEENE & FEFERMAN, 1986] Stephen Cole Kleene and Sol(omon) Feferman. The foundations of mathematics. 1986. In [GOETZ, 1986, Vol. 23, pp. 594–602].

[KLEENE, 1952] Stephen Cole Kleene. *Introduction to Metamathematics*. D. Van Nostrand, 1952.

[KLIEN, 1947] Horst Klien, editor. *Duden. Rechtschreibung der deutschen Sprache und der Fremdwörter*. 13. Auflage. Bibliographisches Institut AG, Leipzig, 1947.

[KLOP, 1992] Jan Willem Klop. Term rewriting systems. 1992. In [ABRAMSKY &AL., 1992, Vol. II, pp. 1–116].

[KNESER, 1921/22] Hellmuth Kneser. Grundlagen der Mathematik. Private notes of [HILBERT, 1921/22]. Archive of HELLMUTH KNESER's papers, Niedersächsische Staats- und Universitätsbibliothek Göttingen, Handschriftenabteilung, Cod. Ms. H. Kneser, 1921/22.

[KNESER, 1922/23] Hellmuth Kneser. Logische Grundlagen der Mathematik. Private notes of [HILBERT, 1922/23]. Archive of HELLMUTH KNESER's papers, Niedersächsische Staats- und Universitätsbibliothek Göttingen, Handschriftenabteilung, Cod. Ms. H. Kneser, 1922/23.

[KOHLENBACH & OLIVA, 2003] Ulrich Kohlenbach and Paolo B. Oliva. Proof mining: a systematic way of analysing proofs in mathematics. *Proc. Steklov Inst. Math.*, 242:136–164, 2003.

[KREISEL, 1951] Georg Kreisel. On the interpretation of non-finitist proofs — Part I. *J. Symbolic Logic*, 16:241–267, 1951.

[KREISEL, 1952] Georg Kreisel. On the interpretation of non-finitist proofs — Part II. *J. Symbolic Logic*, 17:43–58, iv, 1952.

[KREISEL, 1958] Georg Kreisel. Mathematical significance of consistency proofs. *J. Symbolic Logic*, 23:155–182, 1958.

[KREISEL, 1965] Georg Kreisel. Mathematical logic. 1965. In [SAATY, 1965, Vol. III, pp. 95–195].

[KREISEL, 1968] Georg Kreisel. Survey of proof theory. *J. Symbolic Logic*, 33:321–388, 1968.

[KREISEL, 1982] Georg Kreisel. Finiteness theorems in arithmetic: an application of HERBRAND's theorem for Σ_2-formulas. 1982. In [STERN, 1982, pp. 39–55].

[LANDSMANN, 2006] N. P. Landsmann. Between classical and quantum. 2006. In [BUTTERFIELD &AL., 2006, pp. 417–554]. philsci-archive.pitt.edu/archive/00002328/01/handbook.pdf.

[LEEUWEN, 1990] Jan van Leeuwen, editor. *Handbook of Theoretical Computer Sci.*. MIT Press, 1990.

[LEŚNIEWSKI, 1929] Stanisław Leśniewski. Grundzüge eines neuen Systems der Grundlagen der Mathematik. *Fundamenta Informaticae*, 15:1–81, 1929.

[LEWIS, 1918] Clarence Irving Lewis. *A Survey of Symbolic Logic*. Univ. of California Press, Berkeley and Los Angeles, 1918. http://www.archive.org/details/asurveyofsymboli00lewiuoft.

[LÖWENHEIM, 1915] Leopold Löwenheim. Über Möglichkeiten im Relativkalkül. *Mathematische Annalen*, 76:228–251, 1915. English translation *On Possibilities in the Calculus of Relatives* by STEFAN BAUER-MENGELBERG with an introduction by JEAN VAN HEIJENOORT in [HEIJENOORT, 1971, pp. 228–251].

[ŁUKASIEWICZ & TARSKI, 1930] Jan Łukasiewicz and Alfred Tarski. Untersuchungen über den Aussagenkalkül. *Sprawozdania z posiedzeń Towarzystwa Naukowego Warszawskiego, Wydział III (Comptes Rendus Séances Société des Sciences et Lettres Varsovie, Cl. III)*, 23:30–50, 1930. English translation in [ŁUKASIEWICZ, 1970, pp. 130–152].

References

[ŁUKASIEWICZ, 1970] Jan Łukasiewicz. *Selected Works*. North-Holland (Elsevier), 1970. Ed. by L. BORKOWSKI.

[MACLANE, 1995] Sounders MacLane. Mathematics at Göttingen under the Nazis. *Notices of the American Math. Soc.*, 42:1134–1138, 1995.

[MAHONEY, 1994] Michael Sean Mahoney. *The Mathematical Career of* PIERRE *de* FERMAT *1601–1665*. Princeton Univ. Press, 1994. 2nd rev. ed. (1st ed. 1973).

[MAJER, 2006] Ulrich Majer. HILBERT's axiomatic approach to the foundations of science — a failed research program? 2006. In [HENDRICKS &AL., 2006, pp. 155–184].

[MANCOSU, 1998] Paolo Mancosu, editor. *From* BROUWER *to* HILBERT — *The Debate on the Foundations of Mathematics in the 1920s*. Oxford Univ. Press, 1998.

[MANCOSU, 1999a] Paolo Mancosu. Between RUSSELL and HILBERT: BEHMANN on the foundations of mathematics. *Bulletin of Symbolic Logic*, 5:303–330, 1999.

[MANCOSU, 1999b] Paolo Mancosu. Between Vienna and Berlin: the immediate reception of GÖDEL's incompleteness theorems. *History and Philosophy of Logic*, 20:33–45, 1999.

[MARTIN-LÖF, 1984] Per Martin-Löf. *Intuitionistic Type Theory*. Bibliopolis, Napoli, 1984.

[MEINONG, 1904a] Alexius Meinong. Über Gegenstandstheorie. 1904. In [MEINONG, 1904b, pp. 1–50].

[MEINONG, 1904b] Alexius Meinong, editor. *Untersuchungen zur Gegenstandstheorie und Psychologie*. Barth, Leipzig, 1904.

[MOORE, 1997] Gregory H. Moore. HILBERT and the emergence of modern mathematical logic. *Theoria: Revista de Teoría, Historia y Fundamentos de la Ciencia*, 12:65–90, 1997.

[MUELLER, 2006] Ian Mueller. English translation of [HILBERT & BERNAYS, 1968, §§ 1–2]. Web only: http://www.phil.cmu.edu/projects/bernays/b-translations.html, 2006.

[MYHILL, 1975] John Myhill. Constructive set theory. *J. Symbolic Logic*, 40:347–382, 1975.

[NEUMANN, 1927] John von Neumann. Zur HILBERTschen Beweistheorie. *Mathematische Zeitschrift*, 26:1–46, 1927. Received July 29, 1925. *(First occurrence of the notion of basic typus (Grundtypus), the cardinal notion in the theory of ε-substitution. Clarification that the consitency proof of [*ACKERMANN, 1925a*] does not include second order with comprehension.)*

[NICOD, 1917] Jean Nicod. A reduction in the number of primitive propositions of logic. *Proc. Cambridge Philosophical Soc.*, 19:32–44, 1917. Received Oct. 30, 1916. http://commons.wikimedia.org/wiki/Category:Jean_Nicod_texts_(English).

[ODIFREDDI, 1996] Piergiorgio Odifreddi, editor. *Kreiseliana — About and Around* GEORG KREISEL. A K Peters, Wellesley (MA), 1996.

[PARSONS, 1998] Charles Parsons. Finitism and intuitive knowledge. 1998. In [SCHIRN, 1998, pp. 249–270].

[PARSONS, 2008] Charles Parsons. *Mathematical thought and its objects*. Cambridge Univ. Press, 2008.

[PECKHAUS & KAHLE, 2002] Volker Peckhaus and Reinhard Kahle. HILBERT's paradox. *Historia Mathematica*, 29:157–175, 2002.

[PECKHAUS, 1990] Volker Peckhaus. HILBERT*programm und kritische Philosophie*. Vandenhoeck & Ruprecht, Göttingen, 1990.

[PECKHAUS, 2001] Volker Peckhaus. Umfangslogik / Inhaltslogik. 2001. In [RITTER &AL., 1971–2007, Vol. 11, p. 83f.].

[PECKHAUS, 2003] Volker Peckhaus. The pragmatism of HILBERT's program. *Synthese*, 137:141–146, 2003.

[PEIRCE, 1880a] Charles S. Peirce. A boolean algebra with one constant. 1880. Unpublished Note MS 378: Lectures winter term 1880/81. In [PEIRCE, 1933, Vol. IV, pp. 13–18]. Also in [PEIRCE, 1989, pp. 218–221].

[PEIRCE, 1880b] Charles S. Peirce. On the algebra of logic. *American J. of Mathematics*, 3:15–57, 1880. Also in [PEIRCE, 1989, pp. 163–209].

[PEIRCE, 1885] Charles S. Peirce. On the algebra of logic: A contribution to the philosophy of notation. *American J. of Mathematics*, 7:180–202, 1885. Also in [PEIRCE, 1993, pp. 162–190].

References

[PEIRCE, 1933] Charles S. Peirce. *Collected Papers of* CHARLES S. PEIRCE, *Volumes III and IV, Exact Logic (Published Papers) and The Simplest Mathematics.* Belknap Press, Cambridge (MA), 1933. Ed. by CHARLES HARTSHORNE and PAUL WEISS.

[PEIRCE, 1989] Charles S. Peirce. *Writings of* CHARLES S. PEIRCE — *A Chronological Edition, Vol. 4, 1879–1884.* Indiana Univ. Press, 1989. Ed. by CHRISTIAN J. W. KLOESEL.

[PEIRCE, 1993] Charles S. Peirce. *Writings of* CHARLES S. PEIRCE — *A Chronological Edition, Vol. 5, 1884–1886.* Indiana Univ. Press, 1993. Ed. by CHRISTIAN J. W. KLOESEL.

[POHLERS, 1997] Wolfram Pohlers. *Proof theory — An introduction.* Number 1407 in Lecture Notes in Mathematics. Springer, 1997.

[POINCARÉ, 1902] Henri Poincaré. Review of [HILBERT, 1899]. *Bulletin des sciences mathématiques,* 26:249–272, 1902.

[POINCARÉ, 1905] Henri Poincaré. Les mathématiques et la logique. *Revue de Métaphysique et de Morale,* 13:815–835, 1905. English translation in [EWALD, 1996, Vol. 2, pp. 1021–1038].

[PRAWITZ &AL., 1994] Dag Prawitz, B. Skyrms, and D. Westerståhl, editors. *Logic, Methodology and Philosophy of Science IX.* Number 134 in Studies in Logic and the Foundations of Mathematics. North-Holland (Elsevier), 1994.

[QUINE, 1954] Willard Van O. Quine. Quantification and the empty domain. *J. Symbolic Logic,* 19:177–179, 1954. Received Sept. 28, 1953.

[QUINE, 1981] Willard Van O. Quine. *Mathematical Logic.* Harvard Univ. Press, 1981. 4th rev. ed. (1st ed. 1940).

[QUINE, 1982] Willard Van O. Quine. *Methods of Logic.* Harvard Univ. Press, 1982. 4th rev. ed. (1st ed. 1950).

[RAVAGLIA, 2003] Mark Ravaglia. *Explicating the finitist standpoint.* PhD thesis, Dept. of Philosophy, Carnegie Mellon Univ., 2003.

[REID, 1970] Constance Reid. HILBERT. Springer, 1970.

[RITTER &AL., 1971–2007] Joachim Ritter, Karlfried Gründer, and Gottfried Gabriel, editors. *Historisches Wörterbuch der Philosophie.* Wissenschaftliche Buchgesellschaft, Darmstadt, 1971–2007.

[ROWE, 2000] David Rowe. The calm before the storm: HILBERT's early views on foundations. 2000. In [HENDRICKS &AL., 2000, pp. 55–94].

[RUBIN & RUBIN, 1985] Herman Rubin and Jean E. Rubin. *Equivalents of the Axiom of Choice.* North-Holland (Elsevier), 1985. 2nd rev. ed. (1st ed. 1963).

[RUSSELL, 1905a] Bertrand Russell. On denoting. *Mind,* 14:479–493, 1905.

[RUSSELL, 1905b] Bertrand Russell. Review of [MEINONG, 1904b]. *Mind,* 14:537–537, 1905.

[SAATY, 1965] T. L. Saaty, editor. *Lectures on Modern Mathematics.* John Wiley & Sons, New York, 1965.

[SCHAPER, 1898/99] Hans von Schaper. Elemente der EUKLIDischen Geometrie. Notes of [HILBERT, 1898/99], 175 bound + 2 loose pp.. (Also in HILBERT Nachlass, Georg August Universität Göttingen, Mathematisches Inst., Lesesaal. With an introduction by MICHAEL HALLETT and together with [HILBERT, 1898/99] also in [HILBERT, 2004, pp. 185–407].) HILBERT Nachlass, Niedersächsische Staats- und Universitätsbibliothek Göttingen, Handschriftenabteilung, Cod. Ms. D. Hilbert 552, 1898/99.

[SCHIRN, 1998] Matthias Schirn, editor. *The Philosophy of Mathematics Today.* Oxford Univ. Press, 1998.

[SCHNEIDER, 1958] Hubert H. Schneider. Semantics of the predicate calculus with identity and the validity in the empty individual-domain. *Portugaliae Mathematica,* 17:85–96, 1958. Received April 1958.

[SCHOLZ, 1932/33] Heinrich Scholz. Logistik. Unpublished lecture course, 1932/33.

[SCHÜTTE, 1934] Kurt Schütte. Untersuchungen zum Entscheidungsproblem der mathematischen Logik. *Mathematische Annalen,* 109:572–603, 1934. Received June 30, 1933.

[SCHÜTTE, 1935] Kurt Schütte. Über die Erfüllbarkeit einer Klasse von logischen Formeln. *Mathematische Annalen,* 110:161–194, 1935. Received Nov. 3, 1933.

References

[SCHÜTTE, 1960] Kurt Schütte. *Beweistheorie*. Number 103 in Grundlehren der mathematischen Wissenschaften. Springer, 1960. Thoroughly revised English translation: [SCHÜTTE, 1977].

[SCHÜTTE, 1977] Kurt Schütte. *Proof theory*. Number 225 in Grundlehren der mathematischen Wissenschaften. Springer, 1977. Translated from a thorough revision of [SCHÜTTE, 1960] by JOHN N. CROSSLEY.

[SCHULZ &AL., 1995ff.] Hans Schulz, Otto Basler, and Gerhard Strauß. *Deutsches Fremdwörterbuch*. de Gruyter, Berlin, New York, 1995ff..

[SCHWICHTENBERG, 1977] Helmut Schwichtenberg. Proof theory: Some applications of cut-elimination. 1977. In [BARWISE, 1977, pp. 867–895].

[SHEFFER, 1913] Henry Maurice Sheffer. A set of five independent postulates for boolean algebras, with application to logical constants. *Transactions of the American Math. Soc.*, 14:481–488, 1913. http://www.ams.org/journals/tran/1913-014-04/home.html.

[SIEG & RAVAGLIA, 2004] Wilfried Sieg and Mark Ravaglia. DAVID HILBERT and PAUL BERNAYS, Grundlagen der Mathematik I and II — A Landmark. Tech. Report CMU–PHIL–154, Carnegie Mellon Univ., 2004. www.hss.cmu.edu/philosophy/techreports/154_Sieg.pdf.

[SIEG & SCHLIMM, 2005] Wilfried Sieg and Dirk Schlimm. DEDEKIND's analysis of number: Systems and axioms. *Synthese*, 147:121–170, 2005.

[SIEG, 1988] Wilfried Sieg. HILBERT's program sixty years later. *J. Symbolic Logic*, 53:338–348, 1988.

[SIEG, 1991] Wilfried Sieg. HERBRAND analyses. *Archive for Mathematical Logic*, 30:409–441, 1991.

[SIEG, 1994] Wilfried Sieg. Mechanical procedures and mathematical experience. 1994. In [GEORGE, 1994, pp. 71–117].

[SIEG, 1999] Wilfried Sieg. HILBERT's programs: 1917–1922. *Bulletin of Symbolic Logic*, 5:1–44, 1999.

[SIEG, 2005] Wilfried Sieg. Only two letters: The correspondence between HERBRAND and GÖDEL. *Bulletin of Symbolic Logic*, 11:172–184, 2005.

[SIEG, 2009] Wilfried Sieg. HILBERT's Proof Theory. 2009. In [GABBAY & WOODS, 2004ff., Vol. 5: Logic from RUSSELL to CHURCH, pp. 321–384].

[SKOLEM, 1920] Thoralf Skolem. Logisch-kombinatorische Untersuchungen über die Erfüllbarkeit und Beweisbarkeit mathematischer Sätze nebst einem Theorem über dichte Mengen. *Skrifter, Norske Videnskaps-Akademi i Oslo (= Videnskapsselskapet i Kristiania), Matematisk-Naturvidenskapelig Klasse*, J. Dybwad, Oslo, 1920/4:1–36, 1920. Also in [SKOLEM, 1970, pp. 103–136]. English translation of § 1 *Logico-Combinatorial Investigations in the Satisfiability or Provability of Mathematical Propositions: A simplified proof of a theorem by* LEOPOLD LÖWENHEIM *and generalizations of the theorem* by STEFAN BAUER-MENGELBERG with an introduction by JEAN VAN HEIJENOORT in [HEIJENOORT, 1971, pp. 252–263]. *(LÖWENHEIM–SKOLEM Theorem via SKOLEM normal form and choice of a sub-model)*.

[SKOLEM, 1923] Thoralf Skolem. Einige Bemerkungen zur axiomatischen Begründung der Mengenlehre. In *Proc. 5th Scandinaviska Matematikerkongressen, Helsingfors, July 4–7, 1922*, pages 217–232, Helsingfors, 1923. Akademiska Bokhandeln. Also in [SKOLEM, 1970, pp. 137–152]. English translation *Some remarks on Axiomatized Set Theory* by STEFAN BAUER-MENGELBERG with an introduction by JEAN VAN HEIJENOORT in [HEIJENOORT, 1971, pp. 290–301]. *(The best written and, together with [SKOLEM, 1920], the most relevant publication on the LÖWENHEIM–SKOLEM Theorem; although there is some minor gap in the proof according to WANG [1970], cured in [SKOLEM, 1929]. Includes SKOLEM's Paradox and a proof of the LÖWENHEIM–SKOLEM Theorem which does not require any weak forms of the Axiom of Choice, via SKOLEM normal form and construction of a model without assuming the previous existence of another one)*.

[SKOLEM, 1928] Thoralf Skolem. Über die mathematische Logik (Nach einem Vortrag gehalten im Norwegischen Mathematischen Verein am 22. Oktober 1928). *Nordisk Matematisk Tidskrift*, 10:125–142, 1928. Also in [SKOLEM, 1970, pp. 189–206]. English translation *On Mathematical Logic* by STEFAN BAUER-MENGELBERG and DAGFINN FØLLESDAL with an introduction by BURTON DREBEN and JEAN VAN HEIJENOORT in [HEIJENOORT, 1971, pp. 508–524]. *(First explicit occurrence of SKOLEMization and SKOLEM functions)*.

References

[SKOLEM, 1929] Thoralf Skolem. Über einige Grundlagenfragen der Mathematik. *Skrifter, Norske Videnskaps-Akademi i Oslo (= Videnskapsselskapet i Kristiania), Matematisk-Naturvidenskapelig Klasse*, J. Dybwad, Oslo, 1929/4:1–49, 1929. Also in [SKOLEM, 1970, pp. 227–273]. *(Detailed discussion of* SKOLEM*'s Paradox and the* LÖWENHEIM–SKOLEM *Theorem).*

[SKOLEM, 1970] Thoralf Skolem. *Selected Works in Logic*. Universitetsforlaget Oslo, 1970. Ed. by JENS E. FENSTAD. *(Without index, but with most funny spellings in the newly set titles).*

[STAUDT, 1847] Karl Georg Christian von Staudt. *Geometrie der Lage*. Verlag von Bauer und Raspe, Nürnberg, 1847.

[STERN, 1982] Jacques Stern, editor. *Proc. of the* HERBRAND *Symposium, Logic Colloquium '81, Marseilles, France, July 1981*. North-Holland (Elsevier), 1982.

[TAIT, 1981] William W. Tait. Finitism. *J. of Philosophy*, 78:524–546, 1981.

[TAIT, 2005] William W. Tait. *The provenance of pure reason – Essays in the philosophy of mathematics and its history*. Oxford Univ. Press, 2005.

[TAKEUTI, 1987] Gaisi Takeuti. *Proof Theory*. Elsevier, 1987. 2nd rev. ed. (1st ed. 1975).

[TAPP, 2006] Christian Tapp. *An den Grenzen des Endlichen — erkenntnistheoretische, wissenschaftsphilosophische und logikhistorische Perspektiven auf das* HILBERT*programm*. PhD thesis, Ludwig-Maximilians-Universität, München, 2006.

[TARSKI, 1986] Alfred Tarski. What are logical notions? *History and Philosophy of Logic*, 7:143–154, 1986. Ed. by JOHN CORCORAN.

[TAYLOR & WILES, 1995] Richard Taylor and Andrew Wiles. Ring theoretic properties of certain HECKE algebras. *Annals of Mathematics*, 141:553–572, 1995. Received Oct. 7, 1994. Appendix due to GERD FALTINGS received Jan. 26, 1995.

[TOEPELL, 1986] Michael Toepell. *Über die Entstehung von* DAVID HILBERT*s „Grundlagen der Geometrie"*. Vandenhoeck & Ruprecht, Göttingen, 1986.

[TORRETTI, 1984] Roberto Torretti. *Philosophy of Geometry from* RIEMANN *to* POINCARÉ. Springer, 1984.

[TROELSTRA & SCHWICHTENBERG, 1996] Anne Sjerp Troelstra and Helmut Schwichtenberg. *Basic Proof Theory*, volume 43 of *Cambridge Tracts in Theoretical Computer Science*. Cambridge Univ. Press, 1996.

[TROELSTRA, 1973] Anne Sjerp Troelstra. *Metamathematical investigations of intuitionistic arithmetic and analysis*. Number 344 in Lecture Notes in Mathematics. Springer, 1973.

[TURING, 1936/7] Alan M. Turing. On computable numbers, with an application to the Entscheidungsproblem. *Proceedings of the London Mathematical Society, Ser. 2*, 42:230–265, 1936/7. Received May 28, 1936. Correction in [TURING, 1937].

[TURING, 1937] Alan M. Turing. On computable numbers, with an application to the Entscheidungsproblem. A correction. *Proceedings of the London Mathematical Society, Ser. 2*, 43:544–546, 1937. Correction of [TURING, 1936/7] according to the errors found by PAUL BERNAYS.

[VEBLEN, 1904] Oswald Veblen. A system of axiomatic geometry. *Transactions of the American Math. Soc.*, 5:343–384, 1904.

[VORONKOV, 2002] Andrei Voronkov, editor. *18th Int. Conf. on Automated Deduction, København, 2002*, number 2392 in Lecture Notes in Artificial Intelligence. Springer, 2002.

[WAJSBERG, 1933] Mordechai Wajsberg. Untersuchungen über den Funktionenkalkül für endliche Individuenbereiche. *Mathematische Annalen*, 108:218–228, 1933. Received June 20, 1932.

[WAJSBERG, 1934] Mordechai Wajsberg. Beitrag zur Metamathematik. *Mathematische Annalen*, 109:200–229, 1934. Received March 28, 1933.

[WANG, 1970] Hao Wang. A survey of SKOLEM's work in logic. *In [*SKOLEM, 1970*]*, pages 17–52, 1970.

[WEBER, 1895f.] Heinrich Weber. *Lehrbuch der Algebra*. Vieweg, Braunschweig, 1895f..

[WEYL, 1918] Hermann Weyl. *Das Kontinuum. Kritische Untersuchungen über die Grundlagen der Analysis*. Veit, Leipzig, 1918. Reprinted 1932 by de Grutyer, Berlin and Leipzig. Also in [ANON, 1972]. English translation: [WEYL, 1994].

References

[WEYL, 1944] Hermann Weyl. DAVID HILBERT and his mathematical work. *Bulletin of the AMS*, 50:612–654, 1944.

[WEYL, 1994] Hermann Weyl. *The Continuum: A Critical Examination of the Foundation of Analysis*. Dover Publications, 1994. Translation of [WEYL, 1918] by STEPHEN POLLARD and THOMAS BOLE.

[WHITEHEAD & RUSSELL, 1910–1913] Alfred North Whitehead and Bertrand Russell. *Principia Mathematica*. Cambridge Univ. Press, 1910–1913. 1st ed..

[WILES, 1995] Andrew Wiles. Modular elliptic curves and FERMAT's Last Theorem. *Annals of Mathematics*, 141:443–551, 1995. Received Oct. 14, 1994.

[WIRTH &AL., 2009] Claus-Peter Wirth, Jörg Siekmann, Christoph Benzmüller, and Serge Autexier. JACQUES HERBRAND: Life, logic, and automated deduction. 2009. In [GABBAY & WOODS, 2004ff., Vol. 5: Logic from RUSSELL to CHURCH, pp. 195–254].

[WIRTH, 2004] Claus-Peter Wirth. Descente Infinie + Deduction. *Logic J. of the IGPL*, 12:1–96, 2004. http://www.ags.uni-sb.de/~cp/p/d.

[WIRTH, 2009] Claus-Peter Wirth. Shallow confluence of conditional term rewriting systems. *J. Symbolic Computation*, 44:69–98, 2009. http://dx.doi.org/10.1016/j.jsc.2008.05.005.

[WIRTH, 2010] Claus-Peter Wirth. *A Self-Contained and Easily Accessible Discussion of the Method of Descente Infinie and FERMAT's Only Explicitly Known Proof by Descente Infinie*. SEKI-Working-Paper SWP–2006–02 (ISSN 1860–5931). SEKI Publications, DFKI Bremen GmbH, Safe and Secure Cognitive Systems, Cartesium, Enrique Schmidt Str. 5, D–28359 Bremen, Germany, 2010. 2nd ed. (1st ed. 2006). http://arxiv.org/abs/0902.3623.

[WIRTH, 2011] Claus-Peter Wirth. *A Simplified and Improved Free-Variable Framework for HILBERT's epsilon as an Operator of Indefinite Committed Choice*. SEKI Report SR–2011–01 (ISSN 1437–4447). SEKI Publications, DFKI Bremen GmbH, Safe and Secure Cognitive Systems, Cartesium, Enrique Schmidt Str. 5, D–28359 Bremen, Germany, 2011. http://arxiv.org/abs/1104.2444.

[WOLFF, 1719] Christian Wolff. *Vernünftige Gedanken von den Kräften des menschlichen Verstandes und ihrem richtigen Gebrauche in Erkäntniß der Wahrheit*. Rengerische Buchhandlung, Halle an der Saale, 1719. 2nd extd. ed. (1st ed. 1713).

[ZACH, 1999] Richard Zach. Completeness before POST: BERNAYS, HILBERT, and the development of propositional logic. *Bulletin of Symbolic Logic*, 5:331–366, 1999.

[ZACH, 2001] Richard Zach. *HILBERT's Finitism: Historical, Philosophical, and Metamathematical Perspectives*. PhD thesis, Univ. of California, Berkeley, 2001.

[ZACH, 2003] Richard Zach. The practice of finitism: epsilon calculus and consistency proofs in HILBERT's program. *Synthese*, 137:211–259, 2003.

[ZERMELO, 1904] Ernst Zermelo. Beweis, daß jede Menge wohlgeordnet werden kann. *Mathematische Annalen*, 59:514–516, 1904. English translation *Proof that every set can be well-ordered* by STEFAN BAUER-MENGELBERG with an introduction by JEAN VAN HEIJENOORT in [HEIJENOORT, 1971, pp. 139–141].

[ZERMELO, 1908] Ernst Zermelo. Untersuchungen über die Grundlagen der Mengenlehre I. *Mathematische Annalen*, 65:261–281, 1908. English translation "Investigations in the foundations of set theory I" in [HEIJENOORT, 1971, pp. 199–215].

[ZERMELO, 1930] Ernst Zermelo. Über Grenzzahlen und Mengenbereiche. *Fundamenta Mathematicae*, 16:29–47, 1930. English translation in [EWALD, 1996, pp. 1219–1233].

New Index

Ackermann
    function, xlviii
    Wilhelm, v, xxxiii, xl, xlii–xliv, xlvi, l, li, lxi, VII, VIII, 8, 56, 65, 144, References
addition of an arbitrary antecedent to form an implication, *see* Hinzufügen eines beliebigen Implikationsvordergliedes, 83, 84, 107
Allgemeingültigkeit, *see* validity, 8, 9, 143, Glossary
Allzeichen, 4.b, 98, Glossary
anhängen, 22
ansetzen, 22
antecedent, 62, 67, 81–84, 91, 93, 94, 100, 103, 106–108, 114, 115, 135, 150–152, 155, 159, 160, Glossary
Aristotelian logic, *see* logic, Aristotelian
arithmetic
    primitive, lviii
    primitive recursion, xlviii
    recursive, lx
arithmetization, *see* method of arithmetization
atomic formula, *see* formula, atomic
Aussagen-Verknüpfung, 4.b
Aussagenverbindung, 4.b
Aussagenverknüpfung, 4.b
Ausdruck, *see* expression, 5.b
Ausgangsformel, *see* formula, initial, 44
Axiom
    der ebenen Anordnung, *see* plane axiom of order, 6.b
axiom
    of choice, 41
    of order, *see* plane axiom of order
    of parallels, *see* Parallelenaxiom, 4.a, 4.b, 6.a
    in sharper form, 6.a, 6.b, 12.b
axiomatic, 82
axiomatic method, xv
axiomatics
    contentual, 2, 6.a
    existential, *see also* form, existential, xii, xiv, xl, xlv, lv, lx
    formal, 2, 7
    sharpened, lx
Axiome
    der Anordnung, *see* axioms of order, 4.b
    der Verknüpfung, *see* axioms of connection, 4.b
axioms
    of connection, *see* Axiome der Verknüpfung, 4.a, 4.b, 5.a, 5.b
    of incidence, *see* Axiome der Verknüpfung, 4.b
    of order, *see* Axiome der Anordnung, 4.a, 4.b, 6.b

Bauer-Mengelberg, Stefan, 2, References
Behboud, Ali, xxxiv
Behmann, Heinrich, xxxiii, 145
Behnke, Heinrich, xxxiv
Bernays, Paul, iii–ix, xi, xii, xxi, xxxii, xxxix–xlvii, xlix–lii, liv–lxi, lxiii, III, IV, VI–VIII, 1.b, 5.b, 8, 9, 11, 16, 17, 36, 41, 45, 47, 49.b, 50, 55, 56, 67, 70, 71, 75, 82, 86, 87, 90, 93, 104, 114, 118, 123, 128, 131, 132, 144, Glossary, References
Bernstein, Felix, xxxix
Beweisfigur, *see* proof figure, 151
Born, Max, xxv
Bourbaki, Nicolas, vi, References
Brouwer, L. E. J., v, xii, xxxiv, xxxvi, xxxix, 2, 34, 42, 43, References
Buldt, Bernd, iii, viii, References

Cantor
    Georg, xiii, xxiii, xxv, xxvi, xxxix, References
    sche Fundamentalreihe, 36
Cassirer, E., xxxiv
Cauchy
    sche Fundamentalfolge, 36
    sequence, 36
chain inference, 68, 84, 85, 90, 91, 106–108, 111–115, 117, Glossary
Chevalley, Claude, liii
Church, Alonzo, 131
Clausius, Rudolf, 1.a
combining two mutual implications into an equivalence, 85, 107
completeness, X, XI, 63–66, 123, 124, 127–129, 163, Glossary
    ω-, 34
component, 121, Glossary
conjunction, representing, 56, 57, Glossary
consistency, VII
contentual, IX, X, 2, 3, 6.a, 6.b, 7, 20, 22, 23, 29–32, 45, 62, 83, 91–93, 103, 105, 130, 156, 160, Glossary
    axiomatics, *see* axiomatics, contentual
    history of the English word, 2
    inference, *see* inference, contentual
    mathematics, *see* mathematics, contentual
course of values, *see* induction, course-of-values, *see* Wertverlauf, 9
Crossley, John N., viii, References

darstellende Konjunktion, *see* Konjunktion, darstellende
Dawson, John W., Jr., References
decidable class w.r.t. derivability, 143, 144, Glossary
decidable class w.r.t. satisfiability, 143, 144, Glossary
decision method, 19, 131, Glossary
decision procedure, X, 55, 129, 131, 143, 145, Glossary
Dedekind
    cut, 36–39
    Richard, ix, xiii, xv, xvii, xix, xx, xxii, xxiii, xxv, xxvi, xxviii, xxix, xxxix, lxii, 37, References
deduction theorem, XI, 154–157, Glossary
deductive, 63, 82
derivability condition, liv
derivation by the predicate calculus, 104, 121, 148, 153, 154, 156, 157, Glossary
derivation in the predicate calculus, 121, 123, 155, Glossary
descente infinie, 35
Diller, Justus, viii
direct proof, xxi
distribution of truth values, 54, 56, 57, Glossary
domain, 127, Glossary
Dreben, Burton, References
duality, 52, 53, 113, A56

ε-theorem
    first, l
    second, l
Ebert, Philip, viii
Einsetzung für (die Nennform), 108
Elsenbroich, Corinna, vii
end formula, *see* Endformel, 44
Endformel, *see* end formula, 44
Engesser, Kurt, vii
Entscheidungsproblem, xi, IX, XI, 8, 11, 129–131, 143, 144, 155, 163, Glossary
ε-substitution method, xliii
equivalence schema, *see* Schema der Äquivalenz, 85, 107, 110, 111, 114, 116, Glossary
Erfüllung, 18, Glossary
Ersetzbarkeit, *see* replaceability, 47
Euclid
    of Alexandria, 1.a, 20, 25
Euler
    Leonhard, 36
existence, *see* form, existential, 20, 37
existence axiom, 20
existential
    form, *see* form, existential
exportation, 82, 83, 90, 103, 107, 115, Glossary
expression, 5.b, 49.a, 88

Feferman, Sol(omon), vii, xii, lxi, 1.b, References
Fermat

's Last Theorem, 36
's descente infinie, *see* descente infinie
Pierre, 28, 35, 36
Ferreira, Fernando, lxii
finit, VII
finitism, x, xxxix, li, lix, lxii, 32–38
finitistic, VII
finitistisch, VII
form
    existential, *see also* axiomatics, existential, 1.a, 1.b, 2, 20, 37
    normal, 31
        conjunctive, 53
        disjunctive, 53
        full conjunctive, *see* Normalform, ausgezeichnete konjunktive, 57
        full disjunctive, *see* Normalform, ausgezeichnete disjunktive, 56
        prenex, 143, 158, Glossary
formal axiomatics, *see* axiomatics, formal
Formel, *see* formula
formula, 5.a, 87, 88, 95
    atomic, 88, 95, 119–122, 142, 143, 145, 146, Glossary
    initial, 44, 62–66, 90, 104, 106, 107, 150
    numeric, xli
    prime, XI, 145–148, Glossary
Formula (1), 109, 117, 118
Formula (2), 109, 110, 114, 118, 139, 147
Formula (2′), 109
Formula (3), 110, 114
Formula (3′), 110, 139
Formula (4), 110, 111
Formula (5), 111, 113, 114, 116, 118, 138
Formula (6), 111, 113, 116
Formula (7), 112, 113, 118, 136
Formula (8), 113, 116, 137, 138
Formula (9), 113, 136
Formula (10), 113, 116
Formula (11), 114, 117, 118, 159
Formula (12), 115, 117
Formula (13), 115, 116, 138, 139
Formula (13′), 116, 138
Formula (14), 116, 118
Formula (14′), 116
Formula (15a), 116
Formula (16a), 117
Formula (16b), 117
Formula (17a), 117
Formula (17b), 117
Formula (18), 117
Formula (19), 117
Formula (20), 117, 118
Formula (a), 99, 103–106, 108–115, 120, 134, 148, 152
Formula (b), 100, 103–106, 108, 109, 113, 115, 120, 133
formula variables, 88
Formulas (15), 116

formulas of the predicate calculus, XI, 104, 109, 118, 120, 121, 123, 126–128, 131, 140, 143, 153–155, 158, Glossary
Fraenkel, Adolf Abraham, 42
Frege, Gottlob, xxvi, xxviii, 9, 10, 15, 18, 37, 63, 71, 87, References
function
    calculable, lviii
function symbol, 163, Glossary
fundamental sequence, see Cantorsche Fundamentalreihe, Cauchysche Fundamentalfolge, 36
Fundamentalfolge, see Cauchysche Fundamentalfolge
Fundamentalreihe, see Cantorsche Fundamentalreihe
Funktion
    ganze rationale, see polynomials

Gabbay
    Dov, iii, iv, viii
    Michael, iii, viii
    Murdoch J., viii
Gegenstand, 86, Glossary
Gentzen
    Gerhard, vi, References
Gentzen, Gerhard, xi, xxxix, xli, lviii, lix, lxi
genus, 2, 41
Geometrie
    der Lage, see geometry of position, 4.b
geometry
    of position, see Geometrie der Lage, 4.a, 4.b, 6.a, 6.b
    plane, 4.a
Gödel
    Kurt, vi, xi, xxxix, xlv, xlix–liii, lviii, lxi, lxii, VII, VIII, 127, 128, 144, 151, 158, References
Goldbach
    's Conjecture, 36
    Christian, 36
Goldfarb, Warren, References
graph, 9, 35, 87

Hallett, Michael, xxiv
Hasenjaeger, Gisbert, V, VI, 123
Haubrich, Ralf, xxiii
Heidgegger, Martin, 1.b
Heijenoort
    Jean van, xx, xxix–xxxi, xlii, xliv, 2, 151, References
Helmholtz, Hermann von, 29
Hendricks, Vincent, xxxiv
Herbrand
    Jacques, vi, xi, xlv, xlvi, l–liii, lviii, lxii, VIII, 10, 128, 130, 131, 145, References
Hertz
    Heinrich, xvi
    Paul, 47, 68, References

Hessenberg, Gerhard, xxvii
Heyting, Arend, v, lix, 70, References
Hilbert
    's "Foundations of Geometry", VIII, 1.a, 1.b, 4.a, 4.b, 5.a, 5.b, 6.b, 12.b, 65
    's "Grundlagen der Geometrie", 1.a, 5.a, 6.a, 6.b, 12.a, 12.b
    David, iii–xxxi, xxxiii–xlvii, xlix–lii, liv–lix, lxi, lxiii, III, VII, VIII, 1.a, 1.b, 4.b, 5.b, 6.b, 8, 9, 11, 12.b, 16, 17, 35–37, 45, 49.b, 50, 55, 56, 67, 70, 71, 75, 82, 86, 87, 90, 93, 104, 114, 118, 123, 128, 131, 132, Glossary, References
Hillenbrand, Thomas, viii, 31, References
homogeneity, xxx
Huntington, Edward V., 13
Hurwitz, A., xxvi
Husserl, Edmund, 53
Huygens, Christiaan, 36
hypothetical connective, 5.a, 47
hypothetical inference, 62, 67
hypothetical sentence, 5.a

identically false, 55, 121, A57, Glossary
identically true, X, 54, 55, 57, 60–66, 68, 69, 71, 72, 78, 82, 84–87, 90, 119–121, 129, 131, 149, A56, Glossary
implication formula, X, 67–70, 78, Glossary
    regular, X, 67–69
importation, 82, 83, 102, 107, 115, Glossary
impossibility proof, IX, 19, 27, 42, 44, Glossary
incompleteness theorem, li
individual symbol, 88, 89, 104, 120, 129, 150, 153–156, 161, Glossary
individual variable, 8, 86, 88, Glossary
induction
    complete, 23
    course-of-values, 23
    mathematical, see Induktion, vollständige, 23
    structural, 23
Induktion
    vollständige, see induction, mathematical, 23
inference
    contentual, 32
inference schema
    of modus ponens, see Schlussschema, 62, 64, 66–69, 72, 74, 82, 84–86, 90, 91, 94, 104, 106, 108, 110, 112, 133, 134, 149, 159, Glossary
Inhaltslogik, see logic of content, 53
initial formula, see Ausgangsformel, see formula, initial
interchange of antecedents, 81–83, 91, 107, 114, 115, 135, 152, 159, Glossary
intuitionism, xii, xxxix, 43
intuitionistic logic, see logic, intuitionistic

Jaśkowski, Stanisław, 106
Jevons, W. S., xxvi

New Index

Kalkul, 45
Kalmár, László, 144
Kant, Immanuel, 1.b, 2, 14, 17, 42, References
Kettenschluss, 68, 84
Kleene, Stephen C., lvii, References
Kolmogorov, Andrey, lix
Konjunktion
    darstellende, 56, 57, Glossary
Kowalewski, Arnold, xxxiv
Kreisel, Georg, lxi, lxii, V, VI, References
Kronecker, Leopold, v, xii–xiv, xxvi, xxvii, xxix, xxx, xxxvii, xli, lxii, 42

lattice
    distributive, 49.b
law
    of the excluded middle, 34, 42
Leśniewski, Stanisław, 64, References
Lewis, Clarence Irving, 67, References
Löwenheim
    -Skolem Theorem, xliv, li, 128, References
    Leopold, 128, References
logic
    Aristotelian, 9, 106, 117
    intuitionistic, 34
    of content, see Inhaltslogik, 53, Glossary
    of extension, see Umfangslogik, 53, Glossary
Łukasiewicz, Jan, 45, 64, 69, 71, 77, References

manifold, 14, 16, 17, 37, Glossary
Mannigfaltigkeit, 14, Glossary
Martin-Löf, Per, lxi
mathematics
    contentual, 2
Melis, Erica, viii
metamathematics, 44
method
    axiomatic, xiv, xxi
    genetic, xiv, xxi
    of arithmetization, xviii, xxii, 3, 18, 41
    of exhibition, see Methode der Aufweisung, 12.a, 12.b, 13–17
    of progressive inference, 13
Methode
    der Aufweisung, see method of exhibition
methodical, 1.a
methodisch, 1.a, 20, 32, 44
methodischer Standpunkt, 1.a
methodological, 1.a
methodological standpoint, 1.a
model, see Erfüllung, see Wertung, 12.a, 18, 73, 128, Glossary
    standard, see Wertung, normale, 73
modus
    ponens, see inference schema of modus ponens, 62
monadic predicate calculus, XI, 122, 124, 131, 143, 145, 146, 148, 163, Glossary
Moschner, Markus, viii

Mueller, Ian, 14, References

Nachbereich, 127, Glossary
Nagel, Ernest, xvii
negation
    finitistic interpretation of, 33
    sharpened, 33
Nelson, Leonard, xxvii
Nennform, 89
Neumann, John von, vi, xl, xliv, l, lxii, References
Newton, Isaac, 1.a
Nicod, Jean, 48, 64, References
nominal form, 89, 97, 108, 115, 139, 152, 159, Glossary
normal disjunction, XI, 158, 160–163, Glossary
normal form, see form, normal
Normalform
    ausgezeichnete disjunktive, 56
    ausgezeichnete konjunktive, 57
    pränexe, Glossary
number
    cardinal, 28–29
    ordinal, 28
number theory
    elementary, 20–28
    recursive, xlviii

omission of multiple antecedents, 83, 107

Parallelenaxiom, see axiom of parallels, 4.b
    in erweiterter Fassung, 6.b
    in schärferer Fassung, 6.b
Parsons, Charles, iii, viii, lvii, References
Peano
    Guiseppe, 18
Peckhaus, Volker, iii, viii, xii, xxv, 53, References
Peirce
    Charles S., 10, 48, 56, References
plane axiom of order, see Axiom der ebenen Anordnung, 6.a, 6.b, 12.a, 12.b
Poincaré, Henri, xxvi, xxxii, xxxiv, xxxix
polynomial functions, elementary theory of, 29
polynomials, elementary theory of, 32
positive logic, 67
positive-identical, 68
predicate system, 10, 11, 14, 128, Glossary
predicate variable, 8, 10, 88
prenex formula, 140
prenex normal form, see form, normal, prenex
Presburger
    Mojżesz, xlvi
Primärformel, see formula, prime
prime formula, see formula, prime
Primformel, see formula, atomic
principle
    least-number, 35
    of choice, 41
proof figure, 151
proof theory, lv, V, VII–IX, 44, 128–130, 158, 160, Glossary

Hilbert's, x
propositional combination, 4.b
propositional connective, 4.b
pure predicate calculus, 150, Glossary
pure predicate logic, 7, 8, Glossary

quantifier, 4.b, 98
quantifier symbols, 4.a, 4.b
Quine, Willard Van O., VIII, 7, 15, 106, References

range, 127, Glossary
recursion, 26
recursiveness conditions, lviii
refutability, 128–131, 143, 144, 155, Glossary
Regel
  der Einsetzung, see rule of substitution, 62
regular implication formula, see implication formula, regular
renaming (of) bound variables, X, 97, 98, 101, 104, 105, 107, 115, 116, 136, 140, 141, 150, 151, Glossary
replaceability, see Ersetzbarkeit, 47
replacement rules, 49.a, 82
  for the equivalence, 83, 107, 109, Glossary
representability condition, liii, lv
representing conjunction, see conjunction, representing
rewrite systems, 49.b
Rezaei, Marianeh, viii
rule
  of substitution, see Regel der Einsetzung, VIII, X, 62, 63, 67, 68, 72, 74, 82, 86, 87, 89–93, 95, 97, 99, 104, 106, 107, 112, 114, 127, 149, Glossary
Rule (δ), 108, 109, 111–113, 115–117, 135
Rule (δ'), 109–111, 113, 114, 116, 135, 138, 139
Rule (ε), 133, 149
Rule (ε'), 134, 149, 155
Rule (η), 135, 136, 138–142, 146, 160
Rule (γ), 107, 110, 113, 114, 151
Rule (γ'), 108, 133, 148
Rule (ι), 137, 140–142, 146, 147, 158, 160
Rule ($\varkappa$), 138, 141
Rule (λ), 139, 140, 142, 143, 160
Rule $S1$, 49.a
Rule $S2$, 49.a
Rule (ϑ), 136, 137, 141, 142, 146, 158
Rule (ζ), 134–136, 138, 156
Rules
  "1", 49.a
  "2", 49.a
  "3", 49.a
  "4", 50
Russell
  's Paradox, 15
  Bertrand, vi, x, xxviii, xxxiii, xxxviii, xliv, 15, 18, 63, 86, References

Sartre, Jean Paul, 1.b

satisfiability, IX, 3, 8–11, 12.a, 13, 14, 17–19, 123, 124, 127–131, 143–145, 158, Glossary
  finite, 144, Glossary
Satzverbindung, see sentential combination, 4.b, Glossary
schärfer, VII
Schaper, Hans von, xvi, xxii
Schema (α), 100–103, 105, 107–110, 112, 113, 115, 121, 149–153, 155
Schema (β), 100, 103, 105, 107–109, 113, 115, 121, 149–153, 155, 157
Schema der Äquivalenz, 85
Schema ($\tilde{s}$) (for conjunction), 79–81, 84, 100, 102, 107, 111–113
Schema ($\tilde{s}'$), 81, 102
Schema ($\mathfrak{t}$) (for disjunction), 81, 84, 107, 113
Schiller, Marvin, viii
Schlussschema, see inference schema (of modus ponens), 62
Schneider, Hubert H., 106
Schönfinkel, Moses, 69, 144
Schröder, Ernst, 18, 53
Schütte, Kurt, VIII, 123, 144, 145
Seinszeichen, 4.b, 98, Glossary
sentential combination, 4.a, 4.b, 47, Glossary
Sheffer
  Henry Maurice, 48
  stroke, 48, 49.b, 64, Glossary
Sieg, Wilfried, iii, vi–ix, xii, xviii, xxxiii, xxxix, liii, lvii, VIII, 14, References
Siekmann, Jörg, iii, viii
singulary, 7
Skolem
  's Theorem, 158, 161, 163, Glossary
  normal form, XI, 158, 162, 163, Glossary, References
  Thoralf, VI, 128, 144, 158, References
Sobociński, Bolesław, 64
substitution, see rule of substitution
substitution for (the nominal form), 108–112, 114–117, 133, 134, 137–139, 152, 159, 160, Glossary
substitution rules, 49.a
suffix, 21–24, Glossary
Suppes, Patrick, xvii
system
  of values for the variables, 55
  simply infinite, xviii

Tait, William W., iii, vii, viii, xii, xviii, xix, References
Tapp, Christian, iii, viii
Tarski, Alfred, 7, 45, 64, 69, References
Term, 5.b
tertium non datur, xxxiv, xxxvi, xl, lix, lxi, 35, 37, 40, 41, 124, 127, 128
  for integers, 34–36, 38, 128
thought experiment, xlvi, 20, 32
Torretti, Roberto, xv

New Index

Umfangslogik, *see* logic of extension, 53

Vaihinger, H., xxxiv
validity, IX, 8–11, 17, 34, 124, 127–129, 131, 143, 160, 161, Glossary
valuation, 53, 55, 74–76, 78, A56, Glossary
Veblen, Oswald, 4.a, 4.b, 12.a, References
Verbindung, 4.b
verifiability, xlviii
Verteilung von Wahrheitswerten, 55, 56
Voigt, Andreas Heinrich, 53
Vorbereich, 127, Glossary
Vorderglied, *see* antecedent, 62

Wajsberg, Mordechai, 69, 120, 123
Waldmeister, 31, References
Wang, Hao, References
Weber, Heinrich, xiv
Weierstraß, Karl, xxiii, xxvii, xxviii, 37
Wertung, *see* model, 73
    normale, *see* model, standard, 73
Wertverlauf, *see* course of values, *see* graph, 9, 87
Wertverteilung der Variablen, 54, 56, 57, Glossary
Weyl, Hermann, v, xi, xxxiv, xxxvi, xxxix, xliv, 39, References
Whitehead, Alfred North, vi, x, xxviii, xxxiii, xxxviii, 63, References
Widerspruchsfreiheit, VII
Wirth, Claus-Peter, iii, vii, viii, 140

Zach, Richard, iii, viii
Zeno
    's Paradox, 16, 17
    of Elea, 16
Zermelo, Ernst, xxvi–xxviii, 15, 41, References

Namenverzeichnis.

Ackermann 144, 335, 337, 346, 430
Aristoteles 9, 106, 117

Behmann 145, 199
Bernays 144
Brouwer 2, 34, 42, 43

Cantor 335
Carnap 432
Church 131
Clausius 1

Dedekind 36, 37, 39, 212, 301, 422, 423

Euklid 1, 20, 25, 332

Fermat 371
Fraenkel 42
Frege 15, 18, 63, 70, 87, 165

Gödel 127, 128, 144, 158, 424, 431

Hasenjaeger 123, 272, 386, 437, 450
Helmholtz 29
Herbrand 128, 130, 131, 145, 233, 463

Heyting 70
Hilbert 1, 65, 248, 335
Huntington 13

Jaśkowski 106

Kalmár 144, 389
Kleene 451
Kreisel 308
Kronecker 42

Leśniewski 64
Lewis 67
Löwenheim 127, 128, 199
Łukasiewicz 64, 69, 71, 77

Mendelson 451

Neumann, von 244, 423, 431
Newton 1

Peano 18, 218, 219, 286, 355, 382, 398
Peirce 48
Péter 331, 337, 349
Politzer vgl. Péter
Presburger 233, 364

Quine 106

Raggio 255
Rosser 435, 436, 437, 450
Russel 15, 18, 63, 165, 393

Schneider 106
Schönfinkel 69, 144
Schröder 18
Schröter 451
Schütte 123, 144, 145, 450
Sheffer 48, 64
Skolem 128, 144, 157, 158, 161, 162, 163, 199, 309, 349
Sobociński 64

Tarski 64, 69

Veblen 4, 12

Wajsberg 69, 120, 123
Weyl 39
Whitehead 63, 393

Zeno 16
Zermelo 15, 41

Sachverzeichnis.

Ableitbarkeit 121
Ableitung durch den Prädikatenkalkul 104
—, formale 64
Ackermannsche Funktion 335 f., 430
Addition von Ziffern 22
— in der rekursiven Zahlentheorie 304
Algebra der Logik 53
—, formale 29
—, inhaltlicher Standpunkt in der 29
Allgemeingültigkeit 8, 127
Allzeichen 4, 94
Analysis 36
Anzahlbegriff 28 f., 164, 171 f.
Anzahlformeln 164 f., 177, 198
Äquivalenz 47
—, Transitivität der 85
Argumentstelle, ausgezeichnete 288, 293
— einer Formel 102
Arithmetik, axiomatische 19
—, logische Begründung der 19
—, Überschreitung des finiten Standpunkts in der 34
Arithmetisierung der Größenlehre 41
—, Methode der 3, 18
Auflösungsfigur eines Beweises 223
Aufweisung der Unendlichkeit 17
—, Methode der 12
Ausdruck 49, 88
Ausgangsformel 64 f., 104, 107
Aussagenkalkul 48, 52, 82
Aussagenlogik, elementare 45
Ausschaltung der Existenzaussagen 216
— von Formelvariablen 200, 227, 248
— von Funktionszeichen 291
Austausch freier Variablen gegen gebundene 199
Auswahlprinzip 41
Axiomatik 2
— bei Euklid 20
Axiome als Formeln 90
— der Anordnung 6
— der Verknüpfung 5
—, eigentliche 432
Axiomenschemata 247

Axiomensystem (s. a. „System")
— der ersten Stufe 154

barbara 117
Befund 33
Bestimmung variabler Prädikate 8
Beweis 27 f., 220
—, aufgelöster 223
—, —, im weiteren Sinne 227
—, für „Ableitung" 220
Beweisfäden, Auflösung in 220 f.
Beweistheorie 44

darii 117
Dedekindsche Unendlichkeitsdefinition 212
Dedekindscher Schnitt 36 f.
Deduktion 18
Deduktionsgleichheit 148
Deduktionstheorem 154
deduktiv 63
Definition 286
—, explizite 292, 400 f.
—, rekursive 25, 292
Deutung der elementaren Zahlentheorie 30
—, finite 32
—, hypothetische 32
Diagonalverfahren 335
dictum de omni 62, 103
diophantische Gleichung 381
Disjunktion 4, 45 f.
—, erstreckt über einen Individuenbereich 98
—, n-gliedrige 10
—, nullgliedrige 56, 106
distributive Entwicklung 57
distributives Gesetz 49
Division, bei Ziffern 24
— in der rekursiven Zahlentheorie 322
—, rekursiv 322
Dualbruch 37, 40
Dualität 52, 113, 127, 129
Durchschnitt 126

Eindeutigkeitssatz für Reduzierte 238
Einführung, rekursive 287

Einlagerung 397
Einsetzung, Einschränkung für die 97
— für Formelvariablen 227
Einsetzungen, Rückverlegung der 224f., 230, 267
—, Verlegung der — in die Ausgangsformeln 224
Einsetzungsregel, Einschränkung der 97
—, erweiterte 89
— für Formeln im Aussagenkalkul 62
— im Prädikatenkalkul 97
elementare Aussagenlogik 45
elementarer Kalkul mit freien Variablen 295
Elimination der Kennzeichnungen 431f., 448
Endlichkeitsbedingung 122
Entscheidungsbereich 143
Entscheidungsproblem 8, 129f., 142f.
— für den einstelligen Prädikatenkalkul 190
erfüllbar 8
—, im Endlichen 184
—, k-zahlig 184
Erfüllbarkeit 8, 124, 127, 130
Ersetzbarkeit von Aussagenverknüpfungen 47f.
— der Disjunktion durch die Implikation 51
Ersetzungsregel 49
—, Vollständigkeit der 59
Evidenzproblem 20
existentiale Form 1
Existenzaussagen, Ausschaltung der 216
Existenzsatz in finiter Auffassung 32
explizite Definition 292
Exportation 82
Extrapolation, gedankliche 15

Feld einer Funktion 127
Fermatscher Satz 381
finite Deutung 32, 330
finiter Standpunkt 32
—, Erweiterung des — durch den Intuitionismus 43
— in der Theorie der algebraischen Zahlenkörper 42
finites Schließen 20
formaler Standpunkt 45
Formalisierung des hypothetischen Urteils 91f.
— des Schließens 18, 44, 61
Formalismen, Gleichwertigkeit der — $\mathfrak{S}$ und $\mathfrak{S}^*$ 457

Formalismus 357
— der rekursiven Zahlentheorie 330
— des Systems (B) 357
—, Erweiterung eines 357, 361 f.
Formel 4, 87 f., 95
—, ableitbare 105
—, allgemeingültige 127
—, aussagenlogisch wahre 438
— des Prädikatenkalkuls 104
—, identische 89
—, —, des Aussagenkalkuls 104
—, falsche 228
— $[\iota]$ 433
— $\{\iota\}$ 437
— $[\iota, B]$ 435
—, Komponenten einer 121
—, numerische 228, 249
—, pränexe 140
—, rekursive 322
—, wahre 228
— $\mathfrak{D}$ 213
— $\mathfrak{F}$ 122
— $\mathfrak{F}_0$ 211
— $\mathfrak{G}$ 123
— $\mathfrak{H}$ 123
— (1), (2), (2'), (3) 109
— (3'), (4), (5) 110
— (6) 111
— (7) 112
— (8), (9), (10) 113
— (10a), (10b), (10c), (11), (12) 114
— (13), (13a) 115
— (13'), (14), (14'), (15a) 116
— (15) 117
— (16a), (16b), (17a), (17b), (18), (19), (20) 117
Formelbestandteil 5
Formeln $(\mu_1), (\mu_2), (\mu_3)$ 405
—, abgeleitete — des Prädikatenkalkuls 109ff.
Formelschemata 388
Formelsystem $(\mathfrak{D})$ 212
— $(\mathfrak{D}_0)$ 214
— $(\mathfrak{F}), (\mathfrak{G})$ 208
— $(\mathfrak{F}_0)$ 210
— $(\mathfrak{G}^*)$ 213
— $(\mathfrak{H})$ 209
Formelvariablen 88, 155, 163
—, Ausschaltung der 200ff., 248, 267
—, Vermeidung der 387
Fundamentalreihe 36
Funktion 26, 86, 396
—, Darstellung als ι-Term 396
—, logische 125

Funktion, mathematische 189, 288
—, rekursive 322
—, Wert einer 64
Funktionszeichen 163, 186f., 214

Gattung 2, 41, 87f.
genetische Methode 1
Gleichheitsaxiome (J_1), (J_2) 164
—, Ersetzung von (J_2) durch speziellere Axiome 382f.
Gleichwertigkeit der Formalismen $\mathfrak{S}$ und $\mathfrak{S}^*$ 457
Größenbeziehung 22
Grundformeln des Prädikatenkalkuls 104
Grundgleichungen 76
Grundprädikat 88
Grundregeln 106

hypothetische Deutung 32
hypothetischer Satz 4
hypothetisches Urteil 91f.

Idealisierung in der Wissenschaft 3
Identifikation von Variablen 164
identisch falsch 55
—, im Endlichen 120, 184
—, k-zahlig 118, 184
— wahr 54, 61, 122
identische Formel 89
— des Aussagenkalkuls 104
Identität 4, 7, 130, 163
—, Satz der 166
— der elementaren Algebra 30f.
Identitätsaxiome 164
Implikation 4, 45f.
— im Vergleich zur hypothetischen Verknüpfung 92
—, unendliche 100
Implikationsformeln, reguläre 167
implizite Bestimmung der Grundbeziehungen durch Axiome 7
Importation 82
Individuenbereich 2, 8
— der rationalen Zahlen 36
—, leerer 9, 106
—, unendlicher 14, 17
Individuensymbol 88, 217
Individuenvariablen 8, 86f.
—, freie und gebundene 4, 95
Induktion 304
—, vollständige 23, 264
Induktionsaxiom 264, 431
Induktionsschema 264, 298

Induktionsschema, Erweiterung des 348
Infinitesimalrechnung 36
Inhaltslogik 53
Interpretation 71
Intuitionismus 43
ι-Axiom 453, 455, 457
ι-Ausdruck 398
ι-Regel 393
—, Erweiterung der 433
ι-Symbol 393
ι-Term 397
—, eigentlicher 435

Kalkul, elementarer mit freien Variablen 295
— der zweiten Stufe 422
Kennzeichnung 393
Kettenschluß 84
Klammerersparung 5, 107, 310
Kleiner, Definition von $\leq$, $>$, $\geq$ durch 293
—, — von —, durch $\delta(a, b)$ 303
—, — von —, durch $+$ 366
Kleinste Zahl, Prinzip der 284
Kollision zwischen gebundenen Variablen 394
Komplementärmenge 125
Komponenten einer Formel 121
Kongruenz, explizit definierte 368, 410f.
Kongruenzen, simultane 330
Konjunktion 4, 45
—, erstreckt über einen Individuenbereich 98
—, n-gliedrige 10
—, nullgliedrige 106
—, unendliche 98
Konsequenz 61
konstruktive Methode 1
Kontinuum 17
Kontraposition 83

Logisches Schließen, Formalisierung des 61

Maximum einer endlichen Zahlenfolge 421
— zweier und dreier Zahlen 342
Mehrfachbindungen 96f.
Menge 125
mengentheoretische Prädikatenlogik 124
Metamathematik 44
Methode, axiomatische 1
— der Arithmetisierung 3, 18

Methode der Aufweisung 12, 14
— der Zurückführung auf die Arithmetik 19
— des progressiven Schließens 14
—, genetische 1
—, konstruktive 1
— zur Ausführung von Unabhängigkeitsbeweisen 73 ff.
Modell 18
modus ponens 62
Multiplikation von Ziffern 24
μ-Formeln 405

Nachbereich 127
Negation 4, 45 f.
— bei elementaren Urteilen 33
—, verschärfte 33
Negationsstrich 4
Nennform 88
Nennvariable 441
Normaldisjunktion 158
Normalform 131, 139
—, ausgezeichnete disjunktive 56
—, — konjunktive 57
—, konjunktive 53
—, disjunktive 53
— für Formeln des erweiterten einstelligen Prädikatenkalkuls 178
— im einstelligen Prädikatenkalkul 190
—, konjunktive 53
—, pränexe 143
—, Skolemsche 158
normierte Rekursion 341
Numerierung der Zahlenpaare 326 f.
numerische Formel 228, 249, 274
Nummern zur Abkürzung von Ziffern 21, 30

obere Grenze, Konstruktion der 38
oder, ausschließendes 52
Ordinalzahl 28
Ordnungsbeziehung, Charakterisierung durch die Axiome ($<_1$), ($<_2$), ($<_3$) 217 f.

Paradoxie des Zeno 16
Paradoxien, logische 15
—, mengentheoretische 15
Parallelenaxiom 6
Parameter 27, 86, 288
— der Rekursion 327, 340
Partialurteil 32
Peanosche Axiome 218, 286

Peanosche Axiome (P_1'), (P_2) 219
Pétersche Funktion 337 f.
Polynom 29 f.
positiv identisch 68
positive Logik 67
Potenz 324
Prädikat 4, 7, 86
—, einzahliges, zweizahliges 165
Prädikatenkalkul 104, 107
—, einstelliger 122, 145
—, erweiterter 214, 431
—, erweiterter einstelliger 178
Prädikatenlogik 87
—, mengentheoretische 124
—, reine 7
Prädikatensymbol 88
Prädikatenvariable 8, 88
Prämissen 61
—, Widerspruch aus 85
pränex 140
Primärformeln 145
—, Zerlegung in 145, 178 ff.
prime Zahlen, zueinander 414
Primformeln 88, 121, 187
primitive Rekursion 331, 421
Primzahl 25
— in der rekursiven Zahlentheorie 323
Primzahlsatz von Euklid 25
Principia Mathematica, Symbolik der 94
Prinzip der kleinsten Zahl 35, 284
Produkt, logisches 53
progressives Schließen 14

Quantität des Urteils 98
Quantor 98

Rechengesetze für Summe, Produkt und Differenz 309 ff.
Reduktion 237
—, modifizierte 249, 275, 280
— nach PRESBURGER 369 ff.
—, partielle 243
Reduzierte 233, 237, 243, 275, 369
— bei der Elimination von ι-Symbolen 439 f.
—, Eindeutigkeit der 238
Reflexivität 166 f.
Regel der Umbenennung 97, 394
— (γ) 107
— (γ'), (δ) 108
— (δ') 109
— (ε) 132
— (ε') 133

Regel (ζ) 134
— (η) 135
— (ϑ) 136
— (ι) 137
— ($\varkappa$) 138
— (λ) 139
Regeln, abgeleitete 102, 106f., 131
— der Äquivalenz 49
— der Einsetzung 62
— der Ersetzung 49
— der Erweiterung 49
— der Kürzung 49
— der Umformung von Aussageverbindungen 48
— des Prädikatenkalkuls (Zusammenstellung) 104f.
— für die Konjunktion und Disjunktion 49
— für die Negation 49
Rekursion 26
—, Erweiterung des Schemas der 330
— mit mehreren Parametern 327
—, normierte 341
—, primitive 331, 421
—, Schema der 287
—, simultane 333
—, verschränkte 336f., 430
—, Wertverlaufs- 331
—, Zurückführbarkeit auf primitive 331
Rekursionsschema 288
Rekursionsverfahren 26
— als abgekürzte Beschreibung einer Anweisung 27
rekursive Definition 25, 292
— Einführung 287
— Formel 322
— Funktion 322
— Zahlentheorie 330
rekursiver Term 321
Relation, formale 17
Rückverlegung der Einsetzungen 224
— bei Anwesenheit gebundener Variablen 230
— bei Einbeziehung des Induktionsschemas 267

Satz von der partiellen Reduktion 243
Satzverbindung 4
Schema der Äquivalenz 85
Schluß 60
—, hypothetischer 62
Schlußschema 62, 104, 162
Schnitt, Dedekindscher 36f.

Seinszeichen 4, 94
Sheffersches Symbol 48
Skolemsche Normalform 158
Skolemscher Satz 157
Spezies 2
Stellen eines Prädikates 9
Strichsymbol 214
Struktur 17
Stufe, Axiomensystem der ersten — 154
—, Kalkul der zweiten 422
Subjekt 86
Subjektsbereich 2
Subjektsstellen 8
substituieren 27
Substitutionsregeln 49
Summe, logische 53
Symbol 0 217
Symmetrie 166f.
System der deduktiven Aussagenlogik 65
— von Frege 63, 71
— von Whitehead und Russell 63
— von Łukasiewicz 70
— (A) 262
— ($A*$) 274
— (B) 273
— (C) 356
— (D) 366
— (D_1) 367
— (Z) 380
— ($Z*$) 460
— ($Z**$) 462

Teilbarkeit im System (Z) 413
— in der rekursiven Zahlentheorie 321
— bei Ziffern 25
Teildisjunktion 56
Teiler, größter gemeinsamer 328
—, kleinster von 1 verschiedener 323
Teilkonjunktion 56
Teilmenge 126
Term 187, 217, 393
—, rekursiver 321
tertium non datur 37, 39
— bei unendlichen Folgen 40
— für ganze Zahlen 128
— für Prädikate 127
— für Zahlen 35
— vom finiten Standpunkt 32
Theorie, axiomatische 87f.
Totalität der Zahlenreihe 15
traditionelle Logik 44
Transitivität 166f.

Übereinstimmung (im Sinne einer Äquivalenzrelation) 167
Überführbarkeit 131 f.
Überordnung 397 f.
—, indirekte 399
Umbenennung von gebundenen Variablen 97, 104, 394
— bei Anwendung der ι-Regel 395
Umfangslogik 53
Umformung 131
unabhängig bei Termen, von einem rekursiv eingeführten Funktionszeichen 287
Unabhängigkeit 13, 70 f.
— der Axiome der Systeme (A), (B) 277 ff.
Unabhängigkeitsbeweise 13
Unendlichkeitsdefinition, Dedekindsche 212
Unerfüllbarkeit 11
Unitätsformeln 393
Unmöglichkeitsbeweis 19, 27
Unwiderlegbarkeit 129
Urteil, allgemeines 9, 32, 93 f.
—, —, Widerlegung eines 34
—, hypothetisches 92
—, kategorisches 105
—, partikuläres 94
—, Quantität des 98

Variablen, Ausschaltung der freien 220
—, Austausch von 191
—, freie 95
—, gebundene 4, 95, 393
—, im Aussagenkalkul 49
vel 4
Vereinigungsmenge 39, 126
verifizierbar 237, 243, 247, 249, 276, 280, 295, 297, 369, 376
Vertretbarkeit von Funktionen 361 f.
— von Funktionszeichen durch Prädikatensymbole 455 f.
— von rekursiven Funktionen in (Z) 451 f.
Vollständigkeit der deduktiven Aussagenlogik 65
— des einstelligen Prädikatenkalkuls 189
— des erweiterten einstelligen Prädikatenkalkuls 204 f.
— des Prädikatenkalkuls 123
— von (A) 262 f.
Vollständigkeitssatz für das Axiomensystem (A) 263

Vorbereich 127
Vorderglied 62

Wahrheitsfunktionen 45 f.
—, Abhängigkeitsbeziehungen unter 47 ff.
—, Beziehung der — zum logischen Schließen 59 ff.
—, Darstellbarkeit aller — durch &, $\vee$, $\overline{}$ 56
Wahrheitswerte 11
Wert, ausgezeichneter 73
Wertbestimmung 11
Wertung, normale 73
— von Łukasiewicz 77
Wertverlauf 9
— eines Prädikates 87
Wertverlaufsrekursion 331
widerlegbar 127
Widerspruch im deduktiven Sinne 18, 85, 129
Widerspruchsfreiheit 129
—, deduktive 18
— der Arithmetik 43 f.
— der Prädikatenlogik 120
— der Zahlenreihe 15
— der Zahlentheorie 15
— des elementaren Kalkuls mit freien Variablen 299
— von (D) 368 ff., 376

Zahl, rationale 36
—, —, Darstellung der — durch ganze Zahlen 37
—, reelle 36
Zahlbegriff, Elimination des — aus den Peano-Axiomen 219
Zahlenfolgen, Abbildung der endlichen — auf die Zahlen 325, 423
Zahlenpaare, Numerierung der 326 f., 333
Zahlentheorie 20, 286
—, Formalisierung der — in (Z) 410
—, offene Probleme der 36
—, rekursive 330
Zeichen, logische 4
— zur Mitteilung 21 f.
Ziffern 10, 21, 218
—, Abbau von 26
— 1 als Ausgangsding 21
— erster, zweiter Art 249, 274, 279
—, unbestimmte als Parameter 27
Ziffernsymbole 292

Glossary: German — English

This glossary of technical terms and critical translations serves as a guide for translators and bilingual readers. It is specific to the English translation of [HILBERT & BERNAYS, 1934; 1939; 1968; 1970], and not necessarily useful in other contexts where the meaning of the same German words may differ significantly.

To avoid conflicts and misunderstandings in our English text, the English translations this glossary provides for a particular German word are strongly suggested for exclusive usage in the English translation.

Because some of the translations may seem wondrous to non-experts on the first sight, and because the expert translators of the field need evidence, we have often added one of following two kinds of reference to the left of the English translation:

Either we give the number of the editors' note to the English translation in which the chosen translation for the given German word is discussed; such references have the format "n. $m.n$" with page number m and note number n.

Or else we give the number of a page of the English translation where the chosen translation for the given German word occurs; this has the different format "p. m". In this case, m is the number of the first page of the English text on which the English translation for the given German word occurs with sufficient significance to justify the chosen translation.

German		English
Abbau (von Konstruktoren)		deconstruction
Abbildung	p. 9	mapping
Ableitbarkeit		derivability
ableiten		to derive
Ableitung		derivation
Ableitung durch den Prädikatenkalkul	p. 104	derivation by the predicate calculus
Ableitung im Prädikatenkalkul	p. 121	derivation in the predicate calculus
abschätzen	p. 145	to bound
abstrahieren von	n. 1.5	to ignore
Allgemeingültigkeit	n. 9.1	validity
Allgemeinheit	p. 124	validity
Allzeichen	n. 4.5	universal quantifier symbol
Alternative		alternative
Analogon		analog
Anfangsgründe	p. 20	foundational elements
anhängen	n. 22.6	to suffix
Annahme		assumption
Anordnung		order
anschaulich		intuitive
ansetzen	n. 22.6	to attach
Anzahl	p. 28	cardinal number
Anzahl von	p. 10	number of
Anzahlenlehre	p. 29	theory of cardinal numbers
Aufbau (von Konstruktoren)		construction
aufbauen (konstruieren)		to construct
auffassen	p. 45	to take the point of view
Auffassung	p. 37	point of view
aufgebaut (von Konstruktoren)		constructed
aufstellen	p. 95	to establish
aufweisen	p. 14	to exhibit
Aufweisung	p. 12.a	exhibition
Ausdruck	n. 5.1	expression
ausführen	p. VI	to carry out
ausführlich	p. VI	elaborate
Ausführung	p. VI	elaboration
Ausgangsformel	n. 44.1	initial formula
ausgezeichnete disjunktive Normalform	n. 56.2	full disjunctive normal form
ausgezeichnete konjunktive Normalform	n. 57.1	full conjunctive normal form

Glossary

Aussage	proposition
Aussagen-Verknüpfung n. 4.7	propositional connective
Aussagenkalkul	propositional calculus
Aussagenlogik	propositional logic
Aussagenverbindung n. 4.7	propositional combination
Aussagenverknüpfung n. 4.7	propositional connective
Auswahlprinzip n. 41.1	principle of choice
Axiom der ebenen Anordnung n. 6.3	Plane Axiom of Order
Axiomatik p. 2	axiomatics
Axiome der Verknüpfung n. 4.3	axioms of connection
Axiomensystem	axiom system
Begriff	notion
Begriffsbildung p. 36	concept formation
Begriffsbildungen p. 24	concept formations
Begründung (einer Theorie) p. 1.a	grounding
Begründung (grundlegende) p. 35	foundational justification
behaupten p. 33	to assert
Behauptung p. 6.a	assertion
Bereich	domain
Bestandteil (einer Formel oder einer Ableitung) p. 5.a	part
bestimmen	to determine
bestimmt p. 21	definite
bestimmt, genauer p. 32	more precisely determined
Bestimmung	determination
Betrachtung (Behandlung) p. 34	treatment
Betrachtung (Erörterung) p. 127	discussion
Betrachtungen (Überlegungen) p. 52	considerations
Beweis	proof
Beweisfigur n. 151.2	proof figure
Beweisführung p. 36	reasoning
Beweistheorie	proof theory
Bezeichnung p. 4.a	designation
Beziehung	relation
Beziehungen	relations, relationship
Bildung (Begriffsbildung, Konstrukt)	formation
Bildungsgesetz p. 119	formation rule
Bildungsprozesse p. 27	formation processes
Bildungsregeln	formation rules
dartun p. 17	to show
darstellende Konjunktion p. 56	representing conjunction
deduktionsgleich	deductively equivalent
Deduktionsgleichheit p. 148	deductive equivalence
Deduktionstheorem	deduction theorem
Denken p. 32	thinking
deutlich p. 64	distinct
Deutung p. 30	interpretation
Ding (Individuum) p. 14	thing
Disjunktionsglied p. 57	element of the disjunction
Durchschnitt (von Mengen)	intersection
einsetzen	to substitute
Einsetzung p. 62	substitution
Einsetzung für (die Nennform) p. 108	substitution for (the nominal form)
Einsetzungen p. 31	substitutions
einstellig n. 7.3	singulary
einstelliger Prädikatenkalkul	monadic predicate calculus
elementenfremd	disjoint
Endformel n. 44.3	end formula
Entscheidungsbereich für die Ableitbarkeit p. 143	decidable class w.r.t. derivability
Entscheidungsbereich für die Erfüllbarkeit p. 143	decidable class w.r.t. satisfiability
Entscheidungsmethode	decision method
Entscheidungsproblem	

Entscheidungsproblem	
Entscheidungsverfahren	decision procedure
Erfahrungskomplexe p. 2	experience-complexes
erfüllbar	satisfiable
erfüllbar, im Endlichen p. 144	finitely satisfiable
erfüllbar, f-zahlig p. 144	f-satisfiable
Erfüllbarkeit	satisfiability
Erfüllbarkeit im Endlichen p. 144	finite satisfiability
Erfüllung n. 18.4	model
Ergänzung, ausreichende p. 13	complement
ersetzbar p. 30	replaceable
Ersetzbarkeit p. 30	replaceability
Ersetzbarkeiten p. 49.a	replaceabilities
ersetzen (an einer Stelle / in einem Kontext) p. 26	to replace (with)
ersetzen (im Gebrauch mittels) p. 43	to replace (by)
Ersetzung	replacement
Ersetzungen p. 153	replacements
Ersetzungsregel p. 49.a	replacement rule
Ersetzungsregeln der Äquivalenz p. 107	replacement rules for the equivalence
Erwägung p. 16	deliberation
erwiesen, ist p. 59	is proved
Evidenz (elementar) p. 43	self-evidence
Evidenz, anschauliche p. 3	intuitive self-evidence
Evidenz, unmittelbare p. 20	immediate self-evidence
Evidenzen n. 2.4	evidences
Existenzbeweis p. 15	existence proof
Existenzsätze p. 3	existence sentences
Exportation p. 82	exportation
Fallunterscheidung p. 72	case analysis
Feld (einer Relation) p. 127	field
fest abgegrenzt p. 86	well-determined and fixed
festsetzen p. 97	to stipulate
feststellbar (durch mechanischen Nachweis) p. 15	verifiable
feststellen (durch mechanischen Nachweis) p. 11	to verify
feststellen (im Allgemeinen) p. 13	to establish
Feststellung (Aussage) p. 160	statement
Feststellung (Erkenntnis) p. 158	finding
festumgrenzt p. 20	well-determined and fixed
Figur n. 151.2	figure
finit n. VII.3	finitistic
Formeln des Prädikatenkalkuls p. 104	formulas of the predicate calculus
Formelsprache p. 45	formula language
Frage der Widerspruchsfreiheit p. 15	issue of consistency
Fragestellung	issue
Funktionszeichen	function symbol
ganze rationale Funktion n. 29.2	polynomial function
Gattung n. 2.5	genus
gebräuchlich	customary
Gegenstände (der (d. h. einer) Theorie) p. 2	objects of a theory
Gegenstände, idelle n. 1.10	intellectual objects
Gegenstände, mathematische n. 1.10	mathematical objects
Gegenstand n. 86.1	thing
Gegenstand der Betrachtung p. 21	object of consideration
Geltung n. 3.5	validity
genetisch (generisch) n. 1.7	genetic
Gesamtheit p. 15	totality

Glossary

Gesamtheit, fertige
p. 15 — completed totality

Gesichtspunkt (vgl. Auffassung)
p. 16 — viewpoint

gewünschte Formel
p. 91 — desired formula

gleichbedeutend
p. 84 — tantamount

Grenzübergang
p. 118 — passage to the limit

Größenbeziehung
p. 22 — relation of magnitude

Größenlehre
p. 41 — theory of magnitudes

Grundbegriffe
p. 1.a — basic notions

Grundlegung
p. 37 — grounding

Gültigkeit
n. 3.5 — validity

Hinterglied
p. 106 — succedent

Hinzufügen eines beliebigen Implikationsvordergliedes
p. 83 — addition of an arbitrary antecedent to form an implication

identisch, im Endlichen
p. 120 — finitely identical

identisch, f-zahlig
p. 118 — f-identical

identisch falsch
p. 55 — identically false

identisch wahr
p. 54 — identically true

identische Formel
p. 90 — identical formula

Implikationsformel
p. 67 — implication formula

Importation
p. 82 — importation

Inbegriff
n. 1.6 — aggregate

Individuen-Variable
n. 8.4 — individual variable

Individuenbereich
p. 2 — domain of indivıuals

Individuensymbol
n. 88.3 — individual symbol

Individuenvariable
n. 88.1 — individual variable

inhaltlich
n. 2.3 — contentual

Inhaltslogik
n. 53.2 — logic of content

Interpretation — interpretation

Kettenschluss (Kettenschluß)
n. 84.2 — chain inference

Klammer
p. 5.a — parenthesis

Klasse — class

Kombination
p. 13 — combination

Komplementärmenge — complement-set

Komponente
p. 121 — component

Konjunktionsglied
p. 116 — element of the conjunction

Lehrsatz
n. 6.5 — theorem

Logistik
p. VIII — formal logic

mannigfach
p. VIII — manifold

Mannigfaltigkeit
n. 14.2 — manifold

Metamathematik — metamathematics

Methode der Aufweisung
n. 12.1 — method of exhibition

Methode der Wertung (vgl. Wertung)
p. X — model construction

Methode des progressiven Schließens
p. 13 — method of progressive inference

methodisch
n. 1.3 — of the method, the treatment of

methodische Anforderung
p. 1.a — methodological requirements

methodische Einstellung
p. 32 — methodological attitude

methodische Umstellung
p. 130 — methodological reorientation

methodischer Hinsicht, in
p. 163 — methodologically

methodischer Standpunkt
p. 7 — methodological standpoint

Modell
p. 12.a — model

Nachbereich (einer Relation)
n. 127.3 — range

nachprüfen
p. 47 — to check

Nachweis
p. 65 — proof

Glossary

Namenverzeichnis p. VI	index of persons
Negationsstrich p. 53	negation bar
Nennform n. 89.1	nominal form
neuere (Logik, die) p. 9	(the) more recent (logic)
Normaldisjunktion p. 158	normal disjunction
normale Wertung n. 73.4	standard model
Normierung p. 55	standardization
Num(m)erierung p. 28	numeration
obere Grenze p. 38	least upper bound
Objekt	object
Paragraph p. 19	section
Parallelenaxiom n. 4.3	Axiom of Parallels
Parallelenaxiom in schärferer Fassung n. 6.4	Axiom of Parallels in Sharper Form
positiv identisch p. 68	positive-identical
Prädikatenkalkul p. 86	predicate calculus
Prädikatenlogik	predicate logic
Prädikatensystem p. 10	predicate system
pränexe Formel p. 140	prenex formula
pränexe Normalform p. 143	prenex normal form
Primformel p. 88	atomic formula
Primärformel p. 145	prime formula
Prinzip der kleinsten Zahl p. 35	least-number principle
Quantor n. 4.5	quantifier
Rechenoperationen p. 21	operations of calculation
Regel der Einsetzung (vgl. Substitutionsregel) p. 62	rule of substitution
reine Prädikaten-Logik p. 7	pure predicate logic
reiner Prädikatenkalkul p. 150	pure predicate calculus
Relation p. 17	relation
Sachgebiet p. 2	subject area
Sachgehalt p. 2	subject matter
Sachlage p. 17	situation
Sachverhalt n. 3.1	state of affairs
Satz	sentence
Satz (Theorem)	sentence, theorem
Satzverbindung n. 4.7	sentential combination
schärfer n. VII.4	sharper
Schema	schema
Schema der Äquivalenz p. 85	equivalence schema
Schemata	schemata
Schließen	inference
Schließen, das gewöhnliche p. 92	the common usage of inference
Schlussfigur (Schlußfigur) p. 117	syllogism
Schlussfolgerung (Schlußfolgerung) p. 45	conclusion
Schlussschema (Schlußschema) n. 62.8	inference schema (of modus ponens)
Schlussweisen (Schlußweisen) p. 19	modes of inference
Seinszeichen n. 4.5	existential quantifier symbol
Sheffersches Symbol n. 48.1	Sheffer stroke
Sinnesqualitäten p. 16	qualia
sinnliche Wahrnehmung p. 16	sense perception
Situation p. 34	situation
Skolemsche Normalform p. 158	Skolem normal form
Skolemscher Satz p. 158	Skolem's Theorem
Sprachgebrauch (gewohnter) p. 21	common usage of language
Standpunkt p. 1.a	standpoint

Glossary

German		English
streng p. 1.a		rigorous
Subjekt p. 2		subject
Substitutionsregel (vgl. Regel der Einsetzung) p. 49.a		substitution rule
Tatsachenkomplex p. 129		body of facts
Tatsächlichkeit n. 2.6		actuality
Teilgesamtheit p. 73		sub-totality
Term		term
Terminus p. 1.a		technical term
Totalität p. 15		totality
transformieren p. 7		transform
überblickbar p. 14		comprehensible
Übereinstimmung p. 29		correspondence
überführbar p. 132		convertible
Überführbarkeit p. 132		convertibility
überführen p. 98		to convert
Überführung p. 140		conversion
Übergang (Schritt) p. 151		step
übergehen (fortschreiten) p. 108		to proceed
übergehen (sich wandeln) p. 10		to turn
Überlegung (pluralisch) p. 22		considerations
Überlegung (singularisch) p. 23		consideration
Übersicht (Abhandlung) p. 148		survey
Übersicht (direkter Zugang) p. 61		easy access
übersichtlich p. 147		easily surveyable
üblich p. VII		customary
Umbennung der gebundenen Variablen p. 97		renaming (of) bound variables
Umfangslogik n. 53.2		logic of extension
umformen p. 103		to transform
Umformung p. 49.a		transformation
umkehrbar eindeutig p. 9		one-to-one
Umstellung der Vorderglieder p. 107		interchange of antecedents
Unmöglichkeitsbeweis p. 19		impossibility proof
Unwiderlegbarkeit		irrefutability
Urteil		judgment
Verbindung (von Formelteilen, Regelanwendungen, etc.) p. 5.a		combination
verbunden p. 53		combined
Vereinigungsmenge		union
Verfahren (Prozedur) n. 3.4		procedure
Verfahren (Vorgehen) n. 3.4		approach
Verfahren der Aufweisung n. 15.1		procedure of exhibition
Verhältnis		relationship
Verifikation		verification
verifizieren		to verify
verifizieren, sich p. 32		to prove true
Verknüpfung p. 45		connective
verschärfen n. VII.4		to sharpen
Verschärfung p. XI		sharpening
Vertauschung der Implikationsvorderglieder p. 81		interchange of antecedents of implications
Vertauschung der Vorderglieder p. 91		interchange of antecedents
Verteilung der Werte α, β p. 74		distribution of the values α, β
Verteilung von Wahrheitswerten p. 56		distribution of truth values
Vertretbarkeit p. VI		representability
vollständige Induktion n. 23.4		mathematical induction
vollständiges System p. 63		complete system
Vollständigkeit, (Frage der) p. 124		(issue of) completeness

German	Page	English
Vollständigkeit von, Frage nach der	p. 123	issue of the completeness of
Vollständigkeitsproblem	p. 124	issue of completeness
Vollständigkeitsproblem von	p. 127	issue of the completeness of
Vollständigkeitssatz	p. 66	completeness theorem
Voraussetzung		presupposition
Vorbereich (einer Relation)	n. 127.2	domain
Vorbetrachtung	p. 24	preliminary consideration
Vorderglied	n. 62.4	antecedent
vorgestellt, anschaulich	p. 32	intuitively conceived
vorliegend	p. 20	present
vorstellbar	p. 32	conceivable
Vorstellung	p. 6.a	conception
Vorstellungen	p. 43	conceptions
Vorwort	p. V	preface
Wertbereich (vgl. Wertevorrat)	p. 86	range of values
Wertbestimmung	p. 11	determination of values
Wertbestimmungen	p. 11	determinations of values
Wertevorrat (vgl. Wertbereich)	p. 63	range of values
Wertsystem der Variablen	p. 55	valuation (of the variables)
Wertung	n. 73.3	model
Wertungsmethode (vgl. Wertung)	p. X	model construction
Wertverlauf	n. 9.3	graph
Wertverteilung der Variablen	p. 54	distribution of truth values (on the variables)
widerlegbar		refutable
Widerlegbarkeit		refutability
Widerspruch		contradiction
widerspruchsfrei	n. VII.1	consistent
Widerspruchsfreiheit	n. VII.1	consistency
Zahlenreihe	p. 15	number series
Zahlentheorie		number theory
Zeichen		symbol
Ziffer	p. 21	numeral
zulässig		admissible
zuordnen	p. 26	to assign
Zuordnung	p. 29	correspondence
Zusammenhang	p. 152	connection

Guidelines for the Typesetting of the English Translation

Line Breaking: The line breaking of the original text will be ignored, even in formulas.

Page Breaking and Original Notes: Regarding the page breaks of the original text, we will admit slight variations of their positions in the English translation, but obey the following rules:

- With very few exceptions, the page breaks will occur in the same paragraph as in the German original.
- The original footnotes and their original marks (simple Arabic numbers, starting from 1 on each page) will go to (the pages of the translations of) the original pages.
- We try to complete sentences without page breaks. If the paragraph is not completed at the page break, we will fill the rest of the line with a horizontal rule, such as shown before the footnote section on Page 1.a.

Although it is not really possible to mark the exact position of the page break in the German original in an English translation, we will roughly indicate this position with the help of the symbol "|".

Editorial Notes: Editorial notes have the form $m.n$, where m is the number of the German original page to whose translation the editorial note refers, and n is the number of the editorial note relative to that page.

Indentation: We will not follow all indentations of the German original. We will remove all indentations that follow a headline or an itemized (i.e. vertical) list (unless this list was introduced in the translation), because these indentations are ugly and carry no information: In the German original they are always present and do not start a new paragraph, but are just a matter of style. We will sometimes replace ugly indentations (such as the ones of paragraphs that have only a single line or that start an itemization at the second line) with some extra vertical space, unless they follow a displayed, non-left allocated text or formula (because then they are relevant for paragraph counting).